NEW DEVELOPMENTS IN RENEWABLE ENERGY

Edited by **Hasan Arman** and **Ibrahim Yuksel**

New Developments in Renewable Energy
http://dx.doi.org/10.5772/45849
Edited by Hasan Arman and Ibrahim Yuksel

Contributors
Tser Chen, Tsai-Lien Yeh, Yi-Hsuan Ko, A Alolah, Ahmed M. Al Salloum, Mashauri Adam Kusekwa, António Cardoso Marques, José Alberto Fuinhas, Rui Flora, Pius Olugbenga Fatona, Oladunjoye Abiodun, Adesanwo Adeola, Adetayo Olumide, Abiodun Abiodun, Mamadou Lamine Doumbia, Gholam Riahy, Sajjad Abedi, Seyed Hossein Hosseinian, Mehdi Farhadkhani, Wan Azlina Wan Ab Karim Ghani, Ehsan Enferad, Daryoush Nazarpour, Jose Pelegri-Sebastia, Miguel Pareja Aparicio, Tomás Sogorb, Vicente Llario, Saeid Eslamian, Masoomeh Fakhri, Mohammad Reza Farzaneh, Ali Eltamaly, Hassan Farh, Robert Peters, Basel I. Ismail, Hasan Arman, Maria Teresa Outeiro, Adriano Carvalho

Published by InTech
Janeza Trdine 9, 51000 Rijeka, Croatia

Publishing Process Manager Iva Lipovic

Technical Editor InTech DTP team

Cover InTech Design team

Additional hard copies can be obtained from orders@intechopen.com

New Developments in Renewable Energy, Edited by Hasan Arman and Ibrahim Yuksel
p. cm.
ISBN 978-953-51-1040-8

For EU product safety concerns:
IN TECH d.o.o., Prolaz Marije Krucifikse Kozulić 3, 51000 Rijeka, Croatia,
info@intechopen.com or visit our website at intechopen.com.

Meet the editors

Dr. Hasan Arman is a Professor at United Arab Emirates University, College of Science, Geology Department since 2008. He received his bachelor degree from Hacettepe University, Turkey in 1984 and Ph.D. degree from University of Arizona, USA in 1992. He worked as a Postdoc at the University of Nevada, Reno, USA from 1992 to 1993. He was as a faculty member at Sakarya University Civil Engineering Department, Turkey between 1993 and 2008. Dr. Arman has taught several different courses in undergraduate and graduate levels related to engineering, environment and geology. His research interests include environmental degradation, water resources, global warming, climate change, renewable and sustainable energy sources. Dr. Arman has published in different peer reviewed national and international scientific journals. In addition to be an editorial board member of several international journals, he also is acting as a scientific reviewer in many others. He was in organization committee of both 1st and 2nd International Conference on Renewable Energy: Generation and Applications (ICREGA) which was held in 2010 and 2012 in Al Ain, UAE.

Dr. Ibrahim Yuksel received his undergraduate degree from Gazi University, Turkey in 1988 and MSc. degree from MidKent College, Chatham, United Kingdom. His Ph.D. is from Karadeniz Technical University, Turkey in 2000. Dr. Yuksel is a civil engineer and has more than 24 years of experience in hydraulic engineering. His study area is renewable and sustainable energy, especially hydro energy. He is a Prof. Dr. at Technology Faculty, Department of Civil Engineering, Sakarya University, Turkey. Also, Dr. Yuksel has been a director of Hendek Vocational School, Sakarya University, Turkey since 2011.

Contents

Preface

Unlike to fossil energy sources, renewable energy sources such as sunlight and wind are existed in widespread geographical areas of the world and provide important opportunities for energy efficiency. Higher growth rate in production of renewable energy and technological diversification of energy sources will contribute a significant energy security and substantial economic benefits to many nations. Consequently, in developing countries, projects related to renewable energy can directly contribute to poverty alleviation via providing the energy needed for setting up businesses and employment. Also, renewable energy technologies can make indirect assistances to poverty alleviation by providing energy for cooking, space heating, and lighting. Renewable energy can also contribute to education by providing electricity to schools.

The recent studies indicate that renewable energy sources have been grown at an average annual rate of 1.7% since 1990s. It is slightly less than the annual growth rate of world's' Total Primary Energy Supply (TPES) which is 1.9%. Especially, the average annual growth rate of wind power is the highest with 25%. However, the production still remains small due to its very low base in 1990s. Most of the production and growth of solar and wind energy are committed by the Organization for Economic Co-operation and Development (OECD) countries. With 10.4% of annual growth rate, renewable municipal waste, biogas and liquid biomass were recorded as the second highest growing energy sources. Having an annual growth rate of 1.2%, the primary solid biomass is the largest contributor to renewable energy in the world and has experienced the slowest growth among the renewable energy sources. Most of the solid biomass is produced in the Nnon-OECD countries, but its growth is comparable forto OECD and non-OECD countries. The annual growth rate of solar photovoltaics and solar thermal is 9.8%. The average annual growth rate of hydropower in non-OECD countries (3.7%) was larger than in OECD countries (0.4%) between 1990 and 2007.

Sustainable energy is the provision of energy that meets the needs of the present without compromising the needs of the future. Renewable energy technologies are important contributors to sustainable energy which utilize renewable energy sources, such as hydroelectricity, solar energy, wind energy, wave power, geothermal energy, and tidal power. They generally contribute to world's energy security by reducing dependence on fossil fuel resources, improving energy efficiency, and providing opportunities for mitigating greenhouse gases.

Sustainable development has evolved to integrate economic, social and environmental aims. Sustainable development has emerged as the key challenge for the 21st Ccentury. Both the opportunities and the lack of progress were highlighted during the Johannesburg World Summit on Sustainable Development in 2002. Decision-makers are looking for sustainable development in order to provide practical approaches for addressing traditional issues as well as the newer challenges.

Energy is important for economical and social development and enhanced quality of life in all nations. However, if technology were to remain stable and if overall quantities were to increase substantially, much of the world's energy could not be sustained under the current production and consumption ways. There is an urgent need to control atmospheric emissions of greenhouse and other gases/substances for efficiency in energy production, transmission, distribution, and consumption in the country. As policymakers and investors all around the world are aware of the electricity's critical role in improving living standards and sustaining economic growth, electricity supply infrastructures in many developing countries are being rapidly expanded.

The book is divided into nineeight sections;: Energy Utilization, Conservation and Social Consideration, Modeling and Analysis, Wind Power, Biomass, Ocean Energy, Geothermal, Bio Fuels, Fuel Cell and Integrated System. Each section has a number of chapters addressing various issues related to renewable energy.

A number of experts have provided progressive contributions for the development of this book. The editor and Cthe co-editor of the book are thankful for their supportive and continuous efforts in completing this book. This book mightwould not be existed if thereit wasn't for was no their remarkable contributions. Finally, the editors would like to kindly thank to all InTech peoplestaff for their invitation and enthusiasm from the first to the final stage of this book.

Editor
Dr. Hasan Arman
Professor, Geology Department
College of Science
United Arab Emirates University
Al Ain, United Arab Emirates

Co-editor
Ibrahim Yuksel
Associate Professor, Civil Engineering
Faculty of Tecnology
Sakarya University
Sakarya, Turkey

Section 1

Energy Utilization, Conservation and Social Consideration

Chapter 1

Present Situation and Future Prospect of Energy Utilization and Climate Change in Turkey

İbrahim Yüksel, Kamil Kaygusuz and Hasan Arman

Additional information is available at the end of the chapter

http://dx.doi.org/10.5772/54319

1. Introduction

Sustainable development has emerged as the key challenge for the 21st Century. The Johannesburg World Summit on Sustainable Development in 2002 highlighted both the opportunities and the lack of progress since the Earth Summit in Rio de Janeiro, a decade previously. Decision-makers are looking to sustainable development to provide practical approaches that could address traditional issues as well as the newer challenges. Although no universally accepted practical definition of sustainable development exists as yet, the concept has evolved to integrate economic, social and environmental aims [1,2]. Recent increases in energy prices are likely to be the precursor of a longer term trend. While they will encourage much needed energy efficiency and stimulate investment, they pose severe difficulties for expanding access to modern energy services to the one third of people who still do not have it, or whose access is inadequate for economic development. An energy system embodying such inequities is neither sustainable nor acceptable [3].

However, developing the remaining hydropower potential offers many challenges and pressures from some environmental action groups over its impact has tended to increase over time. Hydropower throughout the world provides 17% of our electricity from an installed capacity of some 730 GW is currently under construction, making hydropower by far the most important renewable energy for electrical power production. The contribution of hydropower, especially small hydropower (SHP) to the worldwide electrical capacity is more of a similar scale to the other renewable energy sources (1-2% of total capacity), amounting to about 47 GW (53%) of this capacity is in developing countries [3,4].

Affordable energy services are among the essential ingredients of economic development, including eradication of extreme poverty as called for in the United Nations Millennium Development Goal (MDGs). Modern energy services-mainly provided by liquid and gaseous fuels, as well as electricity-are essential. Convenient, affordable energy is also important for improving health and education, and for reducing the human labour required to cook and meet other basic needs [3-6].

Meanwhile, global climate change poses an unprecedented threat to all human beings. While this problem is important in the long-run, most decision-makers recognise (especially in the developing countries), that there are many other critical sustainable development issues that affect human welfare more immediately. However, even in the short term, climate is an essential resource for development. For example, in many countries (especially the poorest ones), existing levels of climatic variability and extreme events pose significant risks for agriculture, economic infrastructure, and vulnerable households. Climatic hazards continue to take their human and economic toll even in wealthy countries. Such climate threats, which undermine development prospects today, need to be better addressed in the context of the long-run evolution of local and regional climates [1,7].

There is a growing concern that long-run sustainable development may be compromised unless measures are taken to achieve balance between economic, environmental and social outcomes. Since the early 1980s, Turkish energy policy has concentrated on market liberalization in an effort to stimulate investment in response to increasing internal energy demand [8]. Turkey's new government has continued this policy despite lower energy demand induced by the 2001 economic crisis. On the other hand, CO_2 and other greenhouse gas emissions of the country are increasing rapidly due to energy and electricity utilization [9].

More generally, climate change and sustainable development interact in a circular fashion. Climate change vulnerability, impacts and adaptation will influence prospects for sustainable development, and in turn, alternative development paths will not only determine greenhouse gas (GHG) emission levels that affect future climate change, but also influence future capacity to adapt to and mitigate climate change. Impacts of climate change are exacerbated by development status, adversely affecting especially the poor and vulnerable socio-economic groups. The capacity to adapt to climate change goes beyond wealth, to other key pre-requisites of good development planning, including institutions, governance, economic management and technology [1,10].

The key to an effective climate change response strategy is a better understanding of relevant policy linkages. Development planners, naturally, place development first, and therefore, climate policies need to be integrated within national sustainable development strategies. In particular, they would like to know whether specific climate change impacts and response measures will make existing development efforts less, or more, sustainable in terms of their economic, social and environmental dimensions [1].

2. Climate change, energy and emission profile in Turkey

Turkey's total carbon dioxide (CO_2) emissions amounted to 239 million tons (Mt) in 2006 (Tables 1-3). Emissions grew by 5% compared to 2001 levels and by just over 50% compared to 1990 levels. Oil has historically been the most important source of emissions, followed by coal and gas. Oil represented 45% of total emissions in 2004, while coal represented 40% and gas 15%. The contribution of each fuel has however changed significantly owing to the increasingly important role of gas in the country's fuel mix starting from the mid-1980s [3,11,12].

	1990	1995	2000	2002	2004	2006
CO_2 sectoral approach (Mt of CO_2)	126.91	152.66	200.56	202.13	2007.25	239.74
CO_2 reference approach (Mt of CO_2)	138.20	157.28	203.48	203.45	209.50	242.61
Total Primary Energy Supply, TPES (Mtoe)	52.94	61.81	76.87	78.73	81.83	94.00
GDP (billion 2000 US$ using PPPs)	309.57	362.57	439.99	464.79	506.30	576.82
Population (millions)	56.20	61.64	67.46	70.71	71.79	72.97
CO_2 / TPES (t CO_2 per TJ)	57.30	59.00	62.30	61.30	60.50	60.90
CO_2 /GDP (kg CO_2 per 2000 US$)	0.91	0.93	1.01	0.96	0.90	0.92
CO_2 / GDP (kg CO_2 per 2000 US$ PPP)	0.41	0.42	0.46	0.43	0.41	0.42
CO_2 / population (t CO_2 per capita)	2.26	2.48	2.97	2.86	2.89	3.29

Table 1. Key indicators in Turkey [3,13].

	CO_2 emissions	Level assessment	Cumulative
IPCC source category	**(Mt of CO_2)**	**(%)**	**total (%)**
Production electricity and heat-coal/peat	42.32	12.6	12.6
Manufacturing industries-coal/peat	42.30	12.6	25.1
Road-oil	36.60	10.9	36.0
Production electricity and heat-gas	27.28	8.1	44.1
Residential-gas	14.45	4.3	48.4
Manufacturing industries-oil	12.35	3.7	52.0
Residential-coal/peat	10.10	3.0	55.0
Non-specified other sectors-oil	9.69	2.9	57.9
Manufacturing industries-gas	8.01	2.4	60.3
Non-specified other sectors-gas	6.51	1.9	62.2
Other transport-oil	5.36	1.6	63.8
Total CO_2 from fuel combustion	239.74	71.1	71.1

Table 2. Key sources for CO_2 emissions from fuel combustion for Turkey in 2006 [3,14].

Years	CO_2	CH_4	N_2O	F gases	Total
1990	139.6	29.2	1.3	0.0	170.1
1992	152.9	36.7	4.0	0.0	193.6
1994	159.1	39.2	2.2	0.0	200.5
1996	190.7	45.0	6.1	0.4	242.1
1998	202.7	47.7	5.6	0.7	256.6
2000	223.8	49.3	5.8	1.1	280.0
2002	216.4	46.9	5.4	1.9	270.6
2004	241.9	46.3	5.5	2.9	296.6
2005	256.3	49.4	3.4	3.2	312.4

Table 3. Greenhouse gas emissions by gas in Turkey (million tons CO_2 eq) [3,13,14].

According to recent projections, total primary energy supply (TPES) will almost double between 2006 and 2020, with coal accounting for an increasingly important share, rising from 24% in 2006 to 36% in 2020, principally replacing oil, which is expected to drop from 40% to 27%. Such trends will lead to a significant rise in CO_2 emissions, which are projected to reach nearly 600 Mt in 2020, over three times 2004 levels [3,12,14,15].

In 2006, public electricity and heat production were the largest contributors of CO_2 emissions, accounting for 30% of the country's total. The industry sector was the second largest, representing 28% of total emissions, followed by transport, which represented 20% and direct fossil fuel use in the residential sector with 8%. Other sectors, including other energy industries, account for 14% of total emissions. Since 1990, emissions from public electricity and heat production have grown more rapidly than in other sectors, increasing by 6%. Simultaneously, the shares of emissions from the residential and transport sectors both dropped by 7% and 3% respectively while the share of emissions from the manufacturing industries and construction sector remained stable [3,11,13,16].

Over 40% of all energy is used by the industrial sector and nearly 35% in the residential sector. The rest is split between transportation and commercial services. Industry in Turkey is energy intensive, especially iron and steel manufacturing and cement production sectors, by far the largest energy users. In the residential and commercial building sector, more than 80% of energy is used for space heating. Use of electrical appliances is rapidly increasing and boosting power demand. Table 4 shows the electric power capacity development in Turkey. Increasing use of air-conditioning, especially in the Mediterranean region, has shifted the peak hours of electricity demand to noon in the summer. Electricity consumption for lighting accounts for 30-40% of power consumption in the residential sector.

Fuel type	2005		2010		2020	
	Installed capacity (MW_e)	Generation (GWh)	Installed capacity (MW_e)	Generation (GWh)	Installed capacity (MW_e)	Generation (GWh)
Coal	14465	48386	16106	104040	26906	174235
Natural gas	10756	66417	18923	125549	34256	225648
Fuel oil	2124	10531	3246	18213	8025	49842
Renewables	14112	50900	25102	86120	30040	104110
Nuclear	0.0	0.0	2000	14000	10000	70000
Total	41457	176234	65377	347922	109227	623835

Table 4. Electric power capacity development in Turkey [11].

On the other hand, the transport sector is dominated by road transport. Vehicle ownership is only seven vehicles per hundred inhabitants compared to the OECD average of fifty. Capacity utilization of available rail lines for passenger transport is low for inter-city traffic and higher for suburban lines [3,17-20].

3. Climate change and greenhouse gas emissions policies in Turkey

Turkey was a member of the OECD when the United Nation Framework Convention on Climate Changes (UNFCCC) was adopted in 1992, and was therefore included among the so-called Annex I and Annex II countries. Under the convention, Annex I countries have to take steps to reduce emissions and Annex II countries have to take steps to provide financial and technical assistance to developing countries. However, in comparison to other countries included in these annexes, Turkey was at a relatively early stage of industrialization and had a lower level of economic development as well as a lower means to assist developing countries. Turkey was not given a quantified emissions reduction or limitation objective in the Kyoto Protocol. Following a number of negotiations, in 2001 Turkey was finally removed from the list of Annex II countries but remained on the list of Annex I countries with an accompanying footnote specifying that Turkey should enjoy favorable conditions considering differentiated responsibilities. This led to an official acceptance of the UNFCCC by the Turkish Grand National Assembly in October 2003, followed by its enactment in May 2004. Turkey has not yet signed the Kyoto Protocol [3,11,14,21].

Throughout this process, the government carried out a number of studies on the implications of climate change and its mitigation. The first efforts were undertaken by the National Climate Coordination Group in preparation for the 1992 Rio Earth Summit. Following this, a National Climate Program was developed in the scope of the UNFCCC. In 1999, a specialized Commission on Climate Change was established by State Planning Organization (DPT) in preparation of the Eighth Five-Year Development Plan (2001-2005). The Five-Year Development Plan was the first planning document to contain proposals for national policies and measures to reduce GHG emissions, and funding for climate-friendly technologies [3,22].

Following the ratification of the UNFCCC, a number of working groups were set up with the objective to define a climate change mitigation strategy and compile the country's first national communication to the UNFCCC. These included a working group on mitigation in the energy sector and a working group on mitigation in the transport sector. However, it remains unclear as to when the strategy and national communication will be completed. The strategy aims to reduce GHG emissions through the implementation of appropriate measures and the development of climate-friendly technologies. Energy efficiency and the development of renewable energy sources are two important components of the strategy. However, the strategy will not include any policies that directly target GHG emissions, such as carbon taxation or emissions trading. It also does not include a specific target for emissions reductions [3,14].

4. Global warming and environmental policy in Turkey

Developing countries, while varying in size and population, political system, economic structure, bear many similarities. They are facing less favourable economic circumstances, worsening environmental degradation and challenges in curbing climate changes. The present paper [1] only focuses on the issues of contradictory objectives, unrealistic standards and limited public participation.

Policy makers in developing countries are well aware of the importance of environmental protection. However, more often than not, they are placed in a dilemma when left to balance between economic growth and environment. Conflicts often rise between social, environmental and economic objectives [1,23]. The headlong pursuit of economic growth is the cornerstone of developing countries. A top Turkish environmental official accepted that economic growth must take precedence over environmental protection for years to come because the former is not only of great importance to maintaining political stability but also to funding the environmental clean-up. This very contradictory objective in developing countries is well materialized in the implementation of "Polluter Pays Principles" (the PPP), the value of which is dramatically belabored. A good example can be found in the way the governments deal with state-owned enterprises (SOEs) in emissions abatement.

On the other hand, for developing countries, great importance should be attached to the acceleration of environmentally responsible development rather than following the past, and arguably the present, path of the industrial world in pursuit of "unrestricted economic growth without considerations to its effects on the natural environment".

http://dx.doi.org/10.5772/54319

Public unawareness of environmental impacts presents a serious impediment in developing countries to effectively implementing environmental policies. Frequently decisions are made in the absence of environmental information in these countries [1,23]. In addition, environmental impacts are normally exposed to the purview of selected environmental departments, and offices in charge, and expert researchers. The public tend to be left in the dark about the seriousness of the worsening environment they are living in, the costs to their health and quality of life, and the opportunity of helping policy-makers to improve the environment. The lack of environmental awareness has resulted in indifference to environmental degradation, an absence of self-regulating motivation and, above all, a lack of enthusiasm to be involved in monitoring polluting operations and enterprises. Public participation could be a cost-effective method of implementing environmental policy, especially for those countries chronically short of funds and trained human resources.

Since possible results of the global warmth gradually started to form the most basic problem on environmental basis, "Framework Convention on Climate Changes" (FCCC) is constituted which was due on March 21, 1994 followed by its approval by 50 countries after being first approved in Rio Environment and Development Conference held in 1992. Aim of the Convention is to keep the concentration of greenhouse gas in the atmosphere at a constant level necessary to prevent its hazardous man caused impact on climate system. On the other hand, international society will come to a common decision in Conference of Parties (COP) held annually where all participating countries are closely involved in decision making process. The countries in Convention's Appendix-1 list decided by Kyoto Protocol to be due between 2008 and 2012 will be forced to reduce total emission level of gases (CO_2, CH_4, N_2O, HFC_s) that have direct greenhouse effect 5% below the level in 1990 [1,24].

5. Electricity generation and CO_2 emission in Turkey

The electricity generation in Turkey is dominated by fossil fuels. As shown in Fig. 1, the share of fossil fuels in total generation has been steadily increasing for last two decades and reached to the peak share of 82,5% in 2008 [25]. The 57,4% of total electricity generation in 2008 was from imported fuels (natural gas, imported coal and liquid fuels). The high level of fossil fuel dependency in the electricity generation is the major cause of increase in the national GHG emissions. Since 1990, the total GHG emission of Turkey has increased more than twofold and reached 366,5 million tons of CO_2e in 2008. Within the same period, the GHG emissions generated upon the electricity generation is increased more than threefold from 30 million tons in 1990 to 101,4 million tons in 2008 [9].

Turkey's GHG emissions were doubled by 2008 and reached to 366,5 million tons CO_2e comparing 1990 level as shown in Fig. 2 [9]. In 2008, around 80% of the total emissions of Turkey were from CO_2 while one third of CO_2 emissions were from electricity generation as shown in Fig. 3 [9]. In other words, more than one quarter of total emissions (27%) are due to electricity generation by fossil fuels. Other important CO_2 sources are industry, road transportation, residential and cement production [9].

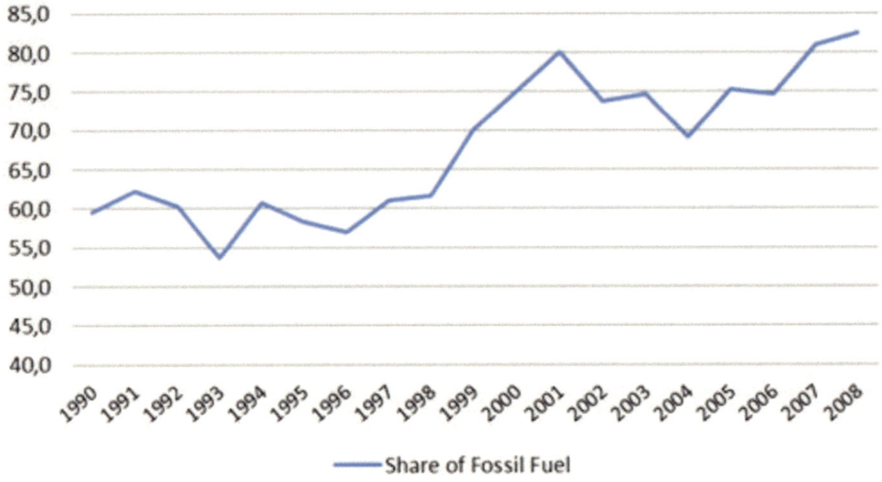

Figure 1. Share of fossil fuels in electricity generation of Turkey by years (%).

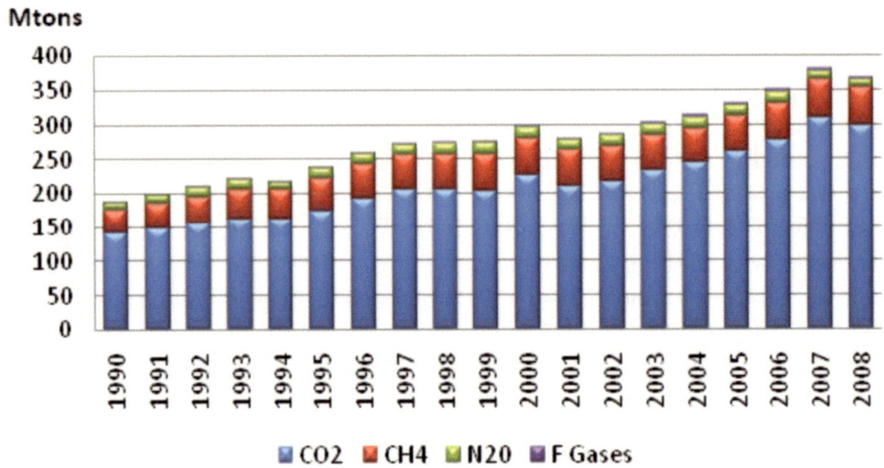

Figure 2. Development of cumulative GHG emissions of Turkey by years.

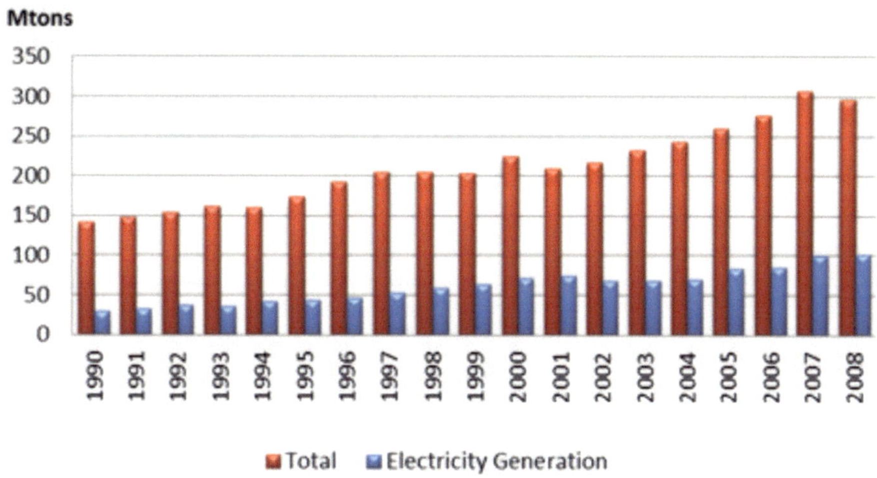

Figure 3. Development of total and electricity generation CO_2 emissions by years in Turkey.

6. Hydropower as a renewable energy in Turkey

Turkey has substantial renewable energy resources. Renewables make the second-largest contribution to domestic energy production after coal. In 2003, energy from renewable sources amounted to 10 Million tons of oil equivalent (Mtoe). More than half of renewables used in Turkey are composed of combustible renewables and waste, the rest being mainly hydro and geothermal as shown in Table 5. Combustible renewables and waste used in Turkey are almost exclusively non-commercial fuels, typically wood and animal products, used in the residential sector for heating. The use of biomass for residential heating, however, has declined owing to replacement of non-commercial fuels by commercial fuels. The contribution of wind and solar is still small but is expected to increase. Electricity generation from renewables totalled 35.5 TWh and contributed 25% to total generation in 2004. In 1990, generation from renewables was 23.2 TWh and their share in power generation was higher, representing 40%. Hydro is the dominant source of renewable electricity, with only 0.15 TWh derived from other sources. Hydro production fluctuates annually depending on the weather [3,12,15,26-28].

Hydropower generation climbed from 2 Mtoe (23.1 TWh) in 1990 to 3.0 Mtoe (35.3 TWh) in 2004, growing on average by 3.8% per year. The economic hydropower potential has been estimated at 128 TWh per year, of which 35% has been exploited. The government has a strategy for developing the hydropower potential and expects a few hundred plants to be constructed over the long term adding more than 19 GW of capacity. Construction costs would be approximately US$ 30 billion. The government expects hydropower capacity to reach about 31000 MW in 2020. Some 500 projects (with a total installed capacity over 20400 MW), which are in different phases of the project cycle, are awaiting realization. On the other hand, Turkey has a lot of potential for small hydropower (< 10 MW), particularly in the eastern part of the country. At present the total installed capacity of small hydropower is 176 MW in 70 locations, with annual generation of 260 GWh. Ten units are under construction with a total installed capacity of 53 MW and estimated annual production of 133 GWh. Furthermore, 210 projects are under planning with a total capacity of 844 MW and annual production of about 3.6 TWh [3,29].

Renewable energy sources	**1990**	**1995**	**2000**	**2002**	**2004**
Primary energy supply					
Hydropower (ktoe)	1991	3057	2656	2897	3038
Geothermal, solar and wind (ktoe)	461	654	978	1142	1215
Biomass and waste (ktoe)	7208	7068	6457	5974	5728
Renewable energy production (ktoe)	9660	10779	10091	10013	10001
Share of total domestic production (%)	38	40	38	40	42
Share of TPES (%)	18	17	12	13	12
Generation					
Hydropower (GWh)	23148	35541	30879	33684	35330
Geothermal, solar and wind (GWh)	80	86	109	153	145
Renewable energy generation (GWh)	23228	35627	30988	33837	35480
Share of total generation (%)	40	41	25	26	25
Total final consumption					
Geothermal, solar and wind (ktoe)	392	580	910	1048	1134
Biomass and waste (ktoe)	7208	7068	6457	5974	5728
Renewable total consumption (ktoe)	7600	7648	7367	7022	6882
Share of total final consumption (%)	18	15	12	12	11

Table 5. Renewable energy supply in Turkey [3,12].

Hydropower is solar energy in a naturally and ideally concentrated form that can be utilized with the help of a mature and familiar technology with unsurpassed rates of efficiency. Moreover, it does not deprive future generations in terms of raw materials, or burdening them with pollutants or waste. Hydroelectric power plants utilize the basic national and renewable resource of the country. Although the initial investment cost of hydropower seems relatively high, the projects have the lowest production costs and do not depend on foreign capital and support, when considering long-term economic evaluation [3,30,31].

7. The role of hydropower and dams for sustainable energy

The generation of hydropower provides an alternative to burning fossil fuels or nuclear power, which allows for the power demand to be met without producing heated water, air emissions, ash, or radioactive waste. Of the two alternatives to hydropower, in the last decade, much attention has been given to thermal power production because of the adverse effect of CO_2 emissions. With the increasing threat of greenhouse gases originating from such anthropogenic activities on the climate, it was decided to take action. Thus the Framework Convention on Climate Change was enacted on 21 March 1994 and has been signed by 174 countries to date [3,31].

Dams that produce electricity by this most productive renewable clean energy source in the world provide an important contribution to the reduction of air pollution. The result of an investigation held in the USA suggests that the productivity of hydroelectric power-plants is higher than 90% of thermal plants and this figure is twice that of thermal plants. In case of Turkey, the public has been wrongly informed. Some people have claimed that hydro plants do not produce as much energy as planned because of irregular hydrological conditions and rapid sedimentation of reservoirs. It is also claimed that the cost of the removal of dams entirely filled by sediment at the end of their physical lives is not considered in the total project cost, and that there are major problems in recovering the cost of investment and environmental issues [3,5,31].

8. Cost of the renewable energy technology

In terms of selection of the capital costs of renewable technologies by 2015; the World Bank Study [32], the market analysis and data tables of the International Energy Agency [33] and report prepared by the Ministry of Environment and Forestry [34] are benefited from. On the other hand, the calculated prices are adjusted for the year 2015 by learning rates for each technology. The learning rates are the decrease in cost of technologies for each doubling of capacity due to technological and operational improvements in these kinds of technologies. The formula used to calculate the future cost of technology is given below as [34]:

$$SCI_F = SCI_P * (C_F/C_P)^{\ln(1-LR/\ln 2)} \tag{1}$$

where SCI_P is the current specific capital investment cost, C_P is the current global installed capacity, C_F is the installed capacity of the technology in a future time, and LR is the learning rate of the technology.

The learning rates for each technology type and estimated 2020 capacities for each technology are taken from the literature [32]. The calculated decrease by 2020 for capital cost of each technology types are given in Table 6.

		Capital cost decrease
Electricity generation technology	**Learning rates**	**rate from 2008 to 2020**
Pvotovoltaics (PV)	24%	13%
Wind-Onshore (good wind)	8%	14%
Wind-Onshore (Moderate wind)	8%	14%
Wind-Offshore	8%	19%
Solar thermal with storage	10%	13%
Geothermal	10%	8%
Biomass Gasifier	5%	8%
MSW/landfill Gas	12%	6%
Biogas	13%	7%
Mini Hydropower	2%	5%

Table 6. Decrease in capital cost by learning rate for each technology types

9. Carbon reduction prices

There are only a few credible studies on future carbon price forecasts. It is assumed that 2010 prices for each generation types which are also given in Table 7 will be applicable in 2020 for voluntary market prices. As given in Table 7, the price projections of some analysts for Phase III period of EU ETS (2010-2020) are used for the CDM/JI and EU ETS prices in 2020.The single projected price of CDM/JI which is 20 €/tCO_2e (25 USD/ tCO_2e) as given in this table, is taken as it is while the average estimation of three different prices (30, 35 and 40 US$) presented is taken as a reference for EU ETS price forecast which is 35 €/tCO_2e (45 USD/ tCO_2e) [35].

As given in Table 8, wind energy potential of Turkey is 50,000 MW. But according to the TEIAS [25], only 8,000 MW of these potential has high level wind speed (i.e. capacity factor is 40%). The entire 40.000 MW potential have a moderate wind speed (ie. capacity factor is 25%-30%).

http://dx.doi.org/10.5772/54319

Project type	Price (USD/tCO_2e)
Pvotovoltaics (PV)	21.50
Wind-Onshore (good wind)	12.40
Wind-Onshore (Moderate wind)	12.40
Wind-Offshore	12.40
Solar thermal with storage	21.50
Geothermal	17.00
Biomass Gasifier	16.20
MSW/Landfill Gas	16.20
Biogas	16.20
Mini Hydropower	5.20

Table 7. Voluntary credit price changes by project type in 2010

Energy sources	Potential (TWh/yr)	Potential (GW)	Utilized (GW)	2030 Target (GW)
Hydropower	180	50	15	180
Wind energy	150	50	1	40
Geothermal	4.2	0.60	0.1	4.2
Solar energy	380	0.1	0.1	10
Biomass	3.6	0.50	0.1	2.2

Table 8. Renewable source potential, utilization by 2010 and target for 2030

10. Results and discussion

According to the result of financial analysis, none of the listed renewable electricity generation technology will be financially attractive without additional carbon finance in 2020. Onshore wind plants in the areas with high level wind speed, landfill gas and biogas power plants will be attractive if they secure emission reduction certification and sell those certificates in the voluntary markets based on the price assumptions. However, the wind projects having smaller capacity factor and geothermal projects can be financially attractive. It is clear that, other than price for mini hydro, at least till 2020, none of this prices are realistic, hence these technologies shall have higher feed-in-tariffs to be more attractive by private investors. The effect of carbon finance as an additional revenue to the renewable electricity generation is analyzed. The renewable electricity generation technologies analyzed are; PV, wind, solar thermal with storage, geothermal, biomass gasifier, MSW/landfill gas, biogas and mini hydropower.

According to the result of financial analysis based on the current VCM conditions, carbon finance opportunities for Turkish renewable projects under the voluntary market would be limited to wind power projects with high speed wind potential and also limited to landfill and biogas projects with financial viability. PV, solar thermal, wind projects with moderate or lower wind speed potential, geothermal, biomass gasifier and mini hydro projects are not projected to be financially attractive even with additional VER revenues based on the VER prices of 2020. The potential electricity generation through those projects is estimated to be around 40,000 MW. The national target for installed capacity for wind projects by 2023 is 20,000 MW but, if the current feed-in-tariff prices are not to be increased, the highest available carbon prices in voluntary market will not be sufficient to enable investments of the wind projects with low speed potential. Hence, the investments would be limited to the 8,000 MW wind power projects which are financially attractive based on their high speed wind potential. The additional 12,000 MW wind capacity are projected to be utilized if CDM/JI like carbon scheme will be applicable by 2020, will result additional reduction of 19 million tCO_2e emissions considering baseline emissions. In addition to the wind, the entire geothermal energy potential for electricity generation (510 MW) and biomass gasifier as well as most of mini hydro (10 MW) potentials can be utilized with any carbon scheme leading emission reduction prices by 2020.

11. Conclusions

Turkey's high rate of energy-related carbon emissions growth is expected to accelerate, with emissions climbing from 57 million tons in 2000 to almost 210 million tons in 2020. Carbon intensity in Turkey is higher than the western developed nation average. Energy-intensive, inefficient industries remain under government control with soft budged constraints, contributing to undisciplined energy use in Turkey.

But the country has made significant progress in reducing local air pollution, particularly in large cities. Nevertheless, significant efforts still need to be made to ensure existing standards are met and to prepare for further reductions in air pollution. The potential long-term impacts of the liberalization process on air pollution and on GHG emissions should be investigated and monitored in order to optimize policy outcomes. The recent construction of a power plant based on fluidized bed combustion technology is laudable. Further adoption of such cleaner coal plants and more efficient technologies would help Turkey meet more stringent air pollution standards. Similar to other industrializing countries, with the increases in energy consumption and economical growth, energy related environmental problems are rapidly growing in Turkey.

Developing countries are likely the most vulnerably to this change because of their less favourable economic circumstances, weaker institutions and more restricted access to capital, technology and information. Given rapid growth of economies and populations, there are a number of implications for developing countries that indicate a need to curb GHGs and thereby

http://dx.doi.org/10.5772/54319

to lessen the impact of climate change. Great efforts have been made in reforming energy pricing, promoting energy efficiency and the use of renewable energy sources. With some possible options, the paper concludes that the reduction of emissions can only be achieved when policies are supportive and well targeted, standards and incentives are realistic and flexible, and the public is actively responsive to environmental degradation.

Author details

İbrahim Yüksel[1*], Kamil Kaygusuz[2] and Hasan Arman[3]

*Address all correspondence to: yukseli2000@yahoo.com

1 Sakarya University, Technology Faculty, Department of Civil Engineering, Sakarya, Turkey

2 Karadeniz Technical University, Department of Chemistry, Trabzon, Turkey

3 United Arab Emirates University, Faculty of Science, Department of Geology, Al-Ain, UAE

References

[1] Yuksel, I., "Energy Utilization, Renewables and Climate Change Mitigation in Turkey", Journal of Energy Exploration & Exploitation, Vol. 26, Number 1, pp. 35 – 51, 2008.

[2] ME, Ministry of Environment, National Report of Turkey (Eds. G. Tüzün and S. Sezer) submitted to "World Summit on Sustainable Development", Johannesburg, 2002.

[3] Yuksel, I. and Sandalci, M., "Climate Change, Energy, and the Environment in Turkey", Journal of Energy Sources, Part A: Recovery, Utilization, and Environmental Effects, Vol. 33, Number 5, pp. 410 – 422, 2011.

[4] Yuksel, I., "Development of Hydropower: A Case Study in Developing Countries", Journal of Energy Sources, Part B: Economics, Planning, and Policy Vol.2, Num. 2, pp.113 – 121, 2007.

[5] UNDP, United Nations Development program, World Energy Assessment Report, New York: United Nations, 2000.

[6] Yuksel, I. and Dorum, A., "The Role of Hydropower in Energy Utilization and Environmental Pollution in Turkey", Journal of Energy Sources, Part A: Recovery, Utilization, and Environmental Effects, Vol. 33, Number 13, pp. 1221 – 1229, 2011.

[7] PEWCLIMATE, "Climate Change Mitigation in Developing Countries: Brazil, China, India, Mexico, South Africa, and Turkey", 2002.

[8] Yüksel, I., Kaygusuz, K. Renewable energy sources for clean and sustainable energy policies in Turkey. Renew Sustain Energy Rev 2011; 15: 4132-44.

[9] TUIK, Turkish Statistical Institute. Turkey Greenhouse gas Inventory, 1990 to 2008, TUIK, Ankara, 2010.

[10] Kaygusuz, K., "Environmental Impacts of Energy Utilization and Renewable Energy Sources in Turkey". Energy Exploration and Exploitation, Vol. 19, pp. 497 – 509, 2001.

[11] MENR, Ministry of Energy and Natural Resources, "Energy report of Turkey", Ankara, Turkey, 2005. (Available from http://www.enerji.gov.tr).

[12] MENR, Ministry of Energy and Natural Resources, "Energy report of Turkey", Ankara, Turkey, 2007. (Available from http://www.enerji.gov.tr).

[13] DPT, State Planning Organization, "Ninth Development Plan 2007-2013", Ankara, Turkey, 2006.

[14] IEA, International Energy Agency, "CO2 Emissions from Fuel Combustion", 2008 Edition, OECD/IEA, 2008. (Available from www.iea.org).

[15] MEF, Ministry of Environment and Forestry, "First National Communication of Turkey on Climate Change", (Eds. Apak, G and Ubay, B) pp. 60 – 150, 2007.

[16] DIE, State Institute of Statistics, "Statistic Yearbook of Turkey in 2005", Prime Ministry, Republic of Turkey, Ankara, 2006.

[17] WECTNC, World Energy Council Turkish National Committee, "Energy report of Turkey in 2002", Ankara, Turkey, 2003.

[18] DIE, State Institute of Statistics, "Statistic Yearbook of Turkey in 2002", Prime Ministry, Republic of Turkey, Ankara, 2003.

[19] Kaygusuz, K., "Climate Change Mitigation in Turkey", Energy Sources Vol. 26, pp. 563-573, 2004.

[20] Yuksel, I., "Global Warming and Renewable Energy Sources for Sustainable Development in Turkey", Journal of Renewable Energy, Vol. 33, Number 4, pp. 802 – 812, 2008.

[21] Kaygusuz, K., "Energy and Environmental Issues Relating to Greenhouse Gas Emissions for Sustainable Development in Turkey", Renewable Sustainable Energy Reviews Vol. 13, pp. 253 – 270, 2009.

[22] ESMAP, Energy Sector Management Assistance Program, "Turkey-Energy and the Environment Review: Synthesis Report", World Bank, 2003.

[23] World Bank, "World Development Report 1992: Development and the Environment", Oxford University Press, 1992.

[24] Say, N. P., "Lignite-fired thermal power plants and SO2 pollution in Turkey", Journal of Energy Policy, Vol. 34, Number 17, pp. 2690 – 2701, 2006.

[25] TEIAS, Turkish Electricity Transmission Company. Turkish electricity statistics in 2010. TEIAS, Ankara, Turkey, 2010, www.teias.gov.tr/ (accessed date 06.09.2011).

[26] Bilgen, S., Kaygusuz, K. and Sari, A., "Renewable Energy for a Clean and Sustainable Future", Energy Sources Vol. 26, pp. 1119 – 1129, 2004.

[27] Kaygusuz, K and Sarı, A., "The Benefits of Renewable in Turkey", Energy sources Vol.1, pp. 23 – 35, 2006.

[28] Kaygusuz, K., "Energy use and Air Pollution Issues in Turkey". Clean Vol. 35, pp. 539-547, 2007.

[29] DSI, State Water Works, "Hydropower Potential in Turkey", Ankara, Turkey, 2005.

[30] Paish, O., "Small Hydro Power: Technology and Current Status", Renewable and Sustainable Energy Reviews, Vol. 6, pp. 537 – 556, 2002.

[31] Yuksel, I., "Hydropower in Turkey for a Clean and Sustainable Energy Future", Journal of Renewable and Sustainable Energy Reviews, Vol. 12, Number 6, pp. 1622 – 1640, 2008.

[32] World Bank. REToolkit: a resource for renewable energy development, Washington, D.C, 2008

[33] IEA, International Energy Agency. World Energy Outlook 2009, IEA, Paris, 2009.

[34] MEF, Ministry of Environment and Forestry. Impact scenarios of carbon finance on the renewable power capacity of Turkey, report prepared by Aslan, R, Ankara, Turkey, 2010.

[35] World Bank. State and trend of the carbon market-2010. World Bank Carbon Finance, May 2010, Washington, DC.

Chapter 2

Energy Savings Resulting from Installation of an Extensive Vegetated Roof System on a Campus Building in the Southeastern United States

Robert W. Peters, Ronald D. Sherrod and
Matt Winslett

Additional information is available at the end of the chapter

http://dx.doi.org/10.5772/55997

1. Introduction

Green building design, as defined by the United States Environmental Protection Agency (USEPA), is "the practice of creating structures and using processes that are environmentally responsible and resource-efficient throughout a building's life-cycle from design, to construction, operation, maintenance, renovation and ultimately deconstruction" [1]. However, according to the United States Green Building Council (USGBC), commercial buildings in the US alone account for:

- 65% of electricity consumption,
- 30% of greenhouse gas emissions,
- 30% of raw materials use,
- 30% of waste output (136 million tons annually), and
- 12% of potable water consumption [2].

The building infrastructure (Residential, Commercial, Institutional, and Industrial) in the United States (US) consumes over two-thirds of the nation's electricity demand and accounts for one-third of all domestic energy consumption [3]. Regrettably, conventional forms of energy production have an adverse impact on natural ecosystems. Collectively, our buildings contribute to 38.9% of the nation's total greenhouse gas emissions. Faced with rising energy costs, diminishing fuel resources and emerging environmental concerns, scientific research has begun to address these challenges by adopting sustainable or green building alternatives.

2. Common roofing systems

The choice and type of urban roofing system is dependent on environmental concerns such as sun exposure and meteorological factors such as temperature, wind and rain. Some roofs may have covenants to determine their usage. Different systems have been developed and designed to perform at its most effective configuration given the exposure conditions of the building location. As a result, many roofing systems exist and are commonly used by the construction community.

There are many roofing systems used today that depend on the building type, whether is for residential or for commercial applications. The most common systems are listed as follows [4]:

- Built-up,
- Preformed metal, and
- Shingles.

A brief description of the three roofing system is provided involving advantages, disadvantages and durability characteristics of each.

2.1. Built-up

Built-up roofing systems are most common for flat roof applications. They consist of several layers of material built-up from the internal structural frame of the building. They are most commonly made of alternating layers of plywood, felt, asphalt, tar and gravel [4]. The major advantage of this system is its waterproofing characteristics, and out of all roofing systems, built-up roofs are considered to have the highest degree of waterproofing. However, the system has poor fire resistance, generally requires a professional to install, and is difficult to locate roof leaks. The system typically lasts for 22.5 years [5]; [6].

2.2. Preformed metal

Another common roofing system is made of preformed metal panels. The panels are generally made from aluminum, steel and copper [4]. They are most commonly found with contemporary designs. They come in flat, ridged, ribbed or corrugated forms. These systems are easy to install and repair, and can be painted any color. They are considered fire resistant. The panels are easily damaged by wind, falling trees and tree limbs, or any other type of contact. They generally last for 25 years [5]; [6].

2.3. Shingles

Shingle roofs come in many forms. Shingles are commonly made of asphalt, slate, wood and clay tiles [4]. Asphalt shingles are the most popular, especially for residential construction. They are available in a variety of sizes, weights, and colors. They require little maintenance and are easy to install. However, they are considered to have poor fire resistive qualities. Asphalt shingles generally last for 16–17 years with proper maintenance [5]. Slate and ceramic tiles are highly expensive due to their aesthetic and durability properties. They are fire resistant and have attractive appearances. Yet, they are extremely heavy and require strong structural support. They are also very brittle, and require increased amount of time to install which often entails specialized tools. Repairs are in most cases difficult. Slate and ceramic tiles are considered the most durable, lasting for 20–100 years with proper maintenance. Wood shingles and shakes are also attractive options. They are easy to install and are a natural insulator. However, they are highly flammable and require treatments for weather and insect protections. They generally last for 10 to 20 years with proper maintenance [5]; [6].

3. Urban heat–island reduction and building energy conservation

For millions of Americans living in and around cities, heat islands are of growing concern. This phenomenon describes urban and suburban temperatures that are 2 to 10°F (1 to 6°C) hotter than nearby rural areas. Elevated temperatures can impact communities by increasing peak energy demand, air conditioning costs, air pollution levels, and heat-related illness and mortality. The Environmental Protection Agency (EPA) recommends installing cool or vegetative green roofs, planting trees and vegetation and switching to cool paving materials as a way of reducing the negative effects of urban heat islands. The EPA says green roofs, if installed widely in a city, can contribute to heat island reduction by replacing heat-absorbing surfaces with plants, shrubs, and small trees. The vegetation cools the air through evapotranspiration (or evaporation of water from leaves). Planted rooftops remain significantly cooler than a rooftop constructed from traditional heat-absorbing materials. Further, green roofs reduce summertime air conditioning demand by lowering heat gain to the building.

Energy modeling (i.e., energy simulation) is a method for predicting the energy consumption of an occupied structure. Building energy analysis must consider numerous thermal characteristics including: wall and roof materials, the size and orientation of the building, how the building is occupied and operated, as well as influences from the local climate.

The surface temperature of a roof exposed to solar radiation, the resulting heat flow into the building, along with associated indoor temperatures and cooling needs depend on the effect of solar radiation, surface absorptivity, ambient air temperature and wind speeds adjacent to the surface [7]. When vegetative roofs are considered, because of added thermal mass, it is also important to take into account hourly heat transfer when determining energy consumption, as the heat flux through a vegetated roof can be quite different from conventional roofing.

R-values and *U*-values have been used for many years as a measurement of a building envelope's thermal performance. However, these attributes do not fully take into account the effects of thermal mass, and by themselves, are inadequate in describing the heat transfer properties of construction assemblies with significant amounts of thermal mass [8]. Vegetated roofs are more dependent on the interaction between the roofing systems' unit weight, density, thermal conductivity, moisture content, vegetal coverage and specific heat. Therefore, it is often necessary to utilize computer software, which incorporates these elements into the analysis of high thermal mass roof structures and associated energy consumption. The steady-state *R*-values traditionally used to measure energy performance will not accurately capture the complex, dynamic thermal behavior of vegetated roof systems.

4. Vegetated (green) roof systems

Green roofs are engineered ecosystems that rely on vegetation to provide benefits such as reduction of roof temperatures and stormwater retention [9]. Green roofs offer benefits of reducing stormwater runoff, improving air and water quality, and providing habitat and biodiversity for urban centers [10]. Hydrologic modeling has demonstrated that widespread green roof implementation can significantly reduce peak runoff rates, particularly for small storm events [11]. By combining the Green-Ampt method with evapotranspiration of green roofs, Roehr and Kong [12] estimated the potential runoff reduction achieved by green roofs is 20%. Green roofs provide an excellent option to improve stormwater runoff [13]. Green roofs are primarily valued based on their increased roof longevity, reduced stormwater runoff, and decreased building energy consumption [14]. Carter and Jackson [11] noted that research studies have primarily been focused on roof-scale processes such as individual roof stormwater retention, plant growth, or growing medium composition. Few studies have examined the impact that widespread green roof application could have on the hydrology of a real-world watershed [11]. A major barrier to increasing the prevalence of green roofs is the lack of scientific data available to evaluate their applicability to local conditions [15].

Green roofs are typically classified as being either an intensive or extensive roof [16]. Intensive green roofs are often used on commercial buildings in order to have large green areas that incorporate all sizes and types of plants. These roofs use grasses, ground covers, flowers, shrubs and even trees. They often include paths and walkways that travel between different architectural features to provide space where people can interact with the natural surroundings. Intensive green roofs, sometimes termed "rooftop gardens", utilize planting mediums that have greater depth than extensive green roofs; the deeper soil allows intensive roofs to accommodate large plants and various plant groupings. Intensive green roofs require more maintenance than extensive green roofs because of the plant varieties they will support.

Extensive green roofs have a planting medium that ranges from 1.6 to 6 inches deep. Typically, drought-tolerant sedums (succulent plants) and grasses are used since they are shallow-rooted and use little water. Plant diversity on these roofs is kept low to simplify care and to be sure all plants have similar moisture requirements.

Extensive green roofs can significantly reduce both the timing and magnitude of stormwater runoff relative to a typical impervious roof [17]. They note, however, that regional climatic conditions such as seasonality in rainfall and potential evapotranspiration can strongly alter the stormwater performance of vegetated roofs. Factors such as type of green roof and its geometrical properties (slope), soil moisture characteristics, season, weather and rainfall characteristics, age of the vegetated roof, and vegetation affect the runoff dynamics from green roofs [18]. Fioretti *et al.* [19] noted that green roofs significantly mitigate storm water runoff generation, as well reducing the daily energy demand. Aitkenhead-Peterson *et al.* [20] note that most studies on runoff quality from green roofs have been conducted in cooler northern climates. Villarreal and Bengtsson [21] recommended the use of a combination of best management practices; additionally, they observed that green roofs are effective at lowering the total runoff from Augestenborg (Sweden) and that detention ponds should successfully attenuate storm peal flows. Niu *et al.* [22] noted that over the lifetime of a green roof (~40 years), the net present value is ~30% to 40% less for a green roofs as compared with conventional roofs (not including green roof maintenance costs). Kirby *et al.* [23] note that extensive vegetated roof systems offer at least 16% enhancement in reducing stormwater runoff as compared to conventional roofs. Clark *et al.* [14] further note that the additional upfront investment of a green roof is recovered at the time when a conventional roof would be replaced. Rosatto *et al.* [24] concluded that green roofs contribute positively in reducing runoff, with greater retention with vegetated plots and thicker substrate.

Vegetated roof systems have a number of advantages over that of conventional roof systems. Benefits associated with green roof systems include [25]:

- Urban greening has long been promoted as an easy and effective strategy for beautifying the built environment and increasing investment opportunity.
- With green roofs, water is stored by the substrate and then taken up by the plants from where it is returned to the atmosphere through transpiration and evaporation.
- Depending on the plants and depth of growing medium, during the summer, green roofs retain 70% to 90% of the precipitation that falls on them; in winter they retain between 25% to 40%.
- Green roofs not only retain stormwater, but also moderate the temperature of the water and act as natural filters for any of the water that runs off.
- Green roofs reduce the amount of stormwater runoff and delay the time at which runoff occurs, resulting in decreased stress on sewer systems at peak flow periods.
- Through the daily dew and evaporation cycle, plants on vertical and horizontal surfaces can cool cities during hot summer months and reduce the Urban Heat Island (UHI) effect. The UHI is also mitigated by the covering some of the hottest surfaces in the urban environment, such as black rooftops.
- Green roofs can also help reduce the amount of dust and particulate matter throughout the city, as well as the production of smog. This plays a role in reducing greenhouse gas emissions and adapting urban areas to a future climate with warmer summers.
- Green roofs help to achieve the principles of smart growth and positively affect the urban environment by increasing amenity and green space.
- The greater insulation offered by green roofs can reduce the amount of energy needed to moderate the temperature of a building, as roofs provide the greatest heat loss in the winter and the hottest temperatures in the summer.
- The presence of a green roof decreases the exposure of waterproofing membranes to large temperature fluctuations, which can cause micro-tearing, and ultraviolet radiation.

- Green roofs have excellent noise attenuation, especially for low frequency sounds. An extensive green roof can reduce sound from outside by 40 decibels, while an intensive one can reduce sound by 46-50 decibels.
- Green roofs can sustain a variety of plants and invertebrates, and provide a habitat for various bird species.

Historically, studies on green roofs have explored their energy performance compared with traditional roofs. Thermal performance indicated a significant reduction (~40%) of a building cooling load during the summer period [26]. Similar results were achieved for a nursery school, with reductions ranging from 6% to 49%, and reduction ranging from 12% to 87% on the last floor of the nursery school [27]. Wong *et al.* [28] note that green roofs tend to experience lower surface temperatures than the original exposed roof, especially in areas well covered by vegetation. When green roofs are well covered by vegetation, the resulting substrate moisture will tend to keep substrate temperature lower than the original exposed bare roof. These studies determined that over 60% of the heat gain was mitigated by vegetated roof systems. Summertime data have indicated significant lower peak roof surface temperature and higher nighttime surface temperature for green roofs as compared to conventional roofs [29]. The maximum average daily temperature seen for the conventional roof surface was 54.4°C (129.9°F) in his study, while the maximum average day green roof surface temperature was 32.8°C (~21.7°C lower than the conventional roof). Green roofs offer cooling potential (~3.02 kWh/day) to maintain an average room air temperature of 25.7°C (78.3°F) [30]. Green roofs help minimize environmental burdens, conserve energy, and extend the life span of the roofing system in overall sustainability [31]. Up to 30% of total rooftop cooling is due to plant transpiration [32]. Bell and Spolek [33] compared different types of plants for use in increasing the thermal resistance (*R*-value) of green roofs, and found that ryegrass delivered the highest effective *R*-value compared with bare soil, *Vinca major, Trifolium repens,* and *Sedum hispanicum.* Also, though increasing the depth of bare soil from 5 to 14 cm (2.0 to 5.5 inches) increased the *R*-value, no difference was found for different depths of planted soil. This implies that the bulk of benefit toward *R*-value is from evapotranspiration and leaf shading, rather than the moist soil [33].

There are several detailed building simulation programs (BSPs) that take into consideration the complete interaction between all thermal-based elements. The most popular BSPs are A Simplified Energy Analysis Method (ASEAM), Building Design Advisor (BDA), Building Load Analysis and Systems Thermodynamics (BLAST), Builder Guide, Bus^{++}, Dynamic Energy Response of Buildings (DEROB), DOE-2, Energy-10, Energy Plus, ENERPASS, ENER-Win, ESP, FEDs, Home Energy Saver, Hot 2000, TRNSYS, and VisualDOE ([34]; [35]; [36]).

UAB has utilized Visual DOE in the past with great success in the analysis of innovative structures designed for energy efficiency. VisualDOE uses the DOE 2 calculating core and provides output in both numerical and graphical forms. This software is a preferred calculation method due to its cost, previous verification/validation success, ease of use, database support and reasonable input/output requirements. We envision that this computer simulation tool will be able to effective capture the differences in roof types being explored in the purposed research.

http://dx.doi.org/10.5772/55997

5. Results and discussion

5.1. Thermal performance of mini-roof structures

5.1.1. Mini-roofs

During this study, 15 mini-roof combinations were observed for trends in internal temperatures. The various 15 mini-roof combinations are summarized in Table 1. Several of the mini-roof structures are depicted in Figure 1. This photo shows the layout of the 15 mini-roofs, and a vegetated roof from which surface temperatures of the mini-roofs were measured periodically using an infrared thermometer (see Figure 1).

Figure 1. a) Layout of the 15 mini-roods; (b) vegetated mini-roof (surface temperatures were measured using an IR thermometer.

The roofing materials used are all standard commercial flat roof materials. Flat roof materials were only looked at during the study, since the primary application for the roofing combinations will be on a commercial flat roof top, and not a slanted roof structure. Each mini-roof is 2.4-m (8.0-ft) long x 1.2-m (4.0-ft) wide x 1.2-m (4.0-ft) deep (see Figure 1). A number of different roofing systems are being examined for their energy performance. All roofs are insulated with 5.1 cm (2.0-in) of extruded polystyrene. Then the particular roofing combination being investigated is applied over the insulation and sealed. The roofs also include a proper drainage spout, to ensure correct water evacuation, such as on a real roof.

The roofing systems being studied using the 15 mini-roofs are listed in Table 1.

Mini-Roof No.	Mini-Roof Description
1 (Sensor B)	White TPO/PVC/ Elvaloy fully adhered, FiberTite Membrane.
2 (Sensor C)	Black 60-mil EPDM fully adhered/coated/white urethane, Mule Hide Membrane.
3 (Sensor D)	Black 60-mil EPDM fully adhered, Mule Hide Membrane.
4 (Sensor E)	Beige TPO/PVC/Elvaloy fully adhered, FiberTite Membrane.
5 (Sensor F)	White granular modified, Firestone SBS Modified Membrane.
6 (Sensor G)	Black granular modified, Firestone SBS Modified Membrane.
7 (Sensor H)	Black granular modified coated/white urethane, Firestone SBS Modified Membrane.
8 (Sensor I)	Bituthene IRMA with lightweight "T Clear" pavers.
9 (Sensor J)	Bituthene IRMA with river rock ballast.
10 (Sensor K)	Bituthene IRMA with vegetative green roof, ½-in. drain mat, 350-lbs dry soil.
11 (Sensor L)	Bituthene IRMA with vegetative green roof, 1-in. drain mat, 350-lbs dry soil.
12 (Sensor M)	Black 60-mil EPDM loose, ballasted with river rock, Mule Hide Membrane.
13 (Sensor N)	Black 60-mil EPDM loose, ballasted with #300 marble chips, Mule Hide Membrane.
14 (Sensor O)	White TPO/PVC/Elvaroy loose laid, ballasted with river rock, FiberTite Membrane.
15 (Sensor P)	Bituthene IRMA with vegetative green roof, ½-in. drain mat, 350-lbs dry soil.
Sensor S	Sensor inside mini-roof No.10 (inside the soil of the green roof).
Sensor T	Sensor under the white TPO/PVC/Elvaloy loose laid, ballasted with river rock.

Notation – TPO: thermoplastic polyolefin; SBS: styrene-butadiene-styrene; PVC: polyvinyl chloride; EPDM: ethylene propylene diene monomer; IRMA: inverted roof membrane assembly.

Table 1. Mini-roof descriptions.

To investigate the thermal properties of the roofing structures an ambient temperature probe was placed inside of each roof (see Figures 2 and 3) recording temperature data every 10 minutes of each day, for more than 3 years. This data was then automatically sent to a data logger and placed into an Excel file for review later. The temperature probe reports the temperature to the nearest hundredth of a degree Centigrade.

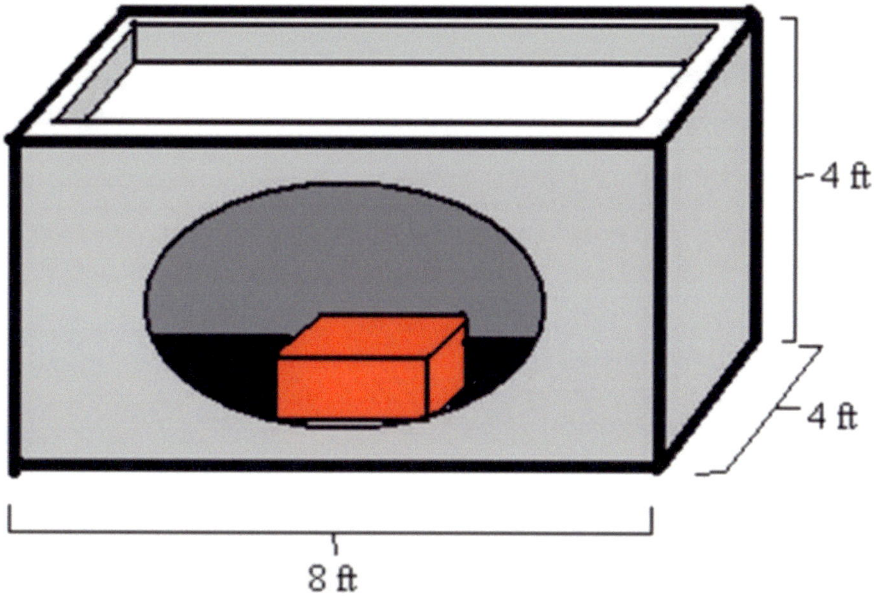

Figure 2. Depiction of a Typical Mini-Roof System.

Figure 3. Mini-roof system showing temperature sensor installed inside a mini-roof.

Figure 4 presents some typical temperature profiles on several different days.

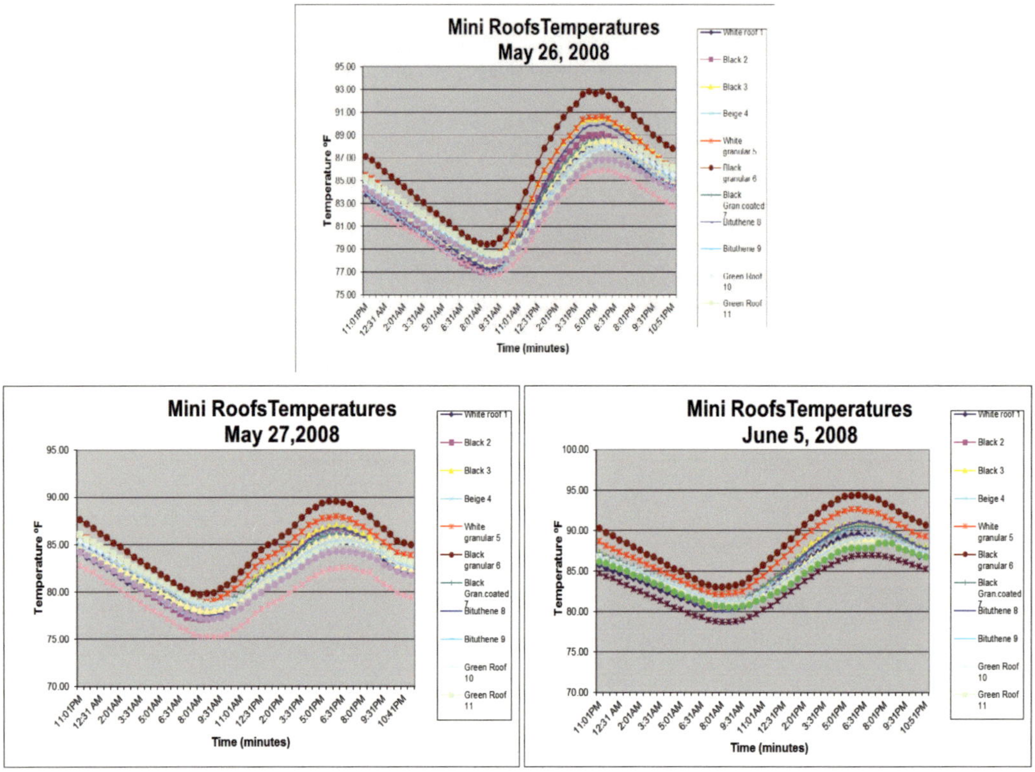

Figure 4. a. Temperature profile inside the various mini-roofs on May 26, 2008. b. Temperature profile inside the various mini-roofs on May 27, 2008 [37]. c. Temperature profile inside the various mini-roofs on June 5, 2008.

The results from these mini-roof structures have shown the following trends [37]:

- Clean white roofs resulted in consistently lower temperatures inside the mini roof than the other roofing materials.
- Black roofs resulted in the highest temperature readings.
- Green roofs resulted in temperatures typically ~1.1-1.7°C (2-3°F) higher than the white roofs; however, they will dampen the drainage of rainfall during a rain storm through the retention of water onto the soil.
- Bituthane (river rock) performs only slightly better than black roofing materials.
- White granular roofing behaved similarly to black granular roofing materials.

- Over time, the reflective (white) roofs become dirty, losing some of their reflectivity, resulting in the roof being less energy efficient.
- White marble chips behaved slightly worse than green roofing materials, but considerably better than black roofing and granular roofing materials.
- While it is too early to come to a definitive conclusion, preliminary evaluations indicate that the "white" and "green" roofs both significantly reduce the roofs surface temperature and therefore, the air temperatures above and around the roof.

Surface temperature measurements during the months of June and July on the various roofing materials used with our mini-roof systems were collected. During this time period, roofing surface temperatures ranging from 20.6°F to 82.2°C were observed. The lower surface temperature values, 20.6°C to 48.3°C, are found in a loose rock or stone roof combination. The higher temperature values occurred on the roofing combinations primarily made of a coating or membrane; these roofs exhibited temperatures roughly between 25.6°C to 82.2°C. The vegetative mini-roofs exhibited surface temperatures ranging between 21.7°C to 52.2°C, while their large counterpart, the pilot roof on top of the University of Alabama at Birmingham (UAB) Hulsey Center, exhibited higher temperatures ranging from 31.7°C to 61.1°C. When looking at the specific roofs contained in each subset of Table 1, it was observed that generally, the lighter colorization of the roof resulted in cooler temperatures. For example, the white Firestone SBS is cooler than the black Firestone SBS on any given day. This thermal property is observed due to the reflectivity of the roof. The darker colored roofs absorb more incoming light radiation than the light colored roofs causing the dark roofs to become hotter. (Since the roofs temperatures observed are taken during the late spring and early summer, it can be inferred that overall roof temperature will increase during late summer).

Photographs of several of these mini-roofs are presented in Figures 5 through 9. Figure 10 presents a typical temperature profile of the 15 mini-roofs during the course of a typical summer week. Series 1 through 5 denoted in the figure refer to the fifteen mini-roofs listed in Table 1. This figure shows a cyclical nature of the temperature readings over each day, generally showing a sinusoidal behavior of temperature; the temperature is cool in the morning, warms up, and is at its hottest during mid-afternoon, and then cools down during evening hours.

Trends seen in statistical comparisons of internal temperatures of similar roofs (using the null hypotheses: $\mu_1 \geq \mu_2$) are summarized below [38]:

- Between the two river rock roofs, Roof 9 will most likely always be hotter or equal to roof 12.
- Between the 3 vegetative roofs, all the roofs are statistically equal to each other in thermal properties.
- Both SBS Firestone roofs are statistically the same, but roof 6 is usually hotter.
- The TPO/PVC/Elvaloy Roofs are statistically the same.
- The 60-mil EPDM roofs are also statistically the same.

Figure 5. Black mini-roof.

Figure 6. White (reflective) mini-roof.

Figure 7. Mini-roof equipped with river rocks.

Figure 8. Mini-roof equipped with crushed marble chips.

Figure 9. Vegetative mini-roof equipped with sedum plants.

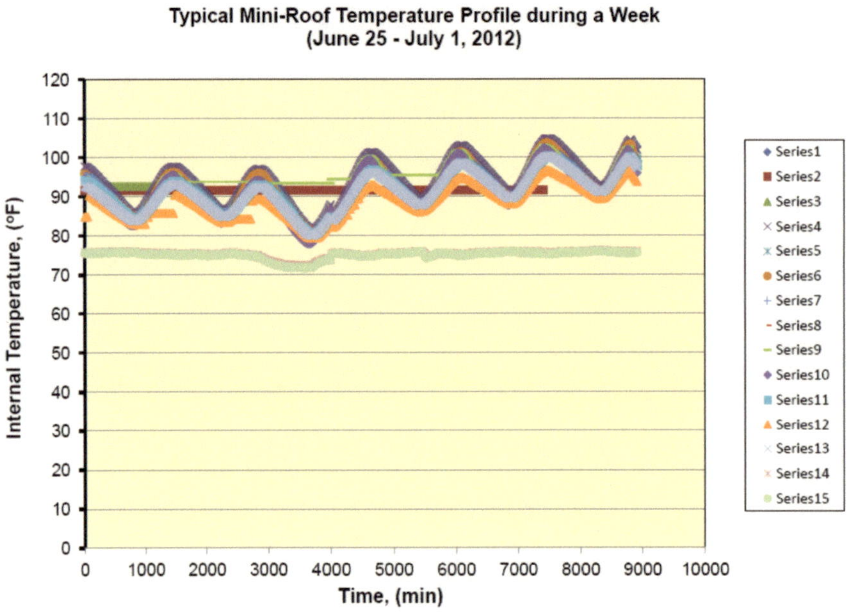

Figure 10. Typical internal mini-roof temperatures during the course of a week.

After acquiring the necessary raw data from the mini-roof sensors, Microsoft Excel (version 2010) was used to plot the data to discover trends in the temperature readings. Subsequent to discovering the trend, a mathematical model was fit to the data. It was noticed that the temperatures cycled in a sinusoidal fashion on yearly and daily time frames, and therefore a general form sine function was utilized as a potential model. Fourier transforms were utilized in order to determine the oscillation frequency of the temperature (ω_0 from the general form $x = A \sin(\omega_0 t + \phi) + C$) by transforming the time domain of the collected raw data into a frequency domain. Two major peaks were discovered from the spectrum: one representing the yearly frequency and the other representing the daily oscillations. After calculating the frequencies, the phase angle (ϕ) and the amplitude (A) of the general form were determined through a regression analysis.

The modeling procedure was applied for all 15 mini-roofs in the study. The developed sine-wave functions indicated that most roofs were statistically different from one another from an amplitude aspect but the phase angles were statistically the same. It was also discovered that almost all roofs had significantly different average mean roof temperatures, but the significance was mostly prevalent in the summer months. During other times of the year, the roofs behaved in a similar fashion (see Tables 2 and 3 for statistics and means of roofs). The fitted sine wave functionalities for the 15 mini-roofs are listed in Table 4.

Statistics of Phase Angles	
Mean	4.0303
Standard Error	0.00343
Median	4.0276
Mode	4.0242
Standard Deviation	0.01329
Sample Variance	0.000177
Kurtosis	12.0174
Skewness	3.3228
Range	0.05512
Minimum	4.0211
Maximum	4.0762
Sum	60.4541
Number (count)	15

Table 2. Summary of statistics for the phase angles from curve fitting.

Statistics of Means	
Mean	66.67
Standard Error	0.18
Median	66.62
Mode	N/A
Standard Deviation	0.68
Sample Variance	0.46
Kurtosis	3.11
Skewness	1.11
Range	3.06
Minimum	65.42
Maximum	68.47
Sum	1000.12
Number (count)	15

Table 3. Summary of statistics for the means from curve fitting.

Roof	Final Equation	Amplitude from sinefind.exe	Phase Angle	r^2	Mean
1	T = 20.1632*sin ((2*π*3.17894 e^{-8} x t) +4.021975)+66.68283	20.163	4.022	0.696	66.683
2	T = 20.5499*sin ((2*π*3.18 e^{-8} x t) +4.02272)+66.63564	20.550	4.023	0.701	66.636
3	T = 20.392*sin ((2*π*3.18 e^{-8} x t) +4.07624)+66.47947	20.392	4.076	0.712	66.479
4	T = 20.482*sin ((2*π*3.18 e^{-8} x t) +4.021125)+66.35505	20.482	4.021	0.698	66.355
5	T = 20.4891*sin ((2*π*3.18 e^{-8} x t) +4.03214)+67.54093	20.349	4.032	0.701	67.541
6	T = 20.7812*sin ((2*π*3.18 e^{-8} x t) +4.03142)+68.46883	20.781	4.031	0.699	68.469
7	T = 20.2886*sin ((2*π*3.18 e^{-8} x t) +4.02415)+67.08382	20.289	4.024	0.701	67.084
8	T = 20.7812*sin ((2*π*3.18 e^{-8} x t) +4.02415)+66.18969	20.781	4.024	0.695	66.190
9	T = 20.3285*sin ((2*π*3.18 e^{-8} x t) +4.024975)+66.91944	20.329	4.025	0.701	66.919
10	T = 19.8608*sin ((2*π*3.18 e^{-8} x t) +4.02758)+66.6981	19.861	4.028	0.697	66.698
11	T = 20.0128*sin ((2*π*3.18 e^{-8} x t) +4.026215)+66.62328	20.013	4.026	0.697	66.623
12	T = 20.0479*sin ((2*π*3.18 e^{-8} x t) +4.027853)+66.27383	20.048	4.028	0.694	66.274
13	T = 19.9914*sin ((2*π*3.18 e^{-8} x t) +4.0325)+66.47272	19.991	4.033	0.696	66.473
14	T = 20.0661*sin ((2*π*3.18 e^{-8} x t) +4.032494)+65.42343	20.066	4.032	0.689	65.423
15	T = 19.4628*sin ((2*π*3.18 e^{-8} x t) +4.028672)+66.27177	19.463	4.029	0.692	66.272

Table 4. Fitted sine-wave functionalities describing the internal temperatures in the mini-roofs.

http://dx.doi.org/10.5772/55997

These results of this study were in agreement with research conducted by Watson [39] addressing the urban heat island effect for the City of Birmingham, Alabama. By studying the effect from different building facades, building materials and seasonal traits. The data indicates amplitudes (A) ranging from 19.5 to 20.7 degrees and phases angles roughly 4.2 to 4.3 radians. The r^2 values ranged from 0.69 to 0.71showing the data modeling was applicable to understanding the trends. The best model developed was for roof 3 (see Figure 11), having the highest correlation coefficient of all 15 mini-roofs.

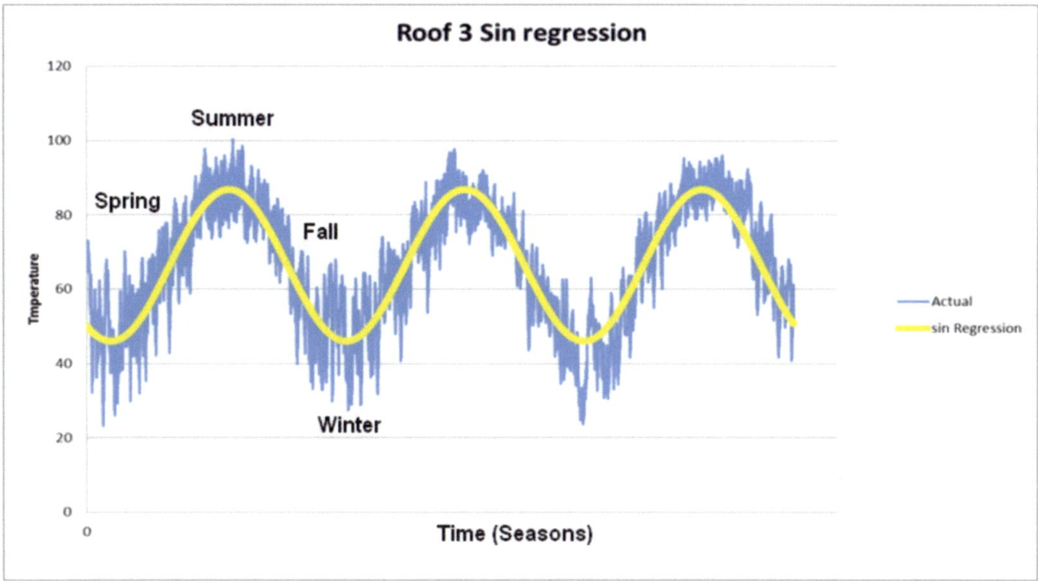

Figure 11. Fitted sine wave regression for a mini-roof system.

Based on the information collected on these mini-roof systems, the UAB Facilities Management Department decided to install an extensive vegetated roof on top of Hulsey Center as a roofing retrofit. UAB wants to obtain more fundamental information and knowledge for establishing green roofs in the southeastern U.S. In the Birmingham, Alabama area, fairly high rainfalls [approximately 132.1 cm/year (52 inches/year) on average] are obtained. However, during the summer months, it is common to have periods of drought with minimal rainfall and very hot and humid days [with temperatures in the 32.2+°C (90+°F range]. Such climatic conditions require plants that can withstand both significant rainfall events and drought conditions.

Photographs of the construction of the pilot green roof system on Hulsey Center are shown in Figures 12 to 14. Photographs of the system taken in June 2009 are shown in Figures 15 to 17.

Figure 12. Construction phase for installing a pilot vegetative roof on top of Hulsey Center.

Figure 13. Construction phase for installing a pilot vegetative roof on top of Hulsey Center.

http://dx.doi.org/10.5772/55997

Figure 14. Initial vegetative roof immediately after installation on top of Hulsey Center.

Figure 15. Vegetative roof on top of Hulsey Center (June 2009).

Figure 16. Vegetative roof on top of Hulsey Center (June 2009).

Figure 17. Close-up of vegetative roof on top of Hulsey Center (June 2009).

Of the roofing area of 1709.4 m ^{2} (18,400 ft ^{2}) for Hulsey Center, an extensive pilot green roof has been installed on ~1388.0 m ^{2} (14,940 ft ^{2}), i.e., occupying ~81.2% of the roofing area, or approximately 1375.9 m ^{2} (0.34 acres). [Prior to installing the pilot green roof, the UAB civil engineering senior design class investigated the loading of a wet vegetative roof on top of Hulsey Center, and found that the current building infrastructure could withstand the loading associated with the vegetative roof. Hulsey Center was originally designed to have two additional floors in the building]. This pilot green roof contains more than 20,000 sedum plants (*Sedum hispanicum*). In the construction of this vegetated roof, the existing roof was removed down to the structural concrete deck. A waterproof roofing membrane was installed directly to the concrete deck. A layer of 7.6-cm (3-inches) of extruded polystyrene roofing insulation was then applied, to which a 226.8-gm (8-oz) non-woven geo-textile scrim sheet was applied. Then, 0.6-m (2-ft) x 0.6-m (2-ft) x 5.1-cm (2-in) prestressed pavers or brick pavers were installed for design, decoration, and access to the vegetative roof. Then, 8.9-cm (3.5-inches) of light weight engineered soil were applied. Sedum plants were planted at the rate of 1615 plants per 100 m ^{2} (150 plants per 100-ft ^{2}). The cost of retrofitting the roof and installing a pilot green roof on top of Hulsey Center was ~$150,000 (USD).

The current pilot green roof on top of Hulsey Center was installed in July 2008. Research conducted at Penn State University shows that green roofs planted with sedum plants reduce the building energy costs [40]. UAB's Facilities Management Department maintains records of the utilities bills (e.g., natural gas, water, and electricity) for each building on campus. The costs of utilities prior to and after implementation of the pilot green roof are shown in Figures 18 through 20 for natural gas, water, and electricity. Since 2006, the costs of natural gas, water, and electricity have increased by 47%, 28%, and 145%, respectively. Due to the increasing rates of these utilities, a more accurate determination is shown with the usage of these utilities, shown in Figures 21 through 23. In the graph symbols, those indicated with a red interior color depict the cost or usage after installation and implementation of the vegetative roof system. The other symbol colors depict the cost or usage prior to installation and implementation of the vegetative roof system. Generally, the usage of natural gas, water, and electricity are lower than that compared to the utility usage prior to installation of the green roof on top of Hulsey Center.

The current pilot green roof on top of Hulsey Center was installed in July 2008. Research previously conducted at Penn State University indicated that green roofs planted with sedum plants reduce the building energy costs [41]. The Facilities Management Department at UAB maintains records of the utilities bills (e.g., natural gas, water, and electricity) for each building on campus. The quantities and costs of utilities prior to (5 years) and after implementation (~3 years) of the pilot green roof have resulted in building energy reductions of 20% to 25%.

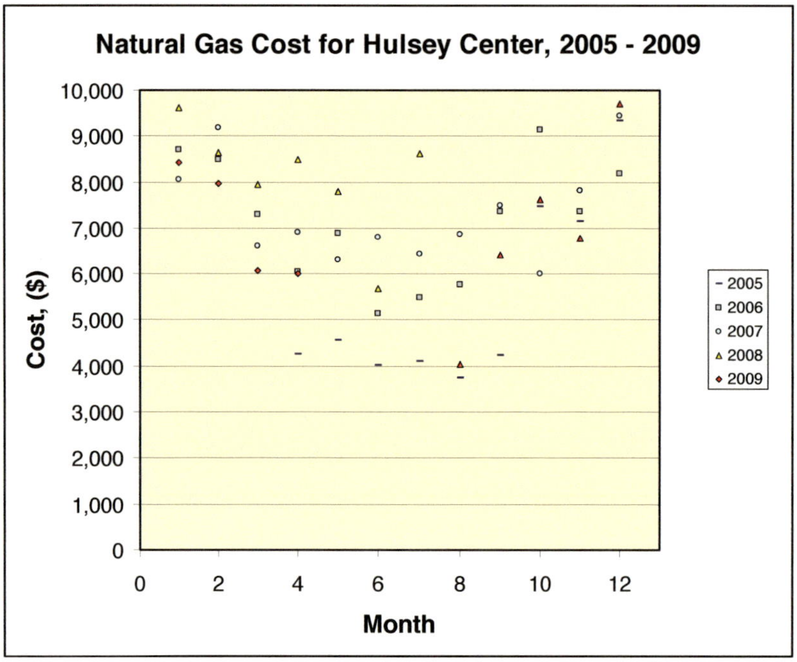

Figure 18. Cost of natural gas for operation of Hulsey Center, 2005 – 2009.

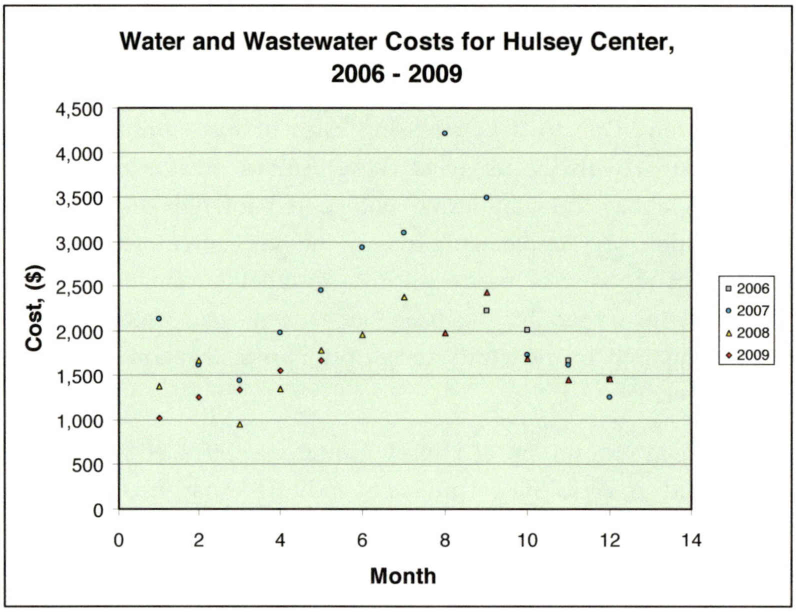

Figure 19. Cost of water/wastewater for operation of Hulsey Center, 2006 – 2009.

http://dx.doi.org/10.5772/55997

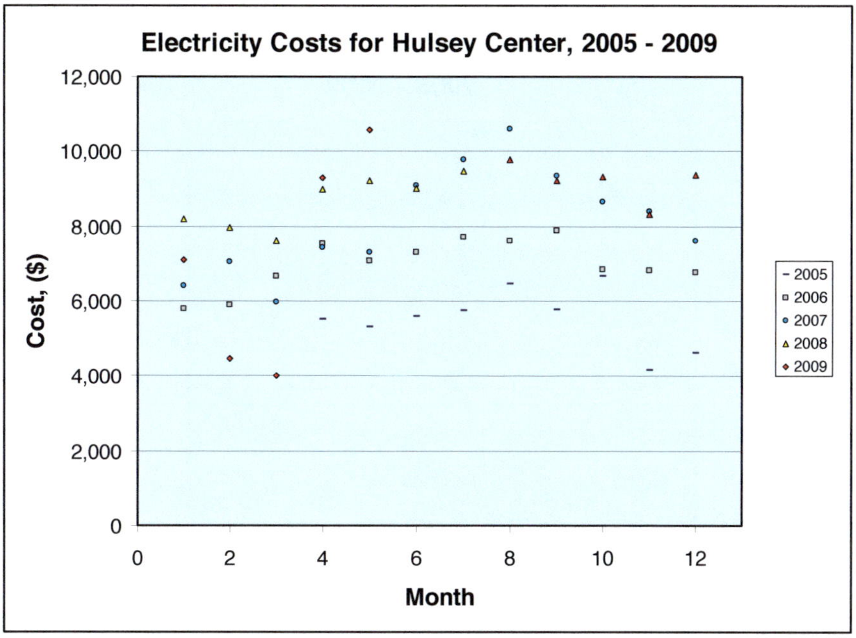

Figure 20. Cost of electricity for operation of Hulsey Center, 2005 – 2009.

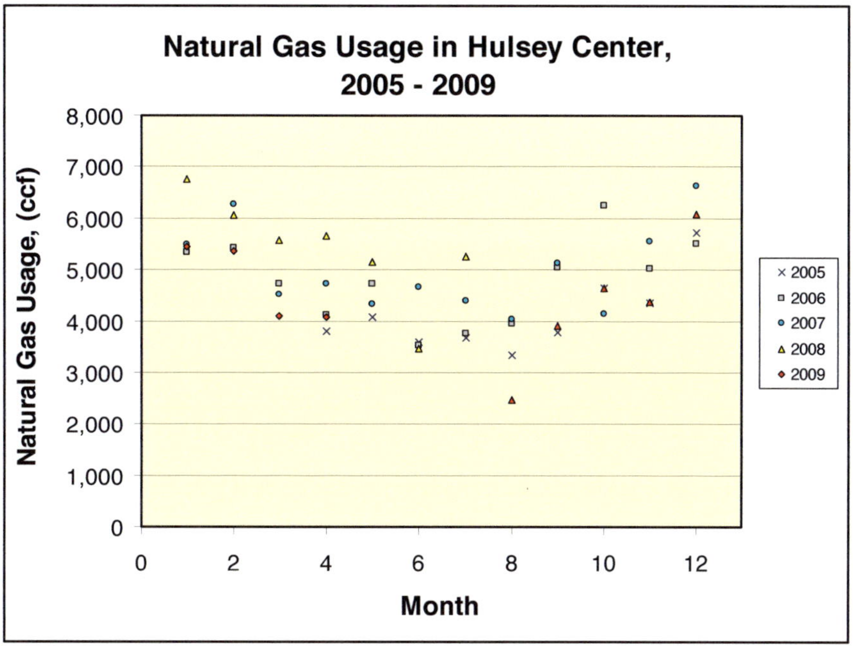

Figure 21. Natural gas usage for operation of Hulsey Center, 2005 – 2009.

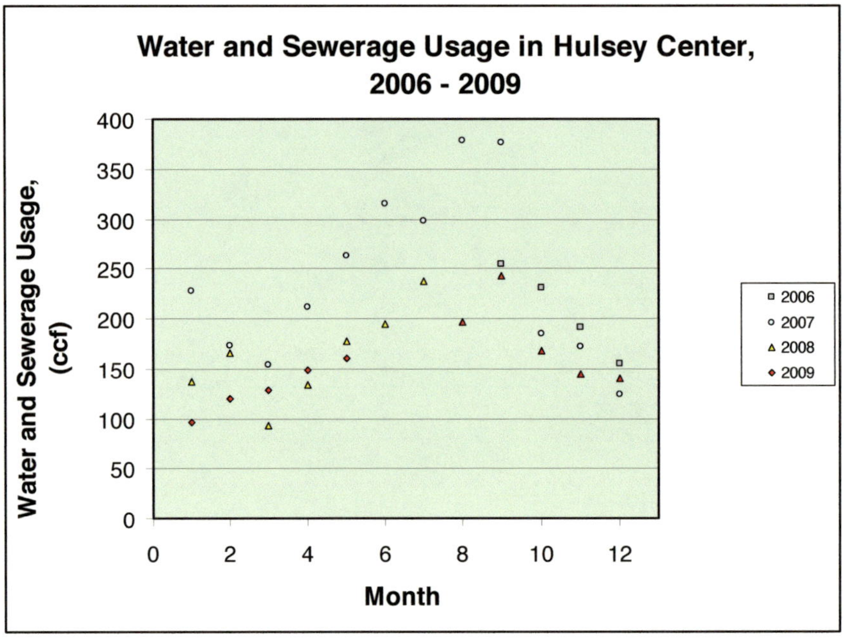

Figure 22. Water/wastewater usage for operation of Hulsey Center, 2006 – 2009.

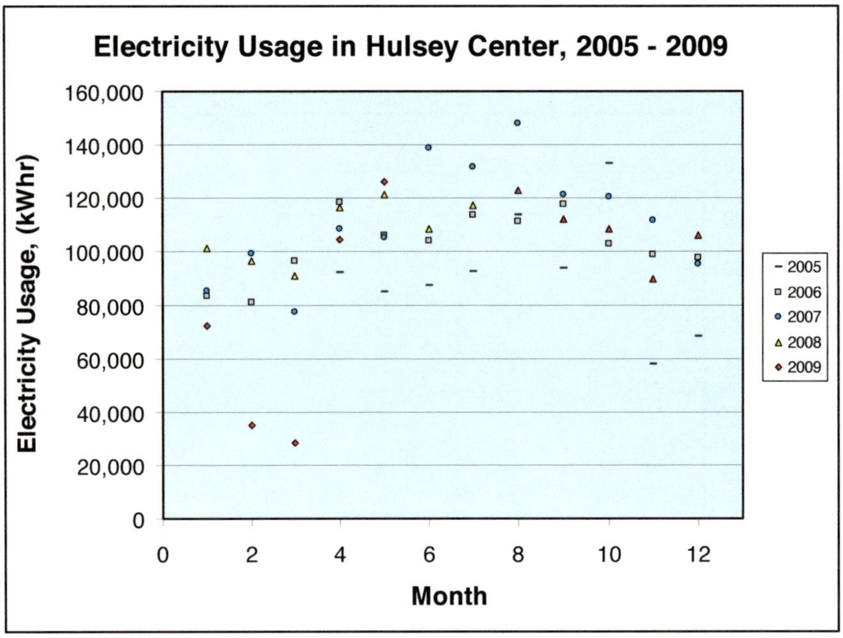

Figure 23. Electricity usage for operation of Hulsey Center, 2005 – 2009.

6. Summary and conclusions

The behavior of different roofing materials affects the heating loads placed upon the buildings and their heating, ventilation, and air conditioning (HVAC) systems. Black roofs resulted in the highest temperature readings. Black roofing materials and bituthane (river rock) perform the poorest of the roofing materials tested, resulting in the highest heating load being placed on the building infrastructure. White granular roofing behaved similarly to black granular roofing materials. Clean white roofs resulted in consistently lower temperatures inside the mini-roof than the other roofing materials; however, over time, the reflective (white) roofs become dirty, losing some of their reflectivity, resulting in the roof being less energy efficient. Green (vegetated) roofs are fairly efficient in terms of energy performance due to evapotranspiration effects. Green roofs resulted in temperatures typically ~1.1-1.7°C (2-3°F) higher than the white roofs; however, they will dampen the drainage of rainfall during a rain storm through the retention of water onto the soil, and thereby lessen the discharge into stormdrains. Green roofs can significantly reduce stormwater runoff, reduce peak flow quantities, and lengthen the time of concentration from roofing structure [23]. "White" and "green" roofs both significantly reduce the roofs surface temperature and therefore, the air temperatures above and around the roof.

Hypothesis testing indicated that, between the 3 vegetative mini-roofs, all the min-roofs are statistically equal to each other in thermal properties. Both SBS Firestone mini-roofs are statistically the same, but mini-roof 6 is usually hotter. The TPO/PVC/Elvaloy mini-roofs are statistically the same. The 60-mil EPDM mini-roofs are also statistically the same.

The temperature varies in a sinusoidal fashion both during the course of the day and on an annual basis. For the various mini-roof structures, the phase angle (ϕ) and the amplitude (A) of the general form were determined through a regression analysis. The developed sine-wave functions indicated that most roofs were statistically different from one another from an amplitude aspect but the phase angles were statistically the same. It was also observed that almost all roofs had significantly different average mean roof temperatures, but the significance was mostly prevalent in the summer months. During other times of the year, the roofs behaved in a similar fashion.

Based on the information collected on these mini-roof systems, the UAB Facilities Management Department installed a vegetated roof on top of Hulsey Center as a roofing retrofit. The vegetated roof is ~1388.0 m^2 (14,940 ft^2) in area, and contains approximately 20,000 sedum plants. Utility bill information both prior to and after implementation of the green roof were gathered for electricity, natural gas, and chilled water were collected. The costs of utilities prior to and after implementation of the pilot green roof indicated utility bill (energy) savings of ~20% to 25% (compared to the case prior to implementation of the vegetated roof system).

Acknowledgements

The financial support and technical support of the UAB's Facilities Management Department is appreciated, including the fabrication and installation of the mini-roofs (on top of the Chemistry Building) and pilot roof on top of Hulsey Center. The input and technical assistance provided by Matt Winslett and Kelly Winnett of the Facilities Management Department in supplying the utility bill information is gratefully acknowledged.

Author details

Robert W. Peters[1*], Ronald D. Sherrod[2] and Matt Winslett[3]

*Address all correspondence to: rwpeters@uab.edu

1 Department of Civil, Construction and Environmental Engineering, University of Alabama at Birmingham, Birmingham, Alabama, USA

2 Department of Chemical Engineering, Worcester Polytechnic Institute, Worcester, Massachusetts, USA

3 Energy Management, Facilities Management Department, University of Alabama at Birmingham, Birmingham, Alabama, USA

References

[1] Environmental Protection Agency (EPA)(2010). *Environmental Protection Agency Green Building Design*. www.epa.gov/greenbuilding/.

[2] United States Green Building Council (USGBC)(2010). United States Green Building Council. www.usgbc.org.

[3] Environmental Protection Agency (EPA)(2009). *Buildings and their Impact on the Environment: A Statistical Summary*. www.epa.gov/greenbuilding/pubs/gbstats.pdf.

[4] Reid, R. N. (2000). *Roofing and Cladding Systems: A Guide for Facility Managers*, Fairmont Press, Lilburn, GA.

[5] Cash, C. G. (1999). The Relative Durability of Low-Slope Roofing, in *Proceedings of the Fourth International Symposium on Roofing Technology*, Gaithersburg, MD, (September 17–19)., 119-124.

[6] Editors of Time Life Books(1979). *Roofs and Siding*, Time-Life Books, Alexandria, VA.

[7] Givoni, B. (1994). *Passive & Low Energy Cooling of Buildings*, John Wiley & Sons, Inc., New York, NY.

[8] Kosny, J, Yarbrough, D, Childs, P, & Mohiuddin, S. (2007). How the Same Wall Can Have Several Different *R*-Values: Relations between Amount of Framing and Overall Thermal Performance in Wood And Steel-Framed Walls, *Oak Ridge National Laboratory*.

[9] Wolf, D, & Lundholm, J. T. (2008). Water Uptake in Green Roof Microcosms: Effects of Plant Species and water Availability, *Ecol. Engrg., 2179186*, 33

[10] Wong, N. H, Tay, S. F, Wong, R, Ong, C. L, & Sia, A. (2003). Life Cycle Cost Analysis of Rooftop Gardens in Singapore, *Building and Environment, 3499509*, 38

[11] Carter, T, & Jackson, C. R. (2007). Vegetated Roofs for Stormwater Management at Multiple Spatial Scales, *Landscape and Urban Planning, 80* (1-2): 84-94.

[12] Roehr, D, & Kong, Y. (2010). Stormwater Runoff Reduction Achieved by Green Roofs: Comparing SWMM Method to TR-55 Method, in *Low Impact Development 2010: Redefining Water in the City, Proceedings of the 2010 International Low Impact Development Conference*, San Francisco, CA, (April 11-14)., 1012-1021.

[13] Hathaway, A. M, Hunt, W. F, & Jennings, G. D. (2008). A Field Study of Green Roof Hydrologic and Water Quality Performance, *Transactions Am. Soc. Agric. & Biolog. Engineers, 13744*, 51

[14] Clark, C, Adriaens, P, & Talbot, F. B. (2008). Green Roof Valuation: A Probabilistic Economic Analysis of Environmental Benefits, *Environ. Sci. Technol., 621552161*, 42

[15] Williams, N. S. G, Nicholas, J. P, & Raynor, K. J. (2010). Green Roofs for a Wide Brown Land: Opportunities and barriers for Rooftop Greening in Australia, *Urban Forestry & Urban Greening, 3245251*, 9

[16] Green Roof Planundated. Website: http://www.greenroofplan.com/intensive-vs-extensive-green-roofs/,accessed August 3, (2012).

[17] Schroll, E, Lambrinos, J, Righetti, T, & Sandrock, D. (2011). The Role of Vegetation in Regulating Stormwater Runoff from Green Roofs in a Winter Rainfall Climate, *Ecol. Engrg., 4595600*, 37

[18] Berndtsson, J. C. (2010). Green Roof Performance towards Management of Runoff Water Quantity and Quality: A Review, *Ecol. Engrg., 4351360*, 36

[19] Fioretti, R, Palla, A, Lanza, L. G, & Principi, P. (2010). Green Roof Energy and Water Related Performance in the Mediterranean Climate, *Build. and Environ., 818901904*, 45

[20] Aitkenhead-Peterson, J. A, Dvorak, B. D, Voider, A, & Stanley, N. C. (2011). Chemistry of Growth Medium and Leachate from Green Roof Systems in South-Central Texas, *Urban Ecosystems, 11733*, 14

[21] Villarreal, E. L, & Bengtsson, A. S. D. L. (2004). Inner City Stormwater Control Using a Combination of Best Management Practices, *Ecol. Engrg., 22* (4-5): 279-298.

[22] Niu, H, Clark, C, Zhou, J, & Adriaens, P. (2010). Scaling of Economic Benefits from Green Roof Implementation in Washington, D.C., *Environ. Sci. Technol., 1143024308,* 44

[23] Kirby, J. T, Peters, R, Watts, S, & Winslett, M. (2011). Experimental Investigation of Runoff Mitigation from Vegetated Roofs, Final report submitted to the Alabama Water Resources Research Institute, Auburn, AL.

[24] Rosatto, H. G, Laureda, D, Perez, D, Nerrera, D, Meyer, M, Gamboa, P, Villalba, G, Friedrich, M, Bargieta, M, Plaza, L. R, Calvo, G, Miranda, M, Inigo, M, & Quaintenne, E. (2010). Water Retention Efficiency of Green Roof Systems, *Revista de la Facultad de Ciencias Agarias, 1213219,* 42

[25] Green Roofs for Healthy Cities (undated). Website: http://www.greenroofs.org/index.php/about-green-roofs/aboutgrnroofs/,accessed August 3, (2012).

[26] Spala, A, Bagiorgas, H. S, Assimakopoulos, M. N, Kalavrouziotis, J, Matthopoulos, D, & Mihalakakou, G. (2008). On the Green Roof System. Selection, State of the Art and Energy Potential Investigation of a System Installed in an Office Building in Athens, Greece, *Renew. Energy, 1173177,* 33

[27] Santamouris, M, Pavlou, C, Doukas, P, Mihalakakou, G, Synneta, A, Hatzibiros, A, & Patargias, P. (2007). Investigating and Analysing the Energy and Environmental Performance of an Experimental Green Roof System Installed in a Nursery School Building in Athens, Greece, *Energy, 91781788,* 32

[28] Wong, N. H, Tan, P. Y, & Yu, C. (2007). Study of Thermal Performance of Extensive Rooftop Greenery Systems in the Tropical Climate, *Build. Environ., 12554,* 42

[29] Sonne, J. (2006). Evaluating Green Roof Energy Performance, *ASHRAE J., 25961,* 48

[30] Kumar, R. (2005). Performance Evaluation of Green Roof and Shading for Thermal Protection of Buildings, *Build. Environ., 1115051511,* 40

[31] Cavanaugh, L. M. (2008). Redefining the Green Roof, *J. Archit. Eng., 146,* 14

[32] Takakura, T, Kitade, S, & Goto, E. (2000). Cooling Effect of Greenery Cover over a Building, *Energy Build., 31*: 1–6.

[33] Bell, H, & Spolek, G. (2009). Measured Energy Performance of Greenroofs, in *Proc. Seventh Annual Greening Rooftops for Sustainable Communities Conference,* The Cardinal Group, Toronto, Canada.

[34] Selim, M. A. (2003). The Effect of Building Envelope Technology on Energy Conservation in Buildings, M.S. Thesis. Cairo University. Cairo, Egypt.

[35] Selim, R. (2008). Investigation of Thermal Performance, Indoor Air Quality and Energy Consumption within Tuskegee Healthy House Demonstrative Model, Ph.D. Dissertation,. University of Alabama at Birmingham, Birmingham, AL.

[36] Mansour, H. M. A. (2001). Sustainable Housing Development with Special Reference to Environmental Aspects in Low Cost Housing in Egyp". M.S. Thesis, Cairo University, Cairo, Egypt. 117 pages.

[37] Nunez, S, & Peters, R. W. (2008). Effect of Roofing Materials on Building Energy Efficiency, Paper included in the conference proceedings of the 2008 Annual AIChE Meeting, Philadelphia, PA, (November 16–21).

[38] Peters, R. W, Price, J. G, Sherrod, R. D, Watts, S. A, Gohlke, J. M, Kirby, J. T, & Winslett, M. (2011). Roofing Types and Their Effects on the Thermal Properties of Runoff, *Proc. Alabama Water Resources Conference*, Orange Beach, AL, (September 8-9).

[39] Watson, C. J. (2011). Sustainable Urban Design Practices that Mitigate Urban Heat Islands and Reduce Energy Consumption, Ph.D. Dissertation, University of Alabama at Birmingham, Birmingham, AL.

[40] Penn State Center for Green Roof Researchundated (a). "Utility Bills Getting You Down? Plant Stuff Up on the Roof, http://aginfo.psu.edu/news/2000/9/roof.html;accessed July 5, (2009).

[41] Penn State Center for Green Roof Researchundated (b). Pennsylvania State University website: http://horticulture.psu.edu/cms/greenroofcenter/about_ctr.html;accessed July 5, (2009).

Chapter 3

On the Public Policies Supporting Renewables and Wind Power Overcapacity: Insights into the European Way Forward

António Cardoso Marques,
José Alberto Fuinhas and Rui Flora

Additional information is available at the end of the chapter

http://dx.doi.org/10.5772/52159

Research supported by: NECE, R&D unit funded by the FCT - Portuguese Foundation for Science and Technology, Ministry of Education and Science.

1. Introduction

The debate about the depth and characteristics of public policy intervention in encouraging renewable energy is more urgent than ever. Some of the literature has started to cast doubt on the strategies being pursued by countries to encourage renewables. These criticisms have been varied with regard to the costs of intervention and the consequences for the economy as a whole. The need to guarantee a continuous supply of electricity requires the existence of backup power, which is essentially based on fossil sources. Marques and Fuinhas (2012a) point out that the increase in renewables has been predominantly based upon direct public subsidies and intervention. Market-driven policies have been deprecated in favor of policy-driven measures. Fossil sources are often identified as beneficiaries of subsidies, so renewable sources should also be stimulated by these policy instruments. However, the way in which this argument is presented can lead to confusion. In fact, a substantial part of these subsidies for fossil fuels is a consequence of the strategy to develop renewables. Fossil fuels are used to provide backup power, such as that from coal plants or combined-cycle gas-fired plants, but they are turned off for long periods. Consequently, overcapacity and economic inefficiency arise, which is the primary reason for the current subsidies for fossil fuels. Some literature, such as Liao et al. (2011), sustains that all incentives/subsidies should be removed, both for fossil fuels and renewables. The authors then propose applying fees to fossil-based products in order to pay for the emission of greenhouse gases. Despite being pragmatic and objective, this perspective is not easy to apply considering the current state of renewable energy technology. In fact, several challenges remain, such as the problem of intermittent generation, which will continue to require the use of fossil sources to offset it.

The deployment of new energy sources is inevitable, which is why alternative energy sources have emerged, resulting from the use of natural resources such as water, wind, sun, tides and heat from the earth. In the meantime, the international community has made several commitments to promoting greater deployment of renewable energy sources. Examples include the Kyoto Protocol in 1997 or the recent European directive for climate and energy measures, known as the 20-20-20 targets. This directive aims to reduce greenhouse gas emissions by 20%, increase the share of energy consumption from renewable energy sources to 20% and promote energy efficiency by reducing primary energy use by 20%. Nevertheless, while the need to develop the use of new energy sources has been consensual, in view of the long-term depletion of fossil fuels and the issue of climate change, there has been no extensive evaluation of the consequences of renewable energy use. The continuous support for renewables, particularly in Europe, has raised the debate about the levels of wind power installed capacity. It is well known that there is no kind of power plant that operates at 100% of its maximum capacity, particularly renewables due to their intermittent nature. Hence, a problem of structural inefficiency arises, which leads to high economic costs associated with the existence of idle capacity.

It is widely accepted that renewables, such as solar and wind, are at the heart of common instruments in reaching European goals of reducing energy dependence, as well as the reduction of greenhouse gases. However, the growth in wind energy magnifies the problem of intermittency. As stated by Holttinen et al. (2009), it is crucial to properly estimate the costs of wind energy in the system as a whole when planning high wind power penetration. Furthermore, the analysis of renewables' intermittent generation is important for policymakers due to the great support for renewables in Europe, in the context of long-term energy goals. Faced with the problem of renewable intermittency, two possible solutions can be considered: (i) energy storage for later use; and (ii) backup electricity generation with fossil fuels. However, the literature (e.g. Beaudin et al., 2010) suggests that energy storage costs are still very high, so upgrading the energy grid in this way is not yet attainable. As a consequence, it would seem more appropriate to combine other energy sources to backup power. Fossil fuel plants can startup and shut down in a short time to keep a secure energy supply and are an effective way of mitigating renewable intermittency (Isla, 1999; and Luickx et al., 2008). A consequence of the divergence between maximum capacity in full-time operation and the electricity actually generated in a given period of time is the idle capacity phenomenon. Idle capacity is noticeable both for renewables, due to their intermittent nature, and for other energy sources since they are turned off more frequently.

We shall focus only on wind power overcapacity. The scarce research on this subject arouses curiosity about the high levels of idle capacity in wind power. Indeed, Boccard (2009), Fiedler and Bukovsky (2011) and Yang et al. (2012) found that wind energy generation is rarely more than 25% of total capacity. This leads us to believe that there may be overcapacity in wind power. The ratio of the actual electricity output to maximum capacity is referred to as the capacity factor. Electricity demand throughout the day is volatile, especially in off-peak and peak-load periods. This may also influence the amount of idle capacity. This problem can force investment into pumped hydro during periods when there is wind overproduction and low grid consumption and secondly into thermal plants like coal-based or gas-fired to provide backup power for wind power when necessary (Luickx et al., 2008).

The vast literature about renewable intermittency, both theoretical and case studies, has not explored the phenomenon of wind overcapacity in enough detail. Indeed, the empirical assessment of overcapacity in wind power merits much more attention. On the one hand, the phenomenon of overcapacity reflects the path taken by renewables and, on the other, since this phenomenon is a leading indicator, it should back the process of updating public guidance. The aim of this chapter is to provide empirical evidence on the drivers that contribute to explaining wind power overcapacity and, secondly, to identify empirically the causes for a panel of 19 countries. While Boccard (2009) addresses the issue of wind intermittency from the perspective of the capacity factors, this chapter is focused on the importance of intermittency and possible wind overcapacity, but using a non-used wind capacity approach.

The role played by several energy sources in creating wind power overcapacity is assessed, controlling for socio-economic drivers and public energy policies and measures. On the whole, this approach can be useful in highlighting the relevance of the intermittent nature of renewables for policymakers in order to deal with wind overcapacity. Econometric techniques of panel data were applied to deal with the energy and socio-economic characteristics of an economic bloc with environmental concerns and long-term energy targets. In particular, the contribution of conventional energy sources to the wind power overcapacity in Europe is appraised. Some light is shed on the public policies that might mitigate the economic inefficiency.

2. Renewables' intermittency context: The debate

The expansion of renewables is the subject of hot debate in the literature regarding the implications of these energy sources, namely their advantages, consequences and prospects for growth. The implications of the unpredictability and inconstancy of wind energy generation prove relevant. In fact, this intermittency in generation makes it increasingly important to combine different energy sources, including fossil fuels, to backup energy supply. A relevant role is merited not only for conventional energy sources, but also for the mix of renewables. Moreover, it is crucial to understand the role that public policy and measures have played. Wind power installation has been strongly stimulated by public guidance and highly subsidized, namely by guaranteed prices under feed-in tariffs which will last for more than 25 years, as stated by Moreno and Martínez-Val (2011). Together with other drivers that promote renewables on a large scale, this creates distortions and increased costs for consumers (Gómez et al., 2011).

2.1. Intermittency and wind power overcapacity

Although the issue of renewable intermittency is far from new in the literature, the relevance of this topic together with the phenomenon of overcapacity requires much more research. The main reasons and impacts of non-constant generation of wind energy are analyzed by authors such as Albadi and El-Saadany (2010), and Green and Vasilakos (2010). Gonzalez et al. (2004) focus on Ireland, Gül and Stenzel (2005) on Scandinavia, the United Kingdom and the United States, Caralis et al. (2008) on Greece and the Chinese case is targeted by Yang et al. (2012) and Zhang and Li (2012).

Intermittency in renewables can be analyzed by using the capacity factor. This is the ratio, for a certain period of time, of the energy generated to the energy that would have been generated in operation by operating total continuous power during the same period (Denholm et al., 2005). Boccard (2009) summarizes that capacity factor depends on: (i) wind variability; (ii) the shadowing phenomenon; and (iii) the intensive focus on subsidy policies. The shadowing phenomenon comes from installing too many wind turbines in a limited area to save costs on land use. Moreover, the short distance between wind farms compromises the individual performance of each farm. The vast use of public financial support policies may have led to fast, but inefficient, wind energy deployment.

Acker et al. (2007) noted that a seasonal influence in the capacity factor can be observed. Caralis et al. (2008) analyzed the capacity factors in Greece and suggest that spatial dispersion of wind farms benefits the wind power capacity factor. They concluded that the accumulation of too many wind farms is not always the optimal solution because it may impair the efficiency of each individual wind farm. More recently, Yang et al. (2012) and Zhang and Li (2012) assessed wind power growth in China, which was driven by three main factors: (i) the perception that China benefits from large wind resources; (ii) the adoption of incentives and subsidies that support the investment in wind power; and (iii) the reduction in wind capital costs. The authors note that more attention to the efficiency of wind turbine allocation in China is needed. In fact, one-third of wind turbines were idle, causing a capacity factor of 16.3% between 2007 and 2010 (Yang et al., 2012).

2.2. Backup and energy storage

It is important to seek new ways to deal with wind speed variability, both in the short and long term. Examples could be additional energy sources to backup power in windless periods or energy storage devices (Purvins et al., 2011). To ensure a secure energy supply, it is necessary to mix wind power with other energy sources, including fossil fuels. Pearce (2009) suggests a solar photovoltaic system mixed with combined heat and power to overcome intermittency in California without resorting to energy storage. Moreno and Martínez-Val (2011) argue that thermal power plants are no longer so important in base load energy generation, turning them into backup sources to substitute renewables. These authors support that by 2020, backup with combined cycle gas turbine plants needs to grow to 8 or 9 Gigawatts. The literature (e.g. Archer and Jacobson, 2007) also mentions another method to smooth wind variability. These authors found that by interconnecting multiple wind parks

http://dx.doi.org/10.5772/52159

through the electricity transmission grid, wind farms behave more similarly over time as a single wind farm with constant wind speed, providing a constant and secure energy supply.

As regards the impact of energy policies and measures on the deployment of renewables, a few studies have provided empirical evidence. Carley (2009) uses the fixed-effect vector decomposition, which is a variant of the fixed-effects model, and finds that the sum total of United States energy policies does not contribute significantly to more electricity from renewables. However, the growth of renewables is promoted by each additional year that a State maintains a policy. A positive relationship between the expansion of wind energy and the adoption of energy policies that promote investment and subsidies is found by Menz and Vachon (2006). Regarding European countries, Marques and Fuinhas (2012a) prove that policies subsidizing the promotion of renewables have been effective in doing so. Overall, they argued that this process is driven by political willingness rather than by economic rationality.

3. Wind capacity, energy sources and European public policies

Wind energy growth in the last decade in Europe was mainly driven by several factors such as: energy demand growth; the commitments made to greenhouse gas reduction under the Kyoto protocol directives; improvements in renewable energy technology; and the reduction of the marginal cost of wind power generation over the past 15 years, approaching the cost of conventional energy sources (Pechak et al., 2011). For these reasons, wind power has registered a strong impulse since the late 1990s and early 2000s. As a consequence, due to the lack of data before 1998 for almost all European countries, this study uses panel data for the time span 1998-2009, for the following countries: Austria, Belgium, the Czech Republic, Denmark, Finland, France, Germany, Greece, Hungary, Ireland, Italy, Latvia, the Netherlands, Norway, Poland, Portugal, Spain, Sweden and the United Kingdom. These countries are part of a group that is driven by long-term energy goals under European directives (EU directive, 2009). Not all countries have the same number of observations due to sporadic missing values, which leads to an unbalanced panel. The remaining countries of the EU27 did not provide available data for wind power installed capacity in the considered time span.

Panel data techniques have several advantages, such as: (i) they allow a more accurate statistical inference; (ii) they provide more informative data and variability; (iii) they increase the number of observations and degrees of freedom; and (iv) they allow for controlling individual heterogeneity and unobserved characteristics of errors which are not detectable in time-series or cross-sectional models (Baltagi, 2005 and Hsiao, 2006).

3.1. Wind capacity

For a better approach to the issue of intermittency and overcapacity, it proved necessary to make the concept of overcapacity operational. To do so, a variable which emulates wind overcapacity ($WOCAP_{c,t}$) was created.

$WOCAP_{c,t}$ is the dependent variable and represents the ratio of idle capacity in a year to the hypothetical maximum energy that could be produced in a year, in a continuous full-power operation. This ratio was computed from raw data, and can be done in two different ways: (i) through idle capacity; and (ii) through capacity factor. Accordingly, for option (i) the result is:

$$WOCAP_{c,t} = \frac{IDCAP_{c,t}}{TOTALCAP_{c,t}}. \tag{1}$$

In equation (1) $TOTALCAP_{c,t}$ is the total of wind installed capacity. $IDCAP_{c,t}$ denotes the idle capacity of wind power in a year. In other words, $IDCAP_{c,t}$ represents the difference between maximum possible wind electricity generation during the year (8760 hours) and the amount of electricity actually generated. $TOTALCAP_{c,t}$ and $IDCAP_{c,t}$ are expressed in Megawatts (MW) and this last one is computed as follows:

$$IDCAP_{c,t} = \frac{(WINDCAP_{c,t}*8760) - (TOTELECGEN_{c,t}*1000)}{8760}, \tag{2}$$

where $TOTELECGEN_{c,t}$ is the total electricity generated in a year, in Gigawatts *per* hour (GWh). $TOTELECGEN_{c,t}$ is multiplied by 1000 to convert to same units.

Regarding option (ii) $WOCAP_{c,t}$ can be computed as the difference between 1 and the capacity factor ($CF_{c,t}$) as follows:

$$WOCAP_{c,t} = 1 - CF_{c,t}. \tag{3}$$

The capacity factor is computed as follows:

$$CF_{c,t} = \frac{TOTELECGEN_{c,t}*1000}{TOTCAP_{c,t}*8760}. \tag{4}$$

In expressions (3) and (4) $CF_{c,t}$ is the ratio of actual wind power to maximum capacity in a year. For example, for Germany and Spain, which are the leader countries in terms of wind installed capacity, in 2009 the total installed capacity was respectively 25777 MW and 18988 MW. Electricity output was 38637 GWh and 36851 GWh. From here, following equation (2) for Germany and Spain, $IDCAP_{c,t}$ in 2009 was:

$$\frac{(25777*8760) - (38637*1000)}{8760} \approx 21366.3836\,MW, \tag{5}$$

$$\frac{(18988*8760) - (36851*1000)}{8760} \approx 14781.2648\,MW. \tag{6}$$

In accordance with equation (1), wind overcapacity ratios ($WOCAP_{c,t}$) for Germany and Spain are given as:

$$\frac{21366.3836}{25777} \approx 0.8289, \tag{7}$$

$$\frac{14781.2648}{18988} \approx 0.7785. \tag{8}$$

For Finland and Latvia, which are the countries with the lowest wind installed capacity (147 MW and 29 MW respectively) with electricity output of 277 GWh and 49 GWh, in 2009 $IDCAP_{c,t}$ was:

$$\frac{(147*8760) - (277*1000)}{8760} \approx 115.3790\,MW, \tag{9}$$

$$\frac{(29*8760) - (49*1000)}{8760} \approx 23.4064\,MW. \tag{10}$$

Hence, wind overcapacity ratios ($WOCAP_{c,t}$) for these two countries respectively are given as:

$$\frac{115.3790}{147} \approx 0.7849, \tag{11}$$

$$\frac{23.4064}{29} \approx 0.\ 8071. \tag{12}$$

Our computations indicate that 82.89 %, 77.85 %, 78.49 % and 80.71 % of the wind installed capacity was idle during the year, i.e., a capacity factor of 17.11 %, 22.15 %, 21.51 % and 19.29 % respectively for Germany, Spain, Finland and Latvia. These values are relatively high. Indeed, it is surprising that this issue has not been addressed earlier with more emphasis in the literature. Average $WOCAP_{c,t}$ values for all countries of our panel for the time span 1998-2009 are presented in Figure 1. Wind overcapacity average values are in line with other authors who addressed capacity factors, like Boccard (2009) and Yang et al. (2012). For example, in Denmark and Portugal, the average $WOCAP_{c,t}$ is 0.7790 and 0.7840 respectively, and according to (4) the capacity factor is 0.2210 and 0.2160. It denotes that Nordic countries (e.g. Norway, Sweden, Finland, Denmark, the United Kingdom and Ireland) as well as southern Europe (e.g. Portugal, Spain and Greece) have less idle capacity and therefore more capacity factors than continental countries. This may be because of higher wind speeds in these regions.

3.2. Variables

Several causes for idle capacity are suggested by the normative literature. Following this closely, the impact of variables with different natures is controlled for, such as: conventional energy sources; other renewable sources; socio-economic drivers; and energy efficiency measures and public policies as follows.

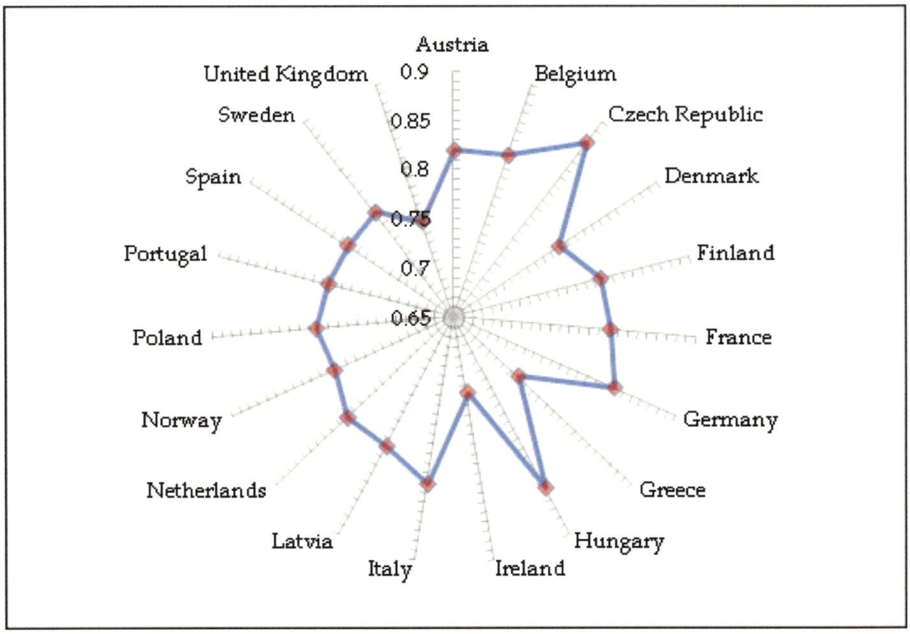

Figure 1. Average $WOCAP_{c,t}$ for the time span 1998 – 2009

3.2.1. Conventional energy sources

To assess the impact of conventional energy sources on wind overcapacity, the shares of fossil energy sources in total electricity generation across European countries were used (see Figure 2 for average values). The variables are for coal-based power plants ($COALSH_{c,t}$), gas-fired ($GASSH_{c,t}$) and oil power plants ($OILSH_{c,t}$). The literature (e.g. Luickx et al., 2008; Østergaard, 2008; Larraín et al., 2010; and Purvins et al., 2011) argues that these variables are the main sources used to backup wind power, especially coal and gas. In fact, gas turbines can be used as a backup source for wind power in windless periods because their startup times are in the order of a few minutes while other conventional power plants may take much longer (Kehlhofer et al., 2009). It is expected that these variables will be highly significant in explaining wind overcapacity. Nuclear power is also part of conventional energy sources. The impact of nuclear capacity factor ($CFNUCL_{c,t}$) in wind overcapacity (computed according to (4)) is controlled. Nuclear power still has great importance in Europe, despite its capacity factor reduction by 7.9% between 1998 and 2009. The toxic waste that comes from nuclear power and the fact that it is difficult to treat as well as risk of disaster have recently brought the debate to Germany to reduce its share of nuclear power in electricity generation.

Figure 2 suggests that the 19 countries included in our study still have a large share of conventional energy sources in total electricity generation, except Nordic countries and Portugal, which have been at the forefront of the support in renewables, namely wind and solar energy.

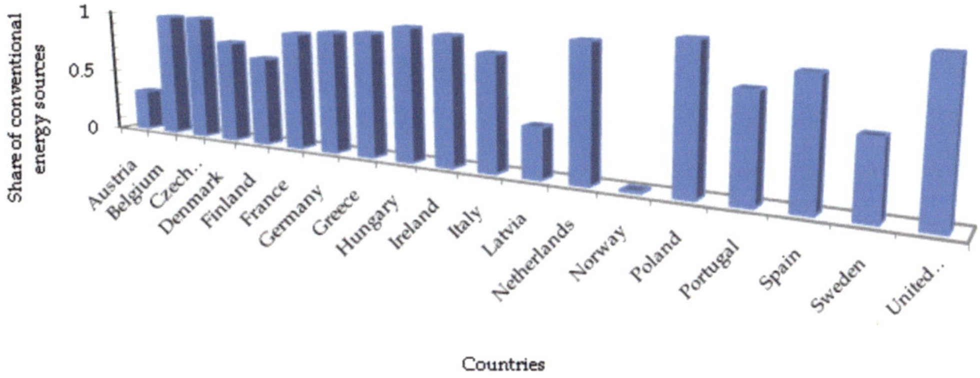

Figure 2. Average conventional energy sources share for the time span 1998-2009

3.2.2. Other renewable energy sources

To assess the impact of renewables on wind overcapacity, our option is to use variables representing the most common renewable energy sources such as hydropower, renewable waste and solar energy. Figure 3 presents the growth rate of renewables' share in the 19 countries under analysis. Regarding hydropower, the effect of the capacity factor of hydropower ($CFHYD_{c,t}$) is controlled. This variable was computed similarly to the nuclear capacity factor, according to expression (4) to avoid multicollinearity problems. In the context of the Europe 20-20-20 targets, renewable waste and solar energy have been increasingly used to generate electricity (Münster and Meibom, 2011). To assess the impact of these two energy sources on wind overcapacity, the effect of the share of waste ($WASTSH_{c,t}$) and solar ($SOLSH_{c,t}$) in the total electricity generated is controlled. We also sought to ascertain the impact of installing more wind power over the years through the growth rate of wind power installed capacity ($WINDGR_{c,t}$). It is expected that the overcapacity of wind power will be positively influenced by more wind power plants.

In Figure 3, values suggest that, in Europe, there has generally been large growth in the share of electricity generated from renewable energy sources. Negative values may indicate that in these countries hydro power is becoming less important in the energy portfolio. For example, Portugal, Germany, Denmark and Ireland have average rates of growth of 10.75%, 9.57%, 10.63% and 11.55% respectively, which indicates huge support in electricity generation from renewables in the early 2000s.

3.2.3. Socio-economic drivers

Potential socio-economic drivers such as population density or economic development were controlled for in order to assess their effect on wind overcapacity (see Figures 4 and 5 for average values). According to Caralis et al. (2008) and Boccard (2009), the spatial dispersion of wind farms may be an important driver for a greater or lesser capacity factor. To control for the effect of spatial dispersion of wind farms, a proxy, the variable $POPDENS_{c,t}$ is used.

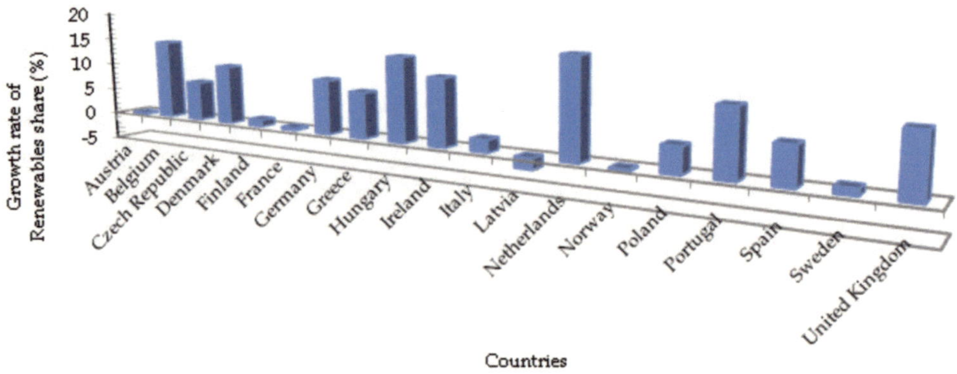

Figure 3. Average rate of growth in renewables' share in total electricity generation (including Hydro) for the time span 1999-2009

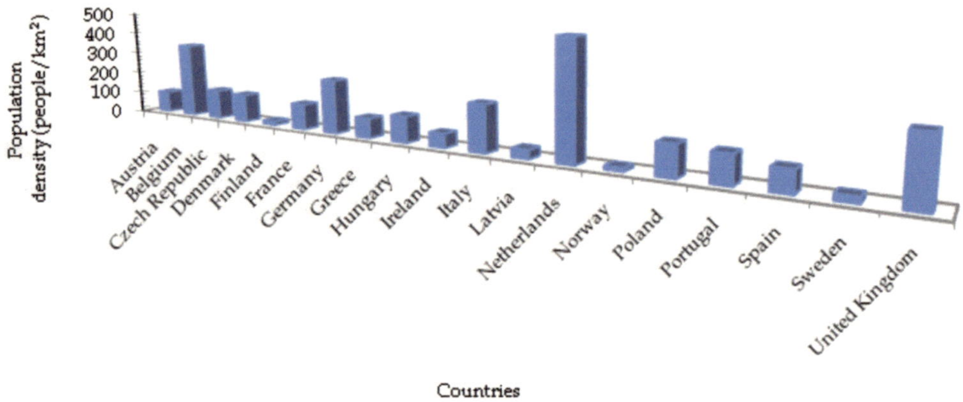

Figure 4. Average population density for the time span 1998-2009

This variable assesses the effect of available and suitable land area for wind park installation in countries with greater or lesser population density. The Netherlands, followed by the United Kingdom, reveal the largest population density in the panel.

Regarding economic development, the natural logarithm of Gross Domestic Product (GDP) *per capita* ($LNGDPPC_{c,t}$) is used to measure the capacity of European countries to invest in more efficient energy generation technologies. It is expected that more developed countries will have greater available financial resources to invest in more efficient energy sources, such as offshore wind parks.

http://dx.doi.org/10.5772/52159

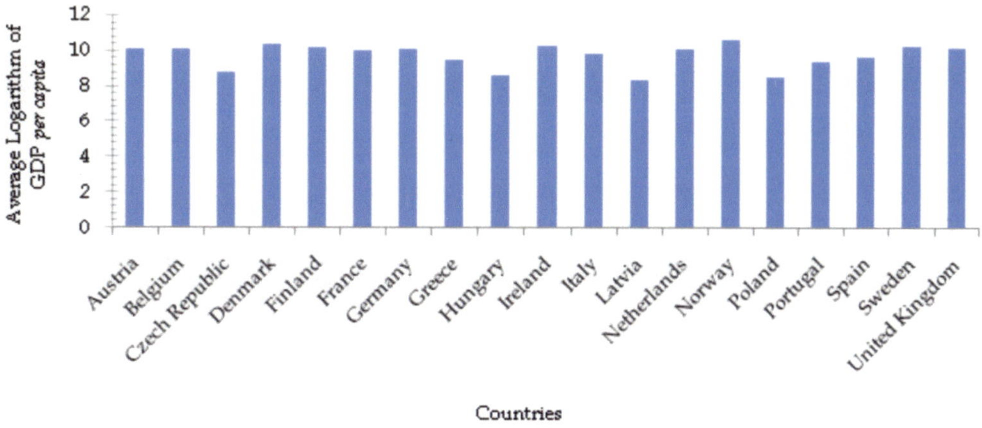

Figure 5. Average natural logarithm of GDP *per capita* for the time span 1998-2009

Latvia is the country with the smallest average natural logarithm of GDP *per capita*, while Norway holds the largest value for this indicator.

3.2.4. *Energy efficiency measures and public policies*

Energy policies and measures have been widely used to promote and support the deployment of renewables. Data from the Mesures d'Utilisation Rationnelle de l'Energie (MURE) database was collected, which provides information concerning the amount and the impact of these measures in order to control for the influence of energy policies on overcapacity. The variables are the cumulative amount of measures taken per household and in the industrial and tertiary sectors, as presented in Figure 6. Firstly, the total energy policies and measures carried out in a year ($ALLPOL_{c,t}$) were considered. It is expected that the total measures may have a positive impact on wind power overcapacity. For a deeper analysis of public policies, the total energy policies are divided into seven individual types to assess the influence of each type individually: (i) Legislative/normative ($NORMPOL_{c,t}$) stands for mandatory standards for buildings, regulations for heating systems and hot water systems, regulations in the field of building and mandatory standards for electrical appliances; (ii) legislative/informative ($INFOPOL_{c,t}$) aims to inform about energy efficiency, mandatory standards in buildings and electrical appliances; (iii) fiscal/tariff ($FISCPOL_{c,t}$) measures include tax exemptions/reductions in retrofitting investments; (iv) incentives/subsidies ($FINPOL_{c,t}$) includes feed-in tariffs, grants and loans. A positive effect of these measures on $WOCAP_{c,t}$ is expected, due to their contribution to renewables' deployment which may positively influence wind overcapacity; (v) information/education ($EDUPOL_{c,t}$) measures aim to provide campaigns by energy agencies and energy suppliers; (vi) co-operative measures ($COOPPOL_{c,t}$) include voluntary programs; and (vii) cross-cutting measures ($CUTPOL_{c,t}$) are the eco-tax on energy consumption or CO2 emissions and other eco-taxes.

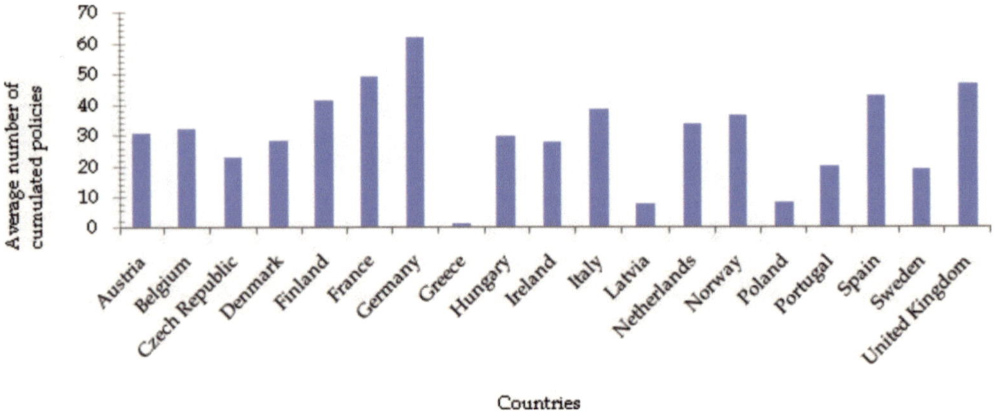

Figure 6. Average number of accumulated energy policies and measures for the time span 1998-2009

Variable	Definition	Source	Obs.	Mean	SD	Min.	Max.
$WOCAP_{c,t}$	Ratio of non-used output to the maximum possible output over a year	EUROSTAT	221	0.7956	0.0543	0.5947	0.9912
$COALSH_{c,t}$	Ratio of elect. gen. to coal (TWh)/total elect. gen. (TWh)	IEA	227	0.2940	0.2483	0	0.9636
$GASSH_{c,t}$	Ratio of elect. gen. to gas (TWh)/total elect. gen. (TWh)	IEA	227	0.2161	0.1761	0.0015	0.6339
$OILSH_{c,t}$	Ratio of elect. gen. to oil (TWh)/total elect. gen. (TWh)	IEA	227	0.0564	0.0770	0.0001	0.4243
$CFNUCL_{c,t}$	Ratio of average plant output to the maximum possible output over a year	IEA	228	0.4324	0.4164	0	0.9659
$CFHYD_{c,t}$	Ratio of average plant output to the maximum possible output over a year	IEA	228	0.2884	0.1231	0.0948	0.6223
$WASTSH_{c,t}$	Ratio of elect. gen. to waste (TWh)/total elect. gen. (TWh)	IEA	227	0.0308	0.0348	0	0.1486
$SOLSH_{c,t}$	Ratio of elect. gen. to solar (TWh)/total elect. gen. (TWh)	IEA	227	0.0004	0.0018	0	0.0210
$WINDGR_{c,t}$	Yearly growth rate of wind installed capacity	EUROSTAT	223	50.5234	98.1664	-7.1429	1000
$POPDENS_{c,t}$	Population density (people/km2)	World Bank, World Development Indicators Database	228	139.6083	115.6765	14.5655	489.6442

Variable	Definition	Source	Obs.	Mean	SD	Min.	Max.
$LNGDPPC_{c,t}$	Logarithm of Gross Domestic Product per Capita	World Bank, World Development Indicator Database	228	9.7194	0.6774	7.9737	10.6431
$ALLPOL_{c,t}$	Total of Accumulated Number of Renewable Energy Policies and Measures	MURE DATABASE	228	29.8553	18.9586	0	82
$NORMPOL_{c,t}$	Accumulated Number of Renewable Energy Policies and Measures – Normative/Legislative	MURE DATABASE	228	6.9035	6.0468	0	36
$FISCPOL_{c,t}$	Accumulated Number of Renewable Energy Policies and Measures – Tariff/ fiscal	MURE DATABASE	221	1.0905	1.8367	0	7
$INFOPOL_{c,t}$	Accumulated Number of Renewable Energy Policies and Measures – Legislative/informative Legislative/ informative	MURE DATABASE	228	3.2807	3.2434	0	13
$FINPOL_{c,t}$	Accumulated Number of Renewable Energy Policies and Measures – Incentives/subsidies	MURE DATABASE	228	8.6754	7.0079	0	26
$EDUPOL_{c,t}$	Accumulated Number of Renewable Energy Policies and Measures – Educational	MURE DATABASE	228	5.5526	4.7159	0	22
$COOPPOL_{c,t}$	Accumulated Number of Renewable Energy Policies and Measures – Co-operative	MURE DATABASE	218	2.7456	3.0511	0	16
$CUTPOL_{c,t}$	Accumulated Number of Renewable Energy Policies and Measures – Cross-cutting	MURE DATABASE	228	1.6404	3.5199	0	16

Notes: MURE DATABASE stands for Mesures d'Utilisation Rationnelle de l'Energie (MURE II Database); co-ordinated by the Institute of Studies for the Integration of Systems and the Fraunhofer Institute for Systems and Innovation Research ISI. IEA stands for International Energy Agency Data Services and EUROSTAT stands for Eurostat Statistics Database available at http://epp.eurostat.ec.europa.eu/portal/page/portal/statistics/search_database with the code nrg_113a.

Table 1. Variables definition, sources and summary statistics

Figure 6 indicates that there has been major support for renewables through energy policies and measures especially in Germany, France, Finland and the United Kingdom.

3.3. Data

In this chapter data from several sources such as the Eurostat database, International Energy Agency (IEA), World Bank and MURE Database is used. Table 1 presents the variables, their definition, sources and summary statistics for the time span 1998-2009.

3.4. Methods

To make a proper empirical analysis the panel dataset structure was analyzed, which generally has a complex nature of term error composition. Several methods were applied: (i) visual analysis of data; (ii) test for first-order autocorrelation in panel data; (iii) test for the presence of groupwise heteroskedasticity; and (iv) test for contemporaneous correlation. Stata v11.2 econometric software was used.

The correlation matrix (Table 2) values suggest that correlation coefficients are low and do not suggest the existence of collinearity among the variables.

	$WOCAP_{c,t}$	$COALSH_{c,t}$	$GASSH_{c,t}$	$OILSH_{c,t}$	$CFNUCL_{c,t}$	$CFHYD_{c,t}$	$WASTSH_{c,t}$	$WINDGR_{c,t}$	$SOLSH_{c,t}$
$WOCAP_{c,t}$	1								
$COALSH_{c,t}$	-0.0365	1							
$GASSH_{c,t}$	-0.1244	-0.1455	1						
$OILSH_{c,t}$	-0.2115	0.071	0.2375	1					
$CFNUCL_{c,t}$	0.2218	-0.1512	-0.0172	-0.4296	1				
$CFHYD_{c,t}$	0.0557	-0.4607	-0.2358	-0.2613	0.1269	1			
$WASTSH_{c,t}$	0.0412	-0.0863	0.0188	-0.2497	0.2778	0.4466	1		
$WINDGR_{c,t}$	0.4853	-0.0079	-0.1081	-0.0593	-0.0462	-0.0219	-0.1532	1	
$SOLSH_{c,t}$	0.0203	-0.0003	0.0637	-0.0357	0.1309	-0.088	0.0184	-0.0723	1
$POPDENS_{c,t}$	0.1344	0.1529	0.5052	-0.0721	0.3865	-0.33	-0.0177	-0.0968	0.0592
$LNGDPPC_{c,t}$	-0.1772	-0.3668	0.0691	-0.0704	0.1725	0.3844	0.301	-0.2477	0.0294
$NORMPOL_{c,t}$	0.0936	-0.2008	0.2193	0.1654	0.0503	-0.1347	-0.0908	-0.109	0.4542
$FISCPOL_{c,t}$	-0.0087	-0.1839	0.2779	-0.2763	0.4015	-0.1103	0.0589	-0.0683	-0.068
$FINPOL_{c,t}$	-0.1705	-0.1179	-0.027	-0.3426	0.5317	0.169	0.088	-0.0178	0.1358
$EDUPOL_{c,t}$	-0.0832	-0.2672	0.0633	-0.3603	0.1555	0.3154	0.252	-0.0749	0.0879
$COOPPOL_{c,t}$	-0.008	-0.1613	0.2452	-0.2366	0.4922	0.3567	0.5535	-0.1681	0.0922
$CUTPOL_{c,t}$	0.0634	0.0865	-0.0564	-0.2202	0.2274	0.175	-0.0224	-0.082	0.249

	$POPDENS_{c,t}$	$LNGDPPC_{c,t}$	$NORMPOL_{c,t}$	$FISCPOL_{c,t}$	$FINPOL_{c,t}$	$EDUPOL_{c,t}$	$COOPPOL_{c,t}$	$CUTPOL_{c,t}$
$POPDENS_{c,t}$	1							
$LNGDPPC_{c,t}$	0.1445	1						
$NORMPOL_{c,t}$	0.1375	0.2511	1					
$FISCPOL_{c,t}$	0.557	0.3284	0.1434	1				
$FINPOL_{c,t}$	0.215	0.2488	0.3877	0.3986	1			
$EDUPOL_{c,t}$	-0.0983	0.3927	0.1509	0.2376	0.4939	1		
$COOPPOL_{c,t}$	0.2422	0.4507	0.1037	0.3808	0.3564	0.3814	1	
$CUTPOL_{c,t}$	0.275	0.3259	0.0897	0.1559	0.4795	0.3185	0.2131	1

Notwithstanding, the Variance Inflation Factor (VIF) test for multicollinearity among variables was performed. Individual values are below 5 for all individual tests and 2.36 for mean VIF (Table 3), which reinforces that multicollinearity among variables is not a problem.

Table 2. Correlation Matrix

Variables	VIF	1/VIF
$COALSH_{c,t}$	2.33	0.4292
$GASSH_{c,t}$	2.42	0.414
$OILSH_{c,t}$	1.81	0.5522
$CFNUCL_{c,t}$	2.76	0.363
$CFHYD_{c,t}$	2.92	0.3423
$WASTSH_{c,t}$	1.95	0.5131
$WINDGR_{c,t}$	1.11	0.8993
$SOLSH_{c,t}$	1.64	0.6092
$POPDENS_{c,t}$	3.47	0.2881
$LNGDPPC_{c,t}$	2.22	0.4514
$NORMPOL_{c,t}$	2.37	0.4226
$FISCPOL_{c,t}$	2.41	0.415
$FINPOL_{c,t}$	3.29	0.304
$EDUPOL_{c,t}$	2.19	0.4556
$COOPPOL_{c,t}$	2.66	0.3758
$CUTPOL_{c,t}$	2.16	0.4621
Mean VIF	2.36	

Table 3. Variance Inflation Factor (VIF)

As part of the empirical research using panel dataset techniques, several tests to detect common panel phenomena in errors structure were performed (see Table 4 for results). The Wooldridge test with the null hypothesis of no first-order autocorrelation to detect serial correlation in the idiosyncratic errors of panel-data (Wooldridge, 2002) was performed. This test follows a normal distribution N(0,1) in Ordinary Least Squares (OLS) estimator. Furthermore, a modified Wald test was applied to search for the presence of groupwise heteroskedasticity in the residuals of a fixed effect (FE) regression model, which assumes homoskedasticity across cross-sections. The modified Wald Test has χ^2 distribution and tests the null hypothesis of: $\sigma_c^2=\sigma^2 \text{for} c=1, \ldots, N$ whereσ^2 is the variance of the $c$country (Greene, 2000). As stated by Marques and Fuinhas (2012b), if one considers that European countries are guided by common energy guidelines, one might expect the presence of contemporaneous correlation in our panel. In order to detect this phenomenon, or rather, test the null hypothesis of cross-section independence, Pesaran (2004), Frees (1995 and 2004), and Friedman (1937) tests were performed. While Pesaran follows a standard normal distribution, the Frees statistic test uses Frees Q-distribution and Friedman uses Friedman's chi-square distributed statistic. Frees and Friedman perform only with available data for all cross-sections. Hausman's statistics test the null hypothesis that the difference of coefficients between fixed-effects and random-effects is not systematic.

	Pooled OLS	**Fixed Effects (FE)**	**Random Effects (RE)**
Wooldridge test F(N(0,1))	3.48		
Modified Wald test (χ^2)		749.41***	
Pesaran test		-1.717	-2.061**
Frees test		0.744	1.191
Friedman test		7.053	4.605

Note: ***, ** denote significance at 1 and 5% significance levels

Table 4. Specification tests and statistics

According to table 4 results, the Wooldridge test value (3.48) does not reject the null hypothesis of no first-order autocorrelation. Accordingly, the autoregressive (AR1) estimator is not suitable. The modified Wald test value (749.41) suggests rejection of the null hypothesis of errors homoskedasticity within cross-sections. Therefore, the presence of groupwise heteroskedasticity is confirmed. As far as the presence of contemporaneous correlation is concerned, with the exception of the Pesaran test for random effects (-2.061), generally the null hypothesis of no contemporaneous correlation was not rejected, suggesting that there is spatial independence across European countries. This is not surprising given the technical nature of our research into the interaction of conventional sources and renewables with wind overcapacity instead of common policy guidelines.

http://dx.doi.org/10.5772/52159

The OLS estimator proves to be consistent when there is no presence of multicollinearity among the explanatory variables and when the regressors are exogenous. It is optimal when there is no serial auto-correlation following $V(\varepsilon)=\sigma_\varepsilon^2 \mid NT$ and when the errors are homoskedastic following $E(\varepsilon)=0$. Therefore, in our case it may be useful to benchmark results of our panel estimation. Moreover, we apply the panel fixed-effects (FE) and random-effects estimators (RE). Using the fixed-effects estimator appears to be appropriate in studying the impact of variables that vary over time. It explores the different variables within groups that have their own characteristics, in our case European countries. The fixed-effects estimator assumes that something time-invariant within groups can affect the dependent variable and cannot be correlated with other groups. In turn, random effects assume that variation across groups is random and not correlated to the dependent and independent variables.

The generic model to estimate is:

$$WOCAP_{c,t} = \alpha + \sum_{k=1}^{k} \beta_k X_{k,c,t} + d_t + \varepsilon_{c,t}\ , \tag{13}$$

where the error term is $\varepsilon_{c,t}=\alpha_c + u_{c,t}$ with α_c uncorrelated with the regressors and $\varepsilon_{c,t}$ homoskedastic with no serial correlation. The dummy for time is denoted by d_t.

4. Empirical evidence of the drivers of wind overcapacity

The estimation results are shown in Table 5. Conventional standard errors (CSE) are provided, as are robust standard errors (RSE) to deal with the presence of heteroskedasticity. Models *I* and *II* represent pooled OLS; models *III* and *IV* are panel fixed-effect estimators (FE); and models *V* and *VI* stand for random-effect estimators (RE). The error term is $\varepsilon_{c,t}=\alpha_c + u_{c,t}$.

A battery of diagnostic tests was applied to test the quality of the estimators. The Breusch-Pagan Lagrange multiplier (LM) is provided to test whether the RE estimator is more suitable than the OLS estimator. The results show that the null hypothesis of variances across groups being equal to zero is rejected, so there is a significant difference across groups. Accordingly, the RE estimator is more suitable than Pooled OLS. The Hausman test to choose the most appropriate estimator between FE and RE was applied. The null hypothesis assumes that the difference in coefficients is not systematic, thus accepting RE over FE estimator (Greene, 2008). The Hausman test accepts the null hypothesis, thus the errors α_c are uncorrelated with the regressors. As a consequence, the discussion will be based on RE estimator with RSE (*VI*). In other words, it seems that differences across countries influence $WOCAP_{c,t}$, so the panel RE estimator is more appropriate than FE to our analysis.

	OLS		FE		RE	
Ind. Variables	***CSE(I)***	***RSE(II)***	***CSE(III)***	***RSE(IV)***	***CSE(V)***	***RSE(VI)***
$COALSH_{c,t}$	-0.0402** (0.0161)	-0.0402** (0.0182)	-0.0696 (0.1264)	-0.0696 (0.0651)	-0.0381** (0.0177)	-0.0381** (0.0192)
$GASSH_{c,t}$	-0.0768*** (0.0234)	-0.0768*** (0.0255)	-0.1174 (0.1330)	-0.1174 (0.0660)	-0.0692*** (0.0253)	-0.0692*** (0.0253)
$OILSH_{c,t}$	-0.0442 (0.0479)	-0.0442 (0.0739)	-0.2819** (0.1382)	-0.2819*** (0.0673)	-0.0646 (0.0508)	-0.0646 (0.0731)
$CFNUCL_{c,t}$	-0.0086 (0.0101)	-0.0086 (0.0137)	0.0893 (0.0785)	0.0893 (0.0619)	-0.0070 (0.0110)	-0.0070 (0.0142)
$CFHYD_{c,t}$	0.0118 (0.0362)	0.0118 (0.0402)	-0.1466 (0.0825)	-0.1466** (0.0621)	0.0049 (0.0386)	0.0049 (0.0415)
$WASTSH_{c,t}$	0.2878*** (0.1084)	0.2878*** (0.0836)	-0.1588 (0.3438)	-0.1588 (0.3102)	0.2666** (0.1174)	0.2666** (0.0849)
$WINDGR_{c,t}$	0.0002*** (0.0000)	0.0002*** (0.0001)	0.0002*** (0.0000)	0.0002*** (0.0001)	0.0002*** (0.0000)	0.0002*** (0.0000)
$SOLSH_{c,t}$	1.2293 (1.8236)	1.2293 (1.0963)	1.5059 (2.0191)	1.5059 (0.8716)	1.1403 (1.8146)	1.1403 (0.9527)
$POPDENS_{c,t}$	0.0002*** (0.0000)	0.0002*** (0.0000)	0.0021 (0.0014)	0.0021 (0.0011)	0.0001*** (0.0000)	0.0001*** (0.0000)
$LNGDPPC_{c,t}$	-0.0300*** (0.0063)	-0.0300*** (0.0076)	0.0878 (0.0563)	0.0878 (0.0511)	-0.0290*** (0.0068)	-0.0290*** (0.0081)
$ALLPOL_{c,t}$	0.0008*** (0.0002)	0.0008*** (0.0003)	0.0009** (0.0004)	0.0009*** (0.0003)	0.0008*** (0.0002)	0.0008*** (0.0002)
CONST	1.0650*** (0.0617)	1.0650*** (0.0746)	-0.2911 (0.6436)	-0.2911 (0.6077)	1.0593*** (0.0666)	1.0593*** (0.0805)
N	218	218	218	218	218	218
R^2	0.4316	0.4316	0.3623	0.3623		
Wald (χ^2)					136.63***	
F (N(0,1))	7.09***		4.82***			
LM (χ^2)					11.76***	
Hausman (χ^2)			30.93			

Notes: OLS - Ordinary Least Squares. RE – Random Effects. FE – Fixed Effects. CSE – Conventional standard errors. The F-test has normal distribution N(0,1) and tests the null hypothesis of non-significance of all estimated parameters. The Wald test has χ^2distribution and tests the null hypothesis of non-significance of all coefficients of independent variables. The LM test has χ^2distribution and tests the null hypothesis of non-relevance of individual effects in the RE model. The Hausman test has χ^2distribution and tests the null hypothesis of the difference in coefficients not being systematic between two selected estimators. Standard errors are reported in brackets. All estimates were controlled to include time effects, although they are not reported for reasons of simplicity. ***, **, denote significance at 1 and 5% significance levels respectively for both coefficient estimators and test statistics.

Table 5. Regression results - Dependent Variable $WOCAP_{c,t}$

Globally, the estimation results provided in Table 5 reveal consistency despite some differences between significance levels. By examining the variables in descending order, for fossil fuels, the effect of $COALSH_{c,t}$ and $GASSH_{c,t}$ proved to be negative and statistically significant at 5% and 1% respectively. On the other hand, variable $OILSH_{c,t}$ does not seem to be significant statistically. This result is in line with expectations, revealing that there is backup for wind power using fossil fuels like coal and gas to overcome intermittency. Oil power plants are not generally used for backup, so these results may reveal our model's robustness.

$CFNUCL_{c,t}$, $CFHYD_{c,t}$ and $SOLSH_{c,t}$ coefficients reveal no statistical relationship between the capacity factor of nuclear and hydro or the share of solar energy and wind power overcapacity. The effects of variables $WINDGR_{c,t}$, $POPDENS_{c,t}$, $LNGDPPC_{c,t}$ and $ALLPOL_{c,t}$ are positive and statistically significant. Therefore, it is assumed that they are important drivers in explaining wind overcapacity. Results from disaggregated policies are presented in table A. 1. None of the individual energy policies proves to be significant in explaining wind overcapacity except $NORMPOL_{c,t}$ and $FISCPOL_{c,t}$. However, it is worth noting that there is no inclusion of the legislative/informative policies ($INFOPOL_{c,t}$) due to their identical nature and the fact that they could create collinearity problems. Contrary to expectations, financial policies have no statistical relation to wind overcapacity.

Exclusion tests were run for the explanatory variables ($FINPOL_{c,t}$, $EDUPOL_{c,t}$, $COOPPOL_{c,t}$ and $CUTPOL_{c,t}$), which do not reveal a statistical significance, following the parsimonious principle. The results are shown in Table A.2. In fact, the models maintain robustness among the estimators for all coefficients with or without these individual policies. This set of variables has no influence either on the ratio of non-used wind capacity ($WOCAP_{c,t}$) or on the remaining model. Nevertheless, given that consistency and robustness are crucial properties, a subsection is opened to provide additional analysis on the reliability of results.

5. Consistency and robustness of empirical evidence

According to Huber (1973), as further evidence of the robustness of results, it is appropriate to apply the robust regression (RREG) estimator to cope with possible outliers from our dataset. These outliers can impair the stability and reliability of results. Such as in Marques and Fuinhas (2012b), robustness is analyzed by providing the robust regression with Huber and Tukey weight functions, as presented in table 6.

As shown from this additional assessment of the robustness of results, the variables maintain their signs, though with small differences in significance levels. In general, the robust regression validates the main results of the estimations.

Ind. Variables	RREG (VII)	RREG (VIII)	RREG (IX)
$COALSH_{c,t}$	-0.0208 (0.0120)	-0.0130 (0.0115)	-0.0229 (0.0125)
$GASSH_{c,t}$	-0.0755*** (0.0176)	-0.0696*** (0.0159)	-0.0700*** (0.0176)
$OILSH_{c,t}$	-0.0540 (0.0358)	-0.1351*** (0.0346)	-0.1424*** (0.0356)
$CFNUCL_{c,t}$	-0.0120 (0.0075)	-0.0028 (0.0065)	-0.0067 (0.0081)
$CFHYD_{c,t}$	0.0084 (0.0270)	0.0270 (0.0253)	0.0015 (0.0284)
$WASTSH_{c,t}$	0.3199*** (0.0808)	0.3410*** (0.0740)	0.3598*** (0.0827)
$WINDGR_{c,t}$	0.0003*** (0.0000)	0.0003*** (0.0000)	0.0003*** (0.0000)
$SOLSH_{c,t}$	1.6675 (1.3598)	-1.7028 (1.3661)	-2.3226 (1.4871)
$POPDENS_{c,t}$	0.0002*** (0.0000)	0.0002*** (0.0000)	0.0002*** (0.0000)
$LNGDPPC_{c,t}$	-0.0316*** (0.0047)	-0.0276*** (0.0042)	-0.0278*** (0.0047)
$ALLPOL_{c,t}$	0.0007*** (0.0002)		
$NORMPOL_{c,t}$		0.0030*** (0.0004)	0.0030*** (0.0005)
$FISCPOL_{c,t}$		-0.0041*** (0.0016)	-0.0042** (0.0017)
$FINPOL_{c,t}$			-0.0001 (0.0005)
$EDUPOL_{c,t}$			-0.0006 (0.0007)
$COOPPOL_{c,t}$			0.0011 (0.0011)
$CUTPOL_{c,t}$			0.0012 (0.0008)
CONST	1.0712*** (0.0464)	1.0238*** (0.0399)	1.0377*** (0.0457)

http://dx.doi.org/10.5772/52159

Ind. Variables	RREG (VII)	RREG (VIII)	RREG (IX)
N	218	218	218
R^2	63.59	0.6920	0.7093
F (N(0,1))	15.48***	19.92***	17.93***

Notes: RREG – Robust Regression. The F-test has normal distribution N(0,1) and tests the null hypothesis of non-significance of all estimated parameters. JST - Joint Significance Test. JST is a Wald χ^2 test with the null hypothesis of $H_0=\beta_1=\beta_2=\beta_3=\beta_4=0$, with β_1, β_2, β_3, β_4 representing the coefficient of $FINPOL_{c,t}$, $EDUPOL_{c,t}$, $COOPPOL_{c,t}$ and $CUTPOL_{c,t}$, respectively. LRT - Linear Restriction Test has the null hypothesis of $H_0=\beta_1+\beta_2+\beta_3+\beta_4=0$. Standard errors are reported in brackets. All estimates were controlled to include time effects, but they are not reported for reasons of simplicity. ***, **, denote significance at 1 and 5% significance levels respectively.

Table 6. Results from Robust Regression – Dependent variable $WOCAP_{c,t}$

6. Conventional energy sources and backup

As stated above, renewable energy sources, particularly wind and solar, suffer from the intermittency phenomenon. This phenomenon could cause overcapacity. There are several factors that may influence overcapacity, such as conventional energy sources and renewable energy sources, socio-economic and energy policies. Our results allow us to explore and discuss them individually and suggest some guidance for energy policy and measures.

As shown in the models, the results for conventional energy sources show a negative effect of fossil fuels on overcapacity, more specifically coal and gas power plants. With greater use of coal and gas, the effect of wind overcapacity is reduced. Two main reasons can be at the origin of this effect: (i) intermittency leads to the uncertainty of energy generation and the need to ensure a continuous electricity supply. It requires the existence of fossil fuels like coal and gas to backup power. With more dependence on these sources in peak-load periods, electricity generation is simultaneously based on renewables and fossil fuels in order to meet electricity demand and this implies a reduction in wind overcapacity; and (ii) in line with Marques et al. (2010) the results for fossil fuels sustain a lobbying effect in the electricity generation industry. This effect promotes the growth of fossil fuels to the detriment of renewables due to more stringent energy policies (Fredriksson et al., 2004). The first sites for installation of wind farms are usually the most efficient ones, and, in some countries, the deployment of renewables is still in its early stages because fossil fuels still have high shares in total electricity generation. Therefore, some countries still benefit from better sites with high wind speeds and from better capacity factor and, as a consequence, wind overcapacity tends to be lower, since wind power is installed in optimum sites.

Nevertheless, assuming that fossil fuels could have a positive effect on overcapacity, an increase in the share of coal, gas and oil would provoke a substitution effect in the electricity generation process due to less use of wind energy. In this case, idle capacity would be greater.

Further regarding conventional energy sources, nuclear power is not statistically significant in wind overcapacity, despite its relevance in Europe. Nuclear power is not an intermittent source and its widespread use in Europe is mainly due to its own characteristics: it is cheaper than oil, with a greater capacity to generate power in a single power plant but is inflexible in providing backup for renewables due to its low startup times. Nuclear power works better in full-power operation than on demand (Dittmar, 2012). It is therefore understandable that nuclear is not significant in explaining wind overcapacity. Ultimately, this result can be seen as additional proof of the robustness of the results.

7. Other renewable energy sources

The contributions that renewable energy sources, other than wind, make to explaining wind overcapacity are also discussed. Similarly to results obtained for nuclear, hydropower is not statistically significant. It is a well-developed renewable energy source, with widespread use in Europe for electricity generation (Balat, 2006), which is relatively stable and mature. Its own characteristics imply its use in base load generation. However, recent developments in this source have involved engaging hydropower with wind power. Indeed, wind power combined with pumped hydro power stations can be useful in meeting electricity demand in peak-load periods. In off-peak, wind overproduction can be used to pump water to an elevated reservoir to later be re-used back in the lower reservoir (Dursun and Alboyaci, 2010). This new technology adds storage capacity, not in terms of energy storage but rather in terms of energy generation storage, thereby giving wind energy a new role as a backup to hydropower. In the future, this could help to mitigate overcapacity effects.

Solar energy merits our attention since it is an increasingly common renewable energy source in Europe. The results show that solar energy does not prove to be statistically significant in explaining wind overcapacity. Nevertheless, a popular and advantageous solution, stated by Nema et al. (2009), is the mix of wind and solar considering that their integration makes them less exposed to intermittency. Supporting such systems, more especially in remote areas, may be a solution to reducing overcapacity. It can be followed by European partners, namely in regions where land space and natural resources allow this investment. A case in point is the recent investment in southern Spain in solar thermal plants with a capacity of 300MW at their completion in 2013. Solar thermal power plants use several available technologies such as power towers, parabolic troughs with heat storage, sterling dishes and concentrated/non-concentrated solar power. Consequently, by combining this mix of technologies, the power plants can operate without sunlight at total capacity for 7.5 hours.

When there is more wind installed capacity, the number of intermittent wind power plants also logically increases. Our results support this assumption since the growth rate of wind

http://dx.doi.org/10.5772/52159

capacity positively influences overcapacity. Regarding the impact of industrial, municipal and renewable waste power, it seems to overlap wind energy due to positive signs in our results. In fact, with more waste processing for energy generation, there appears to be a substitution effect for wind energy. This result is in line with the growing use of waste for electricity generation in recent years, ranging from 2% in 1998 to 5% in 2009.

8. Socio-economic drivers, energy policies and measures

The spatial dispersion and, in other words, the efficient installation of wind parks should be discussed further. Population density was applied as a proxy for spatial dispersion of wind farms to assess the impact of countries with higher or less population density. These countries tend to have less land space to install wind farms properly. Furthermore, continental areas have lower wind speeds than countries with coastlines. Overall, the results support this assumption. Overcapacity is larger in more populated countries. Accordingly, policy makers and players should pay extra attention to this issue in order to support the diversification of wind turbines. To overcome the constraints caused by highly populated countries, offshore technologies can be a better way due to their steadier characteristics mainly driven by higher wind speeds. It also allows the use of higher power generation turbines. In short, offshore wind farms can help to overcome the population density effect on the creation of overcapacity.

In the literature, the role of economic growth as a driver toward renewable energy is far from consensual. Marques et al. (2010) argued that the effect of GDP on renewables depends on the share level of renewables. In their turn, Chang et al. (2009) conclude that economic growth and renewables' development are not directly related. Nonetheless, countries that are in an upward trajectory with high growth rates can support prices of investing in renewables. Despite increased prices for the final consumer, developed countries tend to invest more in renewable energy sources. In this chapter, we focus on the effect of the logarithm of GDP *per capita.* Our results are consistent and reveal that countries with the highest living standards benefit from more advanced and efficient wind power plants which reduce idle capacity.

With the goal of increasing the share of energy from renewable sources, energy policies are stated as an effective instrument for European countries to implement (EU directive, 2009). Several authors (e.g. Gan et al., 2007; and Johnstone et al., 2008) found that incentive taxes, feed-in tariffs, voluntary programs and R&D policy support are the main drivers supporting renewables. Our results allow us to analyze individually and jointly the energy policies adopted in the European context. In our models, the total of accumulated energy policies over the years increases overcapacity, showing that the impact of these policies is sometimes inefficient, taking into account only the players' political will to reach European guidelines. Table A.1 of the appendix shows results from disaggregated energy policies. Normative regulation and efficiency in building policies create wind power overcapacity. Indeed, these policies imply better efficiency and consumption savings. Reducing energy consumption in buildings contributes to aggravating idle capacity. Moreover, fiscal and tariff policies, including tax reduction in retrofitting investments, promote investments in new power plants, replacing old equipment with technologies that generate higher power and are more efficient. Retrofit investments help to upgrade the electricity system and reduce idle capacity.

Overcapacity of renewables is another aspect of the intermittency phenomenon. Wind energy has been a very common and widely accepted instrument in reaching the 20-20-20 targets. It merits a review of the implicit economic consequences. Policy makers should pay more attention to the advantages and consequences of their policies and measures focused on renewables in order to avoid a blindly ill-considered decision-making process. The deployment of wind power installed capacity has implications for the energy grid as a whole and creates economic distortions. To balance conventional energy sources with all renewables is a challenging task that requires enlightened political and scientific intervention. Policy makers should bear in mind that the growth of renewables has to be in line with energy consumption patterns. To mitigate this problem, micro-production incentives seem to be a solution to balancing domestic consumption with network energy supply. Furthermore, installed players should not resist investment in new technologies in order to maximize wind capacity factors. Off-shore sites are a good alternative for countries with coastal areas because in addition to having higher power generation, wind farms can make more efficient use of installed capacity.

Coal-based and gas-fired power plants are actually used to backup wind power. However, this imposes an extra cost on the final consumer since the non-use of conventional power plants is subsidized. Regulatory authorities should be aware and take measures to prevent the price escalation that combines the contribution to investment in renewables with these subsidies for energy industry lobbies. The implementation of mixed systems based on renewable energy in regions with available natural resources can both improve the energy supply economically and supplant the needs of the area (Erdinc and Uzunoglu, 2012). With the opening of energy markets to the private sector, stronger regulation of the market may be an instrument in monitoring immoderate investments.

Generally, there are no incentives to increase the efficiency of renewables' technology. For example, in some countries such as Portugal, there is an incentive based on feed-in tariffs for the solar photovoltaic micro-generation system. The incentive to improve efficiency is non-existent given that the maximum electricity generation that could be sold to the player distributer is bounded. In general, the feed-in tariffs guarantee the price for kWh regardless of whether it is generated by a very efficient device or not. This form of intervention merely ensures income for the players. Policymakers should consider implementing measures that will add competition to the renewables industry, particularly in solar and wind industries, and thus promote patenting and R&D activities.

http://dx.doi.org/10.5772/52159

9. Conclusion

This chapter is centered on a panel dataset of 19 European countries for the time span 1998-2009 in order to understand and analyze the causes of wind overcapacity that may arise from non-constant electricity generation from renewables. To the best of our knowledge, this approach had never been made through panel data techniques and it is a new method in the renewables' intermittency literature. Some light is shed on overcapacity of wind energy and its interaction with conventional energy sources, other renewables, socio-economic drivers and energy policies in the context of an economic bloc with common long-term energy guidelines.

Results from our models reveal that fossil fuel power plants, such as coal-based and gas-fired, are actually used to backup wind power. Oil and nuclear do not appear to be significant in explaining wind overcapacity. These results may highlight the robustness of our model, considering that oil and nuclear power are generally used for base load energy generation and therefore have no direct effect on wind overcapacity. As further robustness assessment, the robust regression estimator was performed to deal with possible outliers from our panel. Overall, the robust regression supports the main results of the estimations.

Renewables such as hydropower and solar photovoltaic seem to make no apparent contribution to explaining wind power overcapacity, unlike industrial and municipal waste. Moreover, the results indicate that population density is a factor in greater wind overcapacity, while countries with a higher standard of living are associated with less overcapacity. The results for public policies and measures suggest that a positive effect on increased overcapacity may be due to inefficient incentives for deployment of wind power. The promotion of renewable energy is a crucial decision because it deals with one of the central inputs of economies and societies in general. In order to gain a full understanding of the appropriate ways in which to promote the paradigm shift from fossil to renewable sources, objectivity is needed in analyzing both the advantages and disadvantages associated with the path that has already been trodden. This cumulative experience should support the intensification of measures that have had a positive impact on the development of renewables and have not added significant distortions to the economy as a whole. Other measures which do not produce the desired effects, or fail to contribute to an egalitarian distribution of benefits, should be reconsidered or even abandoned and replaced.

Policy measures, particularly incentives, should be largely dependent on the level of efficiency achieved by the players. These measures should be oriented towards the market, avoiding distortions between the different players acting in the energy market. Such measures should not result in costs for the economy that endanger the prosperity levels of society in general. In fact, we are dealing with a non-cooperative game played between international players, including countries or economic blocs, where the competitive advantage of this technology domain is more quickly surpassed than the comparative advantage of the possession of fossil resources, such as coal or oil reserves.

Appendix

Ind. Variables	OLS				FE				RE			
	CSE(X)	CSE(XI)	RSE(XII)	RSE(XIII)	CSE(XIV)	CSE(XV)	RSE(XVI)	RSE(XVII)	CSE(XVIII)	CSE(XIX)	RSE(XX)	RSE(XXI)
$COALSH_{c,t}$	-0.0320**	-0.0380**	-0.0320**	-0.0380**	-0.0614	-0.0739	-0.0614	-0.0739	-0.0317	-0.0380**	-0.0317***	-0.0380**
	-0.0162	-0.0179	-0.0111	-0.0158	-0.1289	-0.1299	-0.0789	-0.0719	-0.0167	-0.0179	-0.0114	-0.0158
$GASSH_{c,t}$	-0.0703***	-0.0676***	-0.0703***	-0.0676***	-0.0767	-0.0996	-0.0767	-0.0996	-0.0692***	-0.0676***	-0.0692***	-0.0676***
	-0.0224	-0.0251	-0.0177	-0.0223	-0.1348	-0.1367	-0.0661	-0.0693	-0.0231	-0.0251	-0.0179	-0.0223
$OILSH_{c,t}$	-0.1247**	-0.1219**	-0.1247	-0.1219**	-0.2299	-0.3022**	-0.2299**	-0.3022***	-0.1267**	-0.1219**	-0.1267**	-0.1219**
	-0.0488	-0.0508	-0.06	-0.0577	-0.1359	-0.1411	-0.0877	-0.0747	-0.0497	-0.0508	-0.0606	-0.0577
$CFNUCL_{c,t}$	0.0034	0.002	0.0034	0.002	0.085	0.0837	0.085	0.0837	0.0037	0.002	0.0037	0.002
	-0.0092	-0.0115	-0.0124	-0.0142	-0.0791	-0.0811	-0.0587	-0.0651	-0.0095	-0.0115	-0.0125	-0.0142
$CFHYD_{c,t}$	0.0335	0.0173	0.0335	0.0173	-0.1241	-0.1599	-0.1241	-0.1599**	0.0315	0.0173	0.0315	0.0173
	-0.0356	-0.0404	-0.0387	-0.0457	-0.0818	-0.0834	-0.0648	-0.0632	-0.0364	-0.0404	-0.0392	-0.0457
$WASTSH_{c,t}$	0.3054***	0.3469***	0.3054***	0.3469***	-0.1836	-0.1975	-0.1836	-0.1975	0.3015***	0.3469***	0.3015***	0.3469***
	-0.1044	-0.1179	-0.0682	-0.0899	-0.3496	-0.349	-0.2983	-0.2968	-0.1075	-0.1179	-0.0692	-0.0899
$WINDGR_{c,t}$	0.0002***	0.0002***	0.0002***	0.0002***	0.0002***	0.0002***	0.0002***	0.0002***	0.0002***	0.0002***	0.0002***	0.0002***
	0.000	0.000	-0.0001	-0.0001	0.000	0.000	-0.0001	-0.0001	0.000	0.000	-0.0001	-0.0001
$SOLSH_{c,t}$	-2.2002	-2.8232	-2.2002	-2.8232	1.1918	0.334	1.1918	0.334	-2.0828	-2.8232	-2.0828	-2.8232
	-1.928	-2.1182	-1.6162	-1.9352	-2.1039	-2.2847	-1.0282	-0.922	-1.927	-2.1182	-1.5621	-1.9352
$POPDENS_{c,t}$	0.0002***	0.0002***	0.0002***	0.0002***	0.0015	0.0022	0.0015	0.0022	0.0002***	0.0002***	0.0002***	0.0002***
	0.000	0.000	0.000	0.000	-0.0015	-0.0016	-0.0012	-0.0012	0.000	0.000	0.000	0.000
$LNGDPPC_{c,t}$	-0.0245***	-0.0264***	-0.0245***	-0.0264***	0.0728	0.0851	0.0728	0.0851	-0.0243***	-0.0264***	-0.0243***	-0.0264***
	-0.0059	-0.0067	-0.0059	-0.0072	-0.0559	-0.0563	-0.0524	-0.0512	-0.0061	-0.0067	-0.006	-0.0072
$NORMPOL_{c,t}$	0.0031***	0.0031***	0.0031***	0.0031**	0.0008	0.0009	0.0008	0.0009	0.0031***	0.0031***	0.0031***	0.0031***
	-0.0006	-0.0007	-0.001	-0.0011	-0.0011	-0.0011	-0.0005	-0.0006	-0.0006	-0.0007	-0.001	-0.0011
$FISCPOL_{c,t}$	-0.0050**	-0.0050**	-0.0050**	-0.0050**	0.0066	0.0017	0.0066	0.0017	-0.0049**	-0.0050**	-0.0049**	-0.0050**
	-0.0022	-0.0024	-0.0022	-0.002	-0.0043	-0.0051	-0.0035	-0.0035	-0.0023	-0.0024	-0.0022	-0.002
$FINPOL_{c,t}$		0.0001		0.0001		0.0001		0.0001		0.0001		0.0001
		-0.0007		-0.0007		-0.0011		-0.0008		-0.0007		-0.0007
$EDUPOL_{c,t}$		-0.0001		-0.0001		0.0004		0.0004		-0.0001		-0.0001
		-0.0009		-0.0007		-0.0013		-0.0011		-0.0009		-0.0007
$COOPPOL_{c,t}$		0.0001		0.0001		0.0015		0.0015		0.0001		0.0001
		-0.0015		-0.0012		-0.0028		-0.0016		-0.0015		-0.0012
$CUTPOL_{c,t}$		0.0012		0.0012		0.0050**		0.0050***		0.0012		0.0012
		-0.0012		-0.0008		-0.0024		-0.0008		-0.0012		-0.0008
CONST	1.0006***	1.0251***	1.0006***	1.0251***	-0.0636	-0.2684	-0.0636	-0.2684	1.0003***	1.0251***	1.0003***	1.021***
	-0.0564	-0.0651	-0.0512	-0.0692	-0.6406	-0.6538	-0.6284	-0.6256	-0.058	-0.0651	-0.0525	-0.0692
N	218	218	218	218	218	218	218	218	218	218	218	218
R^2	0.4772	0.4809	0.4772	0.4809	0.3554	0.375	0.3554	0.375				
Wald(χ^2)									171.31***	176.91***		
F(N(0,1))	8.09***	6.80***			4.44***	3.99***						

Notes: OLS - Ordinary Least Squares. RE – Random Effects. FE – Fixed Effects. CSE – Conventional standard errors. RSE – Robust standard errors. The F-test has normal distribution N(0,1) and tests the null hypothesis of non-significance of all estimated parameters. The Wald test has χ^2distribution and tests the null hypothesis of non-significance of all coefficients of independent variables. Standard errors are reported in brackets. All estimates were controlled to include time effects, but they are not reported for reasons of simplicity. ***, **, denote significance at 1 and 5% significance levels respectively

Table A.1 Estimation results with disaggregated variables – Dependent variable $WOCAP_{c,t}$

	OLS	*OLS*	*FE*	*FE*	*RE*	*RE*
	CSE (XI)	*RSE (XIII)*	*CSE (XV)*	*RSE (XVII)*	*CSE (XIX)*	*RSE (XXI)*
Joint Significance test	0.34	0.76	1.36	10.63***	1.34	3.02
Linear restriction test	0.66	0.85	2.08**	3.78***	0.66	0.85

Notes: JST - Joint Significance Test. JST is a Wald χ^2 test with the null hypothesis of$H_0=\beta_1=\beta_2=\beta_3=\beta_4=0$, with β_1, β_2, β_3, β_4 representing the coefficient of $FINPOL_{c,t}$, $EDUPOL_{c,t}$, $COOPPOL_{c,t}$ and $CUTPOL_{c,t}$, respectively. LRT - Linear Restriction Test has the null hypothesis of$H_0=\beta_1+\beta_2+\beta_3+\beta_4=0$. ***, **, denote significance at 1 and 5% significance levels respectively.

Table A.2. Exclusion tests on $FINPOL_{c,t}$, $EDUPOL_{c,t}$, $COOPPOL_{c,t}$ and $CUTPOL_{c,t}$

Acknowledgment

We gratefully acknowledge the generous financial support of the NECE – Research Unit in Business Science and Economics, sponsored by the Portuguese Foundation for Science and Technology.

Author details

António Cardoso Marques[1], José Alberto Fuinhas[1] and Rui Flora[2]

1 NECE and University of Beira Interior, Management and Economics Department, Covilhã, Portugal

2 University of Beira Interior, Management and Economics Department, Estrada do Sineiro, Covilhã, Portugal

References

[1] Acker, T. L., Williams, S. K., & Duque, E. P. N. (2007). Wind resource assessment in the state of Arizona: Inventory, capacity factor, and cost. *Renewable Energy*, 1454-1466.

[2] Albadi, M. H., & El -Saadany, E. F. (2010). Overview of wind power intermittency impacts on power systems. *Electric Power Systems Research*, 627-632.

[3] Archer, C. L., & Jacobson, M. Z. (2007). Supplying Baseload and Reducing Transmission Requirements by Interconnecting Wind Farms. *Journal of Applied Meteorology and Climatology*, 1701-1717.

[4] Balat, M. (2006). Hydropower Systems and Hydropower Potential in the European Union Countries. Energy Sources, Part A: Recovery, Utilization, and Environmental Effects, 28, 10 , 965-978.

[5] Baltagi, B. H. ((2005).). Econometric analysis of panel data- 3rd ed. John Wiley, Chichester.

[6] Beaudin, M., Zareipour, H., Schellenberglabe, A., & Rosehart, W. (2010). Energy storage for mitigating the variability of renewable electricity sources: An updated review. *Energy for sustainable development*, 302-314.

[7] Boccard, N. (2009). Capacity factor of wind power realized vs. estimates. *Energy Policy*, 2679-2688.

[8] Caralis, G., Perivolaris, Y., Rados, K., & Zervos, A. (2008). On the effect of spatial dispersion of wind power plants on the energy capacity credit in Greece. *Environmental Research Letters*, 015003.

[9] Carley, S. (2009). State renewable energy electricity policies: An empirical evaluation of effectiveness. *Energy policy*, 3071-3081.

[10] Chang-H, T., Huang-M, C., & Lee-C, M. (2009). Threefold effect of the economic growth rate on the renewable energy development from a change in energy price: Evidence from OECD countries. Energy Policy, 37, , 5796-5802.

[11] Denholm, P., Kulcinski, G., & Holloway, T. (2005). Emissions and Energy Efficiency Assessment of Baseload Wind Energy Systems. Environment Science & Technology, 39, , 1903-1911.

[12] Dittmar, M. (2012). Nuclear energy: Status and future limitations. *Energy*, 35-40.

[13] Dursun, B., & Alboyaci, B. (2010). The contribution of wind-hydro pumped storage systems in meeting Turkey's electric energy demand. *Renewable and Sustainable Energy Reviews*, 1979-1988.

[14] Erdinc, O., & Uzunoglu, M. (2012). Optimum design of hybrid renewable energy systems: Overview of different approaches. *Renewable and Sustainable Energy Reviews*, 1412-1425.

[15] EU Directive(2009). EC On the promotion of the use of energy from renewable sources and amending and subsequently repealing Directives 2001/77/EC and 2003/30/EC, 2009.

[16] Fiedler, B. H., & Bukovsky, . (2011). The effect of a giant wind farm on precipitation in a regional climate model. *Environmental Research Letters*, 045101.

[17] Fredriksson, P. G., Vollebergh, H. R. J., & Dijkgraaf, E. (2004). Corruption and energy efficiency in OECD countries: theory and evidence. *Journal of Environmental Economics and Management*, 207-231.

[18] Frees, E. (1995). Assessing cross-section correlation in panel data. *Journal of Econometrics*, 393-414.

[19] Frees, E. (2004). Longitudinal and panel data: analysis and applications in the social sciences. Cambridge University Press.10.1017/CBO9780511790928

[20] Friedman, M. (1937). The use of ranks to avoid the assumption of normality implicit in the analysis of variance. Journal of the American StatisticalAssociation, 32, , 675-701.

[21] Gan, L., Eskeland, G. S., & Kolshus, H. H. (2007). Green electricity market development: Lessons from Europe and the US. *Energy Policy*, 144-155.

[22] Gómez, A., Zubizarreta, J., Dopazo, C., & Fueyo, N. (2011). Spanish energy roadmap to 2020: socioeconomic implications of renewable targets. *Energy*, 1973-1978.

[23] Gonzalez, A., Gallachóir, B. Ó., & Mc Keogh, E. (2004). Study of electricity storage technologies and their potential to address wind energy intermittency in Ireland. Sustainable Energy Authority of Ireland.

[24] Onlineat http://www.seai.ie/uploadedfiles/FundedProgrammes/REHC03001FinalReport.pdf

[25] Green, R., & Vasilakos, N. (2010). Market behavior with large amounts of intermittent generation. Energy Policy, 38, , 3211-3220.

[26] Greene, W. (2000). Econometric analysis. New York: Prentice Hall. 2000.

[27] Greene, W. (2008). Econometric analysis. th ed., Upper Saddle River, NJ: Prentice Hall.

[28] Gül, T., & Stenzel, T. (2005). Variability of Wind Power and other Renewables: Management Options and Strategies. IEA. Online at http://www.uwig.org/IEA_Report_on_variability.pdf

[29] Holttinen, H., Lemström, B., Meibom, P., Bindner, H., Orths, A., Van Hulle, F., Ensslin, C., Tiedemann, A., Hofmann, L., Winter, W., Tuohy, A., O'Malley, M., Smith, P., Pierik, J., Olav, Tande. J., Estanqueiro, A., Ricardo, J., Gomez, E., Söder, L., Strbac, G., Shakoor, A., Smith, J., Ch, Parsons. B., Milligan, M., & Wan, Y. (2009). Design and operation of power systems with large amounts of wind power. State-of-the art report. VTT WORKING PAPERS 82.

[30] Onlineat http://www.vtt.fi/inf/pdf/workingpapers/(2007). /W82.pdf

[31] Hsiao, C. (2006). Panel data analysis- Advantages and challenges. IEPR working paper.

[32] Huber, P. J. (1973). Robust Regression: Asymptotics, conjectures and Monte Carlo. Ann. Stat, 1, , 799-821.

[33] Isla, J. (1999). The Gas Turbine: A New Technological Paradigm in Electricity Generation. Technological Forecasting and Social Change, 60, , 129-148.

[34] Johnstone, N., Hascic, I., & Popp, D. (2008). Renewable energy policies and technological innovation: Evidence based on patent counts. NBER Working papers N0. 13760. Cambridge MA.10.3386/w13760

[35] Kehlhofer, R., Hannemann, F., Stirnimann, F., & Rukes, B. (2009). Combined-cycle gas & steam turbine power plants. 3rd ed. USA: Pennwell Cor. poration.

[36] Larraín, T., Escobar, R., & Vergara, J. (2010). Performance model to assist solar thermal power plant siting in northern based on backup fuel consumption. *Renewable Energy*, 1632-1643.

[37] Liao-H, C., Ou-H, H., Chiueh-T, P., & Yu-H, Y. (2011). A challenging approach for renewable energy market development. *Renewable and Sustainable Energy Reviews*, 787-793.

[38] Luickx, P. J., Delarue, E. D., & D'Haeseleer, W. D. (2008). Considerations on the backup of wind power: Operational Backup. *Applied Energy*, 787-799.

[39] Marques, A. C., Fuinhas, J. A., & Manso, J. R. (2010). Motivations driving renewable energy in European countries: a panel data approach. *Energy policy*, 6877-6885.

[40] Marques, A. C., & Fuinhas, J. A. (2012a). Are public policies towards renewables successful? Evidence from European countries. *Renewable Energy*, 109-118.

[41] Marques, A. C., & Fuinhas, J. A. (2012b). Is renewable energy effective in promoting growth? Energy Policy, 46, , 434-442.

[42] Menz, F. C., & Vachon, S. (2006). The effectiveness of different policy regimes for promoting wind power: Experiences from the states. *Energy Policy*, 1786-1796.

[43] Moreno, F., & Martínez-Val, J. M. (2011). Collateral effects of renewable energies deployment in Spain: Impact on thermal power plants performance and management. *Energy Policy*, 6561-6574.

[44] Münster, M., & Meibom, P. (2011). Optimization of use of waste in the future energy system. *Energy*, 1612-1622.

[45] Nema, P., Nema, R. K., & Rangnekar, S. (2009). A current and future state of art development of hybrid energy system using wind and PV-solar: A review. *Renewable and Sustainable Energy Reviews*, 2096-2103.

[46] Østergaard, P. A. (2008). Geographic aggregation and wind power output variance in Denmark. Energy, 33, , 1453-1460.

[47] Pearce, J. M. (2009). Expanding photovoltaic penetration with residential distributed generation from hybrid solar photovoltaic and combined heat and power systems. *Energy*, 1947-1954.

[48] Pechak, O., Mavrotas, G., & Diakoulaki, D. (2011). Role and contribution of the clean development mechanism to the development of wind energy. *Renewable and Sustainable Energy Reviews*, 3380-3387.

[49] Pesaran, M. H. (2004). General diagnostic tests for cross section dependence in panels. Cambridge Working Papers in Economics. Faculty of Economics University of Cambridge 0435.

[50] Purvins, A., Zubaryeva, A., Llorente, M., Tzimas, E., & Mercier, A. (2011). Challenges and options for a large wind power uptake by the European electricity system. *Applied Energy*, 1461-1469.

[51] Wooldridge, J. M. (2002). Econometric Analysis of Cross Section and Panel Data. Cambridge, MA: MIT Press.

[52] Yang, M., Patiño-Echeverri, D., & Yang, F. (2012). Wind power generation in China: Understanding the mismatch between capacity and generation. *Renewable Energy*, 145-151.

[53] Zhang, S., & Li, X. (2012). Large scale wind power integration in China: Analysis from a policy perspective. *Renewable and Sustainable Energy Reviews*, 1110-1115.

Chapter 4

Viewing Energy, Poverty and Sustainability in Developing Countries Through a Gender Lens

Pius Fatona, Abiodun Abiodun, Adetayo Olumide,
Adesanwo Adeola and Oladunjoye Abiodun

Additional information is available at the end of the chapter

http://dx.doi.org/10.5772/51818

1. Introduction

Energy is central to achieving the interrelated economic, social and environmental aims of sustainable development. Energy plays a critical role in poverty reduction. Patterns of energy generation, distribution and utilization directly affect opportunities for income generation, the situation of women, environmental protection and national development. Energy services are an essential engine for growth to enable developing countries to overcome poverty and the conditions of poverty. If man is to realize the sustainable human development goal, the kinds of energy produced and the ways they are used will have to change. Otherwise the environmental damage will accelerate, inequity will increase, and global economic growth will be jeopardized. Yet, worldwide, two billion people rely on traditional fuels including wood, dung and agricultural residues to meet their daily heating and cooking needs (Karlsson and Misana - eds, 2001).

Energy is necessary for meeting basic human needs and a prerequisite for economic development (see figure 1). Increased access to energy resources is the most fundamental requirement to economic growth. Thus, international and national efforts towards less developed countries' poverty reduction and improvements to health and education cannot be achieved without addressing international and national energy policies. Without access to modern energy sources, 40% of the global population relies on traditional biomass fuels in the form of wood, straw, dung, coal and other organic material for meeting their basic household needs (Institute for Development Studies (IDS), 2003). However, the burning of biomass causes many environmental problems, increases global climate change, and is one of the leading causes of premature deaths of women and children in less developed countries.

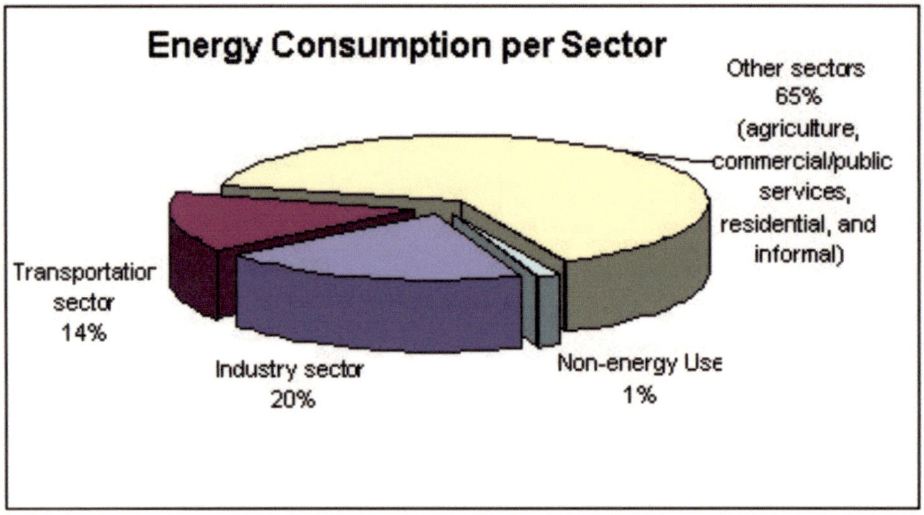

Figure 1. Energy consumption per sector

Worldwide, more than two billion people lack access to sustainable and modern energy services, using traditional solid fuels for cooking and heating. Without access to basic energy services for lighting, cooking, heating, pumping, transportation, communication and other productive purposes, people – most often women - are forced to spend the majority of their time and physical energy on subsistence activities. Lack of energy services is directly correlated with the major elements of poverty, including inadequate healthcare, low education levels and limited employment opportunities.

Development literature has recently embraced the term "feminization of poverty" referring to growing gap between men and women caught in the cycle of poverty – 70% of the 1.5 billion people living on less than a dollar a day are women (United Nations Development Program (UNDP), 1995). It is increasingly evident that gender differentiates the societal processes leading to poverty and the escape routes out. Women living in poverty are consistently without access to key resources such as credit, land and inheritance. Their labor is unrewarded and unrecognized. Their health care and nutritional needs are not given priority, they lack sufficient access to education and support services, and their participation in decision-making at home and in the community are minimal. Caught in the cycle of poverty, women lack access to resources and services to change their situation.

Poverty is one of the world's most fundamental issues, and urgently needs to be addressed. Poverty can be conceptualized in a number of ways, for example, in economic terms (an income of less than $1 a day) or in social terms (lack of access to adequate levels of food, water, clothing, shelter, sanitation, health care and education). It is also possible to identify an energy dimension to poverty: energy poverty. Energy poverty has been defined as the absence of sufficient choice in accessing adequate, affordable, reliable, high quality, safe and environmentally benign energy services to support economic and human development

http://dx.doi.org/10.5772/51818

(Reddy, 2000). Energy poverty interacts with other manifestations of poverty and it is important to explore the issues that surround it, including the gender aspects.

Estimates of the number of people living in poverty put the figure at around 1.3 billion.

Poverty is particularly acute in rural areas in developing countries, and the problem is growing, since the number below the poverty line is increasing in absolute numbers if not as a percentage. Poverty is not static but rather a dynamic process: people can move out of, or into, poverty. This needs to be taken into account when policies and strategies are being designed to help people lift themselves out of poverty. Rural people voice the need for the means to provide themselves with adequate livelihoods. These livelihoods should be sustainable, in the sense that they can withstand stresses and shocks, and they should maintain, or even enhance capabilities and assets without undermining the natural resource base. However, the opportunities for people to support themselves from the land are limited. Already many rural families, who might classify themselves as farmers, have diversified their activities and social support capabilities for survival as well as in order to try and improve their standard of living.

Another option is urban migration, particularly chosen by men, resulting in a large number of female-headed households in rural areas. For example, it is estimated that women make up 80% of the adult rural population in Zimbabwe (Nyoni, 1993). Access to cash incomes through non-agrarian activities might go some way towards stemming the exodus from rural areas (which reduces the total human capital and threatens the sustainability of an area), as well as contributing to the sustainable livelihoods of those who remain.

2. Energy as a poverty issue

Limited availability of modern energy critically impairs socioeconomic development. Industries and productive activities (agriculture, commerce) require energy in various forms to fuel machines, power transformation processes, conserve perishable goods, ensure transport, etc. According to United Nations Industrial Development Organization (UNIDO) (2004), high levels of income per capita tend to be associated with higher levels of industrialization.

Though, not specifically referred to in the targets of the Millennium Development Goals (MDGs), energy supply is an underlining requirement to achieve most of the MDGs. Without access to adequate energy services, the majority of Africans would continue to suffer from deep poverty, since energy is required for most basic household needs, such as cooking and heating. According to World Bank indicators, there is a strong correlation between modern energy consumption and Gross National Product (GNP) per capita. The GNP tends to rapidly increase as per capita commercial energy use increases (figure 2), mainly for low-income countries. When the countries reach a level of per capita energy consumption of around 4.18680e+11 (10.000 TOE), factors such as efficient utilization of energy by industries, energy production and transformation systems and households tend to make the difference for economic growth to continue; therefore, more energy consumption does no longer imply more income for the country.

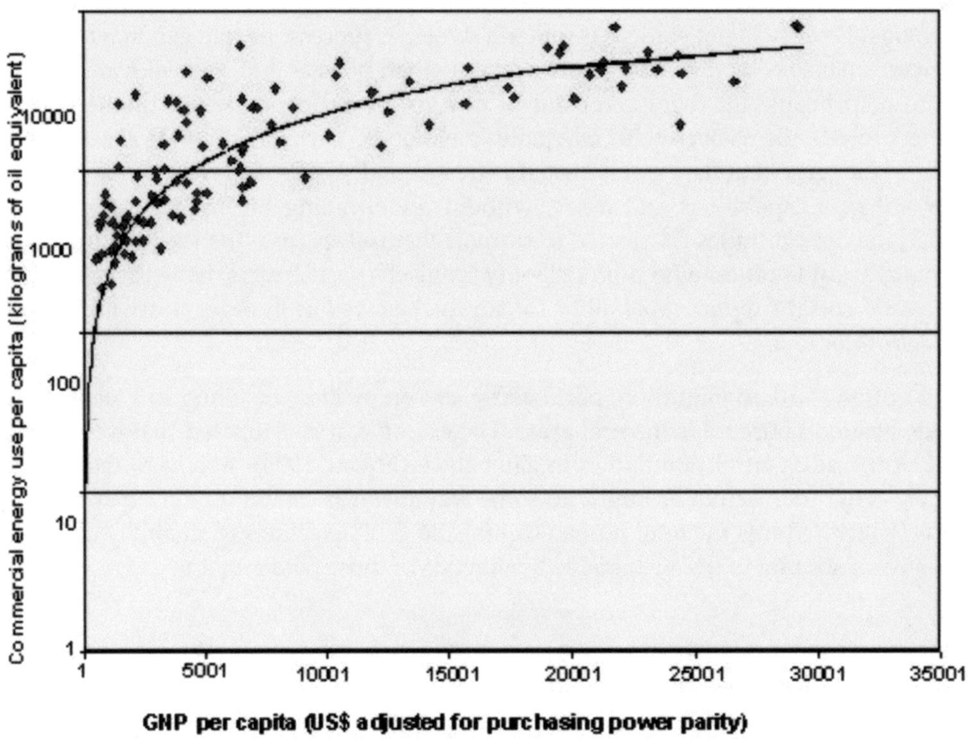

Figure 2. Energy consumption versus GNP. Source: World Bank, World Development Indicators database.

Energy is one of the most essential inputs for sustaining people's livelihoods. At the most basic level, energy provides cooked food, boiled water and warmth. It has long been established that poor people mostly use biomass as their energy carrier and that in many areas there is an increasing shortage in supply, which adds to the burden of the women whose responsibility it is to collect it. This was christened 'the other energy crisis by the World Resources Institute as early as 1975, when the world was still reeling from the international oil crisis (Eckholm, 1975). However, despite the fact that around two billion people still use biomass fuels (World Bank, 1996), and the fact that these are also the two million poorest people on earth, there has been little attempt to analyze the energy-poverty nexus in depth. This can partly be explained by the fact that the biomass in rural areas is collected at zero monetary cost, mainly by women and children, and so it falls outside national energy accounts, the result of which is that the issue renders itself invisible: No data - no visibility; no visibility – no interest (Huyer & Westholm 2001).

The use of biomass has a number of repercussions for poor people. The fuel quality is low, and when burnt it gives off quantities of smoke and particulates that are recognized as having negative effects on health. The several hours a day spent in collecting fuel means that this time cannot be used for other livelihood activities. Although nearly every household in rural areas will use some biomass as an energy carrier, poor households will spend more time searching than those in higher income groups (Reddy, 2000). Wealthier households will also purchase other, higher quality, fuels, which will be used for a greater variety of end-uses than in poor households. In urban areas, poor people have to purchase cooking fuels, and they spend a higher proportion of their income, than higher income households, on fuels (Energy Assessment and Energy Sector Management Assistance Program (ESMAP), 1999). Typically, a poor urban family spends 20% of its income on fuels (Barnes, 1995). In rural areas, poor households will generally restrict fuel purchases to lighting uses (candles and kerosene).

Energy has an equity dimension: poor households use less energy than wealthier ones in absolute terms. Less water is boiled for drinking and other hygiene purposes, increasing the likelihood of water-borne diseases. Illness reduces the ability of poor people to improve their livelihoods and increases their vulnerability, not only preventing adults from working effectively but also negatively effecting children's learning. It is frequently said that more lighting for poor families would allow children to study at night, but the extent to which home lighting really improves educational performance and life chances is unknown. Lighting, in theory, also provides opportunities for extended working hours and thus improved income generation. Further, street lighting and lighting in community centers can open the way for adult education but whether it does so, and whether poor families take advantage of this, are not certain.

Wealthier people are able to exercise some choice in their energy carrier and many opt for the cleaner and more efficient "modern" energy carriers of electricity or gas, including Liquefied Petroleum Gas (LPG) or biogas, although the use of energy carriers is complex. Many better off households use mixtures of modern and traditional fuels, each matched to a specific end purpose, often, for reasons not linked to price. Modern energy carriers do not have the negative health and time effects associated with biomass. Wealthier people are also better able to afford the appliances that make use of these modern energy carriers. In situations where they are reliant on biomass fuels, they are able to purchase more fuel -efficient stoves. In doing so they may be saving a great deal of money per unit of energy consumed. Unfortunately, poor people are often unable to make such investments, opting for lower first cost options, rather than lower life cycle costs, because of their lack of capital (Reddy & Reddy, 1994). The consequences for the poor are that precious cash resources are used on low quality fuels, which are then used at low efficiency, reducing their ability to accumulate the financial resources they need to invest in strategies for improving their livelihoods.

This vicious cycle of energy poverty needs to be broken (figure3 and figure 4). Understanding the decision -making process within households when choosing energy services, which would appear at present to work against sustainable livelihoods, is important for designing effective interventions. A first step towards this should be a widespread acknowledgement among the development community that a lack of access to clean and affordable energy can, and should, be considered a core dimension of poverty. In this respect, it is therefore welcome that the World Summit on Sustainable Development acknowledged that access to energy is needed to help achieve the Millennium Development Goals.

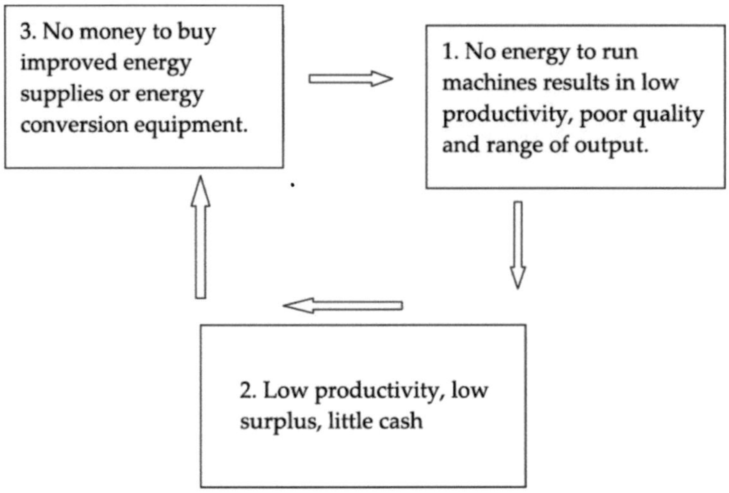

Figure 3. The vicious circle of energy poverty

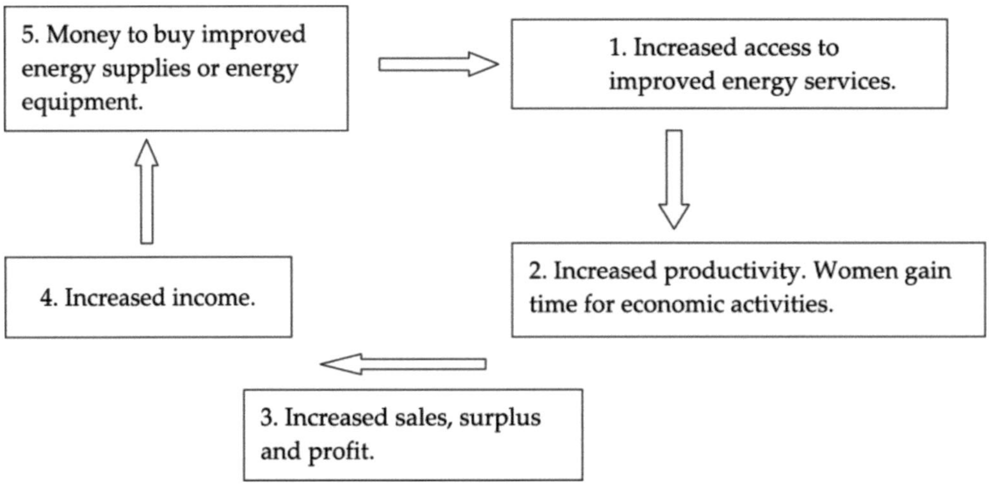

Figure 4. A virtuous circle to break out of energy poverty. Source: Institute for Development Studies (IDS) (2003)

http://dx.doi.org/10.5772/51818

3. Gender and energy

Gender refers to different social roles that women and men play, and the power relations between them. Gender relations influence how communities, households, and institutions are organized, how decisions are made, and how resources are used. To understand how gender shapes activities that affect the environment, it is necessary to examine women's and men's roles and responsibilities, access to and control over resources, knowledge of resources and authority to make decisions about resource use. Therefore understanding women's and men's relationships to the environment plays an important role in developing solutions and meeting challenges for more sustainable use of energy resources. Ignoring gender distorts the understanding of human impacts on the environment as a whole.

From a gender perspective, 70% of the 1.3 people in developing countries living below the poverty threshold are women (Denton, 2001). Of the two billion people without access to modern energy services, most live in rural areas, where women head most of the poor households. Lack of modern fuels impacts more on women and girl child. Hundreds of millions of women and girls spend between three - eight hours carrying fuelwood, dung and other traditional biofuels everyday. The immediate family energy use need is heat; when fuels become scarcer, girl children over boy children are withdrawn from school to support family energy needs. Girls who do not go to school cannot be literate. Illiterate women have more children, larger and poorer families that reinforces the cycle of poverty and underdevelopment. 2 million people die each year due to health and respiratory effects from indoor air pollution (MacDade, 2002).

For poor people the energy problem is "we do not have it". But worse for women, the energy problem is lack of access, control, power relations and social dynamics. Men and women in a given society have different roles, needs and aspirations for energy. From a production point of view, the participation of women in the energy sector has largely been restricted to forestry and biomass management in rural areas. In urban areas women have remained victims of environmental impacts of coal based electricity production that serves middle to high income groups. A few women are involved in the formulation of energy policy, and big energy projects have remained the preserve of men. Both rural and urban dwellers have little say and choice over domestic fuels that they use. In rural areas such say and choices is determined by woodfuel availability, and to some extend availability of cash to purchase fuel such as paraffin.

4. The gender component of energy poverty

The energy-poverty nexus has distinct gender characteristics. Of the approximately 1.3 billion people living in poverty, it is estimated that 70% are women, many of whom live in female-headed households in rural areas (Dutta, 1997). It is important to take note of this fact, not only because men and women have different energy needs and may have different ideas about sustainable livelihoods, but also because women and men have different access to resources and decision-making. Women's access to decision-making within the household and community is restricted, limiting their ability to influence processes and resource allocation on many issues including energy.

Women spend more time than men on basic subsistence activities, such as gathering fuelwood, carrying water, and cooking. According to the World Bank (2001) women of all developing countries spend between 2-9 hours a day collecting fuel and fodder, and performing cooking chores. The opportunity cost of these activities frequently prevents women from undertaking income-generating activities, which deprives poor families of much needed income. When rural women do engage in income-generating activities they are performed together with regular domestic work and are generally home-based micro-enterprise or piece rate projects (sewing, weaving, preparing food to sell, etc.) Home lighting, agro-processing, drinking water pumping and more efficient stoves can reduce women's workloads, provide income earnings and improve women's health.

More than half of the world's household's cook with wood, animal waste, crop residues and untreated coal, exposing primarily women and children to indoor air pollution, which according to the World Health Organization, is responsible for the premature death of over 2 million women and children a year worldwide from respiratory infections. In rural India, shifting from fuel wood to cleaner sources of energy, like kerosene or LPG, halves the mortality rate of children under five (World Bank, 2001).

Gender issues have come to the forefront in many development sectors including agriculture, forestry and water but the energy sector has been slow to acknowledge the links between gender equality, energy and development. Traditional energy policies have inadequately addressed the role of energy as an input to development and have largely ignored the critical role women play in energy systems, particularly in rural areas. Insufficient access to modern energy and existing patterns of energy use, processing, and collection affect women and men differently. Because of their socially determined gender roles, women and girls assume a higher proportion of the burden of unavailable energy services and inefficient energy use.

5. Moving women and their families out of energy poverty

There is no doubt that energy plays a major role in meeting women's practical and reproductive needs (such as cooking, food processing and water hauling), but it can also be seen as a component necessary to meet their productive and strategic needs (lighting to enable evening study, street lighting for safety in attending community meetings, power for women's enterprise development). It is remarkable that the use of gender analysis in energy planning is virtually unknown, whereas it has been successfully used for many years in the health, water and agricultural sectors. This is evidently because the gender component in energy poverty has not been fully recognized. Energy planners have usually equated women's interest in energy with cooking, to the exclusion of other needs, particularly of needs related to productive activities and emancipatory goals. In addition, since the main focus of energy planning has been on fossil fuels to the exclusion of biomass fuels, even women's practical needs have hardly been addressed.

If gender aspects of the energy-poverty nexus are to be adequately dealt with, it is clear that two major transformations have to take place. Firstly, women have to be empowered to make choices about energy. Enabling choice is linked to issues of sustainable livelihoods and poverty alleviation, including access to income generating activities. However, there is more at stake than just improvements in women's financial resources. Women should be able to act upon the energy choices open to them, and their scope for this type of action is linked to decision-making within households. Such a shift in decision -making requires women's social and political empowerment.

Secondly, it also requires changes on the energy supply-side. It will require responsiveness by the energy sector in the provision of equipment using modern energy forms that reduce the drudgery of much of women's labour, and that at affordable prices.

6. Energy and sustainable livelihoods

The World Summit on Sustainable Development (WSSD) in Johannesburg in 2002 recognized the important role of energy for reaching millennium development goals. Access to affordable, reliable and sustainable energy is essential to sustainable development (Hasna, 2007). An adequate solving of energy problems will contribute to achieving progress across all pillars of sustainable development; social, economic and environmental and in meeting the UN millennium goals. Although there are no MDGs on access to energy, WSSD recognized that inadequate access to energy is both a cause and an effect of poverty and recommended the following:

"Take joint actions and improve efforts to work together at all levels to improve access to reliable and affordable energy service for sustainable development sufficient to facilitate the achievement of the Millennium Development Goals, including the goal of halving the proportion of people in poverty by 2015, and as a means to generate other important services that mitigate poverty, bearing in mind that access to energy facilitates the eradication of poverty".

"Sustainable development" has been defined best by the Brundtland Commission as development that meets the needs of the present without compromising the ability of future generations to meet their own needs (Hasna, 2007). Adequate and affordable energy supplies has been key to economic development and the transition from subsistence agricultural economics to modern industrial and service oriented societies. Energy is central to improved social and economic well being and is indispensable to most industrial and commercial wealth organization. It is the key for relieving poverty, improving human welfare and raising living standards. But however essential it may be for development, energy is only a means to an end. The end is good health, high living standards, a sustainable economy and a clean environment (Fatona, 2011).

Much of the current energy supply and use, based as it is, on limited resources of fossil fuels, is deemed to be environmentally unsustainable. There is no energy production or conversion technology without risk or waste. Somewhere along all energy chains - from resource extractions to the provision of energy service – pollutants are produced, emitted or disposed of, often with severe health and environmental impacts (Dasgupta, 2001;). Combustion of fossil fuels is chiefly responsible for urban air pollution, regional acidification and the risk of human – induced climate change (Dasgupta, 2001; Fatona, 2009).

Achieving sustainable economic development on a global scale will requires the judicious use of resources, technology, appropriate economic incentives and strategic policy planning at the local and national levels. It will also require regular monitoring of the impacts of selected policies and strategies to see if they are furthering sustainable development or if they should be adjusted (Arrow et al, 2004).

When choosing energy fuels and associated technologies for the production, delivery and use of energy services, it is essential to take into account economic, social and environmental consequences (Ott, 2003; Wallace, 2005). There is need to determine whether current energy use is sustainable and, if not, how to change it so that it is. This is the purpose of energy indicators, which address important issues within three of the major dimensions of sustainable development: economic, social and environmental.

7. Gender, energy, poverty and sustainable livelihoods

In most developing countries, the majority of informal sector enterprises are owned and operated by women, with women making up the largest proportion of the work force. Despite this, the contribution of women entrepreneurs to national economies is not explicit in national statistics, leading to the development of policies that do not deal with the specific barriers faced by women linked to their gender-defined roles. Their enterprises tend to be concentrated in a relatively narrow range of activities: beer brewing, knitting, dressmaking, crocheting, cane work and retail trading. These activities tend to have disproportionately low rates of return compared to the activities undertaken by men. However, despite the low financial returns, women's enterprises provide important sources of household income, even in male-headed households. Women-headed enterprises are frequently located in the home, and these "cottage industries" tend to be overlooked by agencies because they are in the informal sector, which is diffuse and difficult to reach. When women are forced to close their enterprises, it is often for non -business reasons, and linked to factors associated with working from home. The low rates of return prevent inward investment, hindering innovation and expansion which are regarded as key factors in enterprise sustainability (Grosh & Somolekae, 1996). There is little research to explain what forces drive these start-ups and shutdowns, and how gender influences these processes. Women's access to resources (such as credit, land and education, which are recognised as key factors in microenterprise development) is significantly less than that of men. Generally, research in small and medium-scale enterprise sustainability indicates that a lack of working capital is one of the two most

http://dx.doi.org/10.5772/51818

common causes of enterprise failure (Grosh & Somolekae, 1996). Although there are a number of microcredit programmes targeting women, Bangladesh's Grameen Bank being the most well known, research is increasingly questioning whether women are able to fully utilize the credit, and what degree of control they retain over the loans once disbursed (Baden et al., 1994). Women's access to decision -making within the household and community is also restricted, reducing their ability to influence processes and resource allocation.

The role of energy in the sustainability of women's enterprises is also not well understood. In food processing enterprises it has been estimated that energy costs are 20 - 25% of the total inputs, which would suggest that technological interventions could increase the scale and profitability of these businesses.

The types of enterprises that women are traditionally involved in are energy intensive and rely on biomass fuels. Even in rural areas, women may have to buy fuel wood to run enterprises such as beer making (McCall, 2001). An important issue is what sort of mechanisms can assist women in gaining access to improved energy services. Grain mills, which are very popular with women, since they improve product quality as well as reducing women's labour, are typically only provided by the private sector, and are still absent in many rural locations.

There are positive examples of women taking up energy technologies that have contributed to increasing their incomes. For example, women's groups in Ghana use LPG for fish preservation, giving them a better quality product than when using wood, and enabling them to reach export standards, and considerably improving their income (Mensah, 2001). Another example is the Multi -Functional Platform (Burn & Coche, 2001). The platform consists of a diesel engine mounted on a chassis to which a variety of end-use equipment can be attached, such as grain mills, battery chargers, oil presses, welding machines and carpentry tools. In addition, the engine can be used to generate electricity for sale, which opens up the prospect of women becoming energy entrepreneurs and setting up their own electricity service companies.

8. Conclusion

Energy is central to sustainable development and poverty reduction efforts. It affects all aspects of development -- social, economic, and environmental -- including livelihoods, access to water, agricultural productivity, health, population levels, education, and gender-related issues. None of the Millennium Development Goals (MDGs) can be met without major improvement in the quality and quantity of energy services in developing countries.

Since there is an energy dimension to poverty known simply as energy poverty, which manifests when there is a lack of access to clean, safe, reliable and affordable energy. Energy is opportunity, and imperative to basic activities such as boiling water, storing vaccines, reading at night. Yet, more than three billion people worldwide rely on traditional fuels such as wood, charcoal, dung, and kerosene for cooking and light.

It is well documented that there are more women than men living in poverty. Because of traditional socio-cultural roles, women and girls in developing countries bear the burden / responsibility of collecting fuel for household energy consumption. Women are at the nexus between poverty and energy, so to talk about either it's necessary to apply a gender lens.

Rural women spend long hours collecting fuel and carrying it home over long distances. The time and labour feeds into another dimension of poverty, time poverty, limiting women's ability to engage in productive or income-generating activities. Sometimes referred to as: women's invisible work, reproductive work, women's double burden, women's double day, time poverty cripples opportunity for education and income generation, which as a result often means her family will remain trapped in poverty.

Women's health suffers from carrying heavy loads of firewood long distances. In the home, over a makeshift cookstove and lantern, the exposure to smoke from fuel-based cooking and lighting contributes to over 2 million deaths per year (WHO, 2002). After-dark and without light, women suffer disproportionately from energy poverty from reduced security and gender-based violence.

An estimated 265 million tonnes of greenhouse gases are emitted annually from burning fuel for light globally, accelerating the effects of climate change and deforestation (WHO, 2002). Climate change makes women's long workday even longer due to unpredictable rains causing food, fuel and water scarcity and therefore longer treks to collect the necessities.

A number of relationships between energy, poverty and gender are clear. The poor, in general, pay more for energy than the rich, certainly in relative terms but also often in absolute terms. Those living in poverty cannot afford the upfront cash for appliances which increase the efficiency of fuel use, or enable the transition from traditional energy to modern energy. The options for poor people are in fact very limited indeed, a situation which has been called energy poverty. Biomass fuels (wood, charcoal, dung etc.) are likely to remain for some time the primary fuels for process heat and cooking because electricity is in almost all situations more expensive for such applications. Other options, such as solar cookers and household biogas systems, have not been able to meet the demand for cooking energy, probably because they require significant changes to cooking practices or are too expensive.

Having accepted that the energy sector is central to human development, the first expectation of a gender-aware policy would be that it recognizes the different needs, role, responsibilities and realities of women and men. With this in view, the gender-aware policy would ensure that the policy statement, planning and strategies for intervention promote equality and equity. The key to planning and energy policy that promotes even development of men and women while also providing remediation of past discriminations in the sector is to undertake gender mainstreaming in the sector.

This means that gender analysis must be undertaken to assess the implications of any course of action or intervention on men and women. It also means that budgetary allocations must be sufficient to address the needs of men and women. More women need to come into influential position in the energy sector.

Gender-sensitive energy programmes can ease the double burden of lack of sufficient energy and poverty that women endure as they perform traditional household and community roles. It can also provide opportunities for education and income generation that will allow women to improve their social and economic status and raise the living standards of their families and communities. However, if energy policies are to become more gender-sensitive not only will women themselves have to become more empowered to make choices about energy, but the energy sector will also have to become more responsive to women's energy needs.

Moreover, it can be concluded that the energy dimensions, both of the poverty-gender nexus and the livelihoods analysis, have been poorly understood and, hence, their significance underestimated. It is also noteworthy here that gender analytical tools and frameworks that are in standard use in the agriculture, health and water sectors (the Harvard Matrix, Gender Analysis Matrix etc.) are hardly used in energy planning. This is partly because these tools do not fully meet the needs of energy planners, focusing on general aspects of women and men's relative situations, rather than on the specific energy dimensions of poverty. To an extent the existing gender tools can be adapted to the energy sector to bring to light hidden aspects of the poverty-energy-gender nexus, but there is a clear need to develop tools specific to the energy sector to ensure that all aspects are analyzed. These gender and energy tools need to be 'user friendly' to help social planners incorporate energy dimensions in their work and energy planners to see the gender and poverty aspects of theirs.

The feminization of energy poverty is a vicious cycle impeding sustainable development. Improved access to clean energy is absolutely critical to rural women and girls' development and empowerment. From this year, 2012, the declared year of Sustainable Energy for All, let us remember to keep gender at the centre of discussion and decision-making.

The International Energy Agency (IEA) estimates that demand for energy will increase by 40 per cent in the next 20 years (IEA, 2011). The heavy dependence on fossil fuel to meet present needs, coupled with rising costs, negative environmental impacts and future growth in demand, has placed renewable energy at the forefront of sustainable development issues as a core component of future development, both in developed and developing countries.

A reliable energy source is a prerequisite for sustainable development and poverty alleviation, and the Millennium Development Goals cannot be achieved without first addressing energy issues. For example, health clinics, schools and other vital infrastructure cannot run properly, while access to drinking water and sanitation cannot be guaranteed without a reliable energy source.

Giving the energy poor access to energy is therefore a development imperative. Yet at the same time, successfully mitigating and adapting to climate change will require a global shift toward renewable energy. Unmanageable and unforeseen climate risks threaten to negate development efforts, so action taken to reduce energy poverty must also be compatible with climate protection.

In short, sustainable development requires sustainable energy. A green economy based on fairness, social welfare, and environmental integrity must be powered by green energy; energy from sources that do not unduly impact the environment or people. Green energy means satisfying growing energy needs, empowering local communities and supporting development, while not producing dangerous carbon emissions or causing other negative environmental impacts.

Author details

Pius Fatona, Abiodun Abiodun, Adetayo Olumide, Adesanwo Adeola and Oladunjoye Abiodun

Ogun State College of Health Technology, Ikese-Ijebu, Nigeria

References

[1] Arrow KJ, P. Dasgupta, L. Goulder, G Daily, PR Ehrlich, GM Heal, S Levin, K-G Maler, S Schneider, DA Starrett, B Walker. (2004). Are we consuming too much? Journal of Economic Perspectives, 18(3):147–172

[2] Baden S, Goetz A M, Green C and Guhathakurta M (1994), Background to Gender Issues in Bangladesh. BRIDGE Report No 26. http://www.ids.ac.uk/bridge.

[3] Barnes D (1995), Consequences of Energy Policies for the Urban Poor. FPD Energy Note No7, The World Bank, Washington.

[4] Burn N and Coche L (2001), Multifunctional Platform for Village Power, in "Generating Opportunities: Case Studies on Energy and Women", UNDP, New York. ISBN 92-1-126124-4.

[5] Dasgupta, P. (2001). Human Well-Being and the Natural Environment. Oxford University Press, Oxford.

[6] Denton, F., (2001). Report on Gender and Energy Workshop: Moving Towards Practical Solutions of Meeting Gender Differentiated Energy Needs Within an Integrated Development Approach, 11 –13th June, 2001, Dakar, Senegal.

[7] Dutta S (1997), Role of Women in Rural Energy Programmes: Issues, Problems and Opportunities, ENERGIA News No.4, pp11-14.

[8] Eckholm, E. (1975), The Other Energy Crisis: Firewood. Washington DC: Worldwatch paper no 1

[9] ESMAP (1999), Household Energy Strategies for Urban India: The Case of Hyderabad, Washington.

[10] Fatona, P. Olugbenga (2011). Renewable Energy Use and Energy Efficiency- A Critical Tool for Sustainable Development. In Majid Nayeripour & Mostafa Khesti (eds)

(2011) Sustainable Growth and Applications in Renewable Energy Sources, INTECH, ISBN 978-953-307-573-0

[11] Fatona, P. Olugbenga (2009). Energy Exploitation, Utilization and its Environmental Effects – the Choice to Make and the Decision to take. Toxicological & Environmental Chemistry, 91: 5, 1015-1019

[12] Grosh B and Somolekae G (1996), Mighty Oaks from Little Acorns: Can Microenterprise Serve as the Seedbed of Industrialisation? World Development Vol. 24, No 12, pp 1879-1890. Elsevier Science Ltd.

[13] Hasna, A. M. (2007). "Dimensions of sustainability". Journal of Engineering for Sustainable Development: Energy, Environment, and Health 2 (1): 47–57.

[14] Huyer S and Westholm G (2001), Toolkit on Gender Indicators in Engineering, Science and Technology, prepared for the UNESCO Gender Advisory Board.

[15] http://gstgateway.wigsat.org/TA/data/toolkit.html

[16] Institute for Development Studies (IDS) (2003). Energy and Poverty: A Review of the Evidence and Case Studies in Rural China. World Bank Report; Washington DC.

[17] International Energy Agency, (2011). 'Energy for All: Financing Access for the Poor', early excerpt of the World Energy Outlook 2011 (OECD/IEA, Paris 2011).

[18] Karlsson G and Misana S, eds, 2001. Generating Opportunities: Case Studies on Energy and Women, United Nations Development Programme, (UNPD), New York.

[19] MacDade, S., (2002). Energy for Sustainable Development: Meeting the Rural Energy Challenge. Paper presented at the UNDP/WEC side event during the third United Nations Preparatory Meeting, 25th March – 5th April, 2002, New York.

[20] McCall M (2001), Brewing rural beer should be a hotter issue. Boiling Point No 47, pp23-25. ITDG, Rugby, UK.

[21] Mensah S A (2001), Energy for Rural Women's Enterprises, in "Generating Opportunities: Case Studies on Energy and Women", UNDP, New York. ISBN 92-1-126124-4.

[22] Nyoni S (1993), Women and energy: lessons from the Zimbabwe experience, Working paper 22, Zimbabwe Environment Research Organisation (ZERO), Harare, Zimbabwe.

[23] Ott, K. (2003). "The Case for Strong Sustainability." In: Ott, K. & P. Thapa (eds.) (2003).Greifswald's Environmental Ethics. Greifswald: Steinbecker Verlag Ulrich Rose.

[24] Reddy A K N (2000), Energy and Social Issues, in "World Energy Assessment", UNDP, New York. ISBN 92-1-126-126-0.

[25] Reddy A K N and B S Reddy (1994), Substitution of Energy Carriers for Cooking in Bangalore. Energy, 19(5): 561-71.

[26] UNDP (1995). Human Development Report 1995. UNDP; New York

[27] UNIDO (2004). Industrial Development Report 2004: Industrialization, Environment and the Millennium Development Goals in Sub-Saharan Africa, The new frontier on the fight against poverty, 2004, www.unido.org

[28] Wallace, Bill (2005). Becoming part of the solution: the engineer's guide to sustainable development. Washington, DC: American Council of Engineering Companies. Initiative 62(3): 282–292.

[29] WHO (2002) "Fuel for Life: Household Energy and Health" – report by the World Health Organization).

[30] World Bank (1996). Rural Energy and Development: Improving Energy Supplies for 2 Billion People. Washington, DC, USA. www.worldbank.org

[31] World Bank (2001). World Bank Database on Private Participation in Infrastructure (PPI), www.worldbank.org

Section 2

Modeling and Analysis

Improved Stochastic Modeling: An Essential Tool for Power System Scheduling in the Presence of Uncertain Renewables

Sajjad Abedi, Gholam Hossein Riahy,
Seyed Hossein Hosseinian and Mehdi Farhadkhani

Additional information is available at the end of the chapter

http://dx.doi.org/10.5772/45849

1. Introduction

Nowadays, governments are developing ambitious goals toward the future green and sustainable sources of energy. In the U.S., the penetration level of wind energy is expected to be 20% by the year 2030 [1]. Several European countries already exhibit the adoption level in the range of 5%–20% of the entire annual demand. Also with further developments in the solar cells technology and lower manufacturing costs, the outlook is that the photovoltaic (PV) power will possess a larger share of electric power generation in the near future. Grid-connected PV is ranked as the fastest-growing power generation technology [2]. PV generates pollution-free and very cost-effective power which relies on a free and abundant source of energy.

Due to the increasing wind and solar penetrations in power systems, the impact of system variability has been receiving increasing research focus from market participants, regulators, system operators and planners with the aim to improve the controllability and predictability of the available power from the uncertain resources. The produced power from these resources is often treated as non-dispatchable and takes the highest priority of meeting demand, leaving conventional units to meet the remaining or net demand. This issue makes the optimum scheduling of power plants in power system cumbersome as embeds the stochastic parameters into the problem to be handled. The unpredictability along with potential sudden changes in the net demand, may face operators with technical challenges such as ramp up and down adaptation and reserve requirement problems [3-4].

Several investigations aiming at handling the uncertain nature of wind and solar energy resources have been reported. Basically, the methods found in the literature can be classified into three groups: methods that deal with the prediction of uncertain variables as an input data pre-processing, methods that use stochastic scenario-based approach within the optimization procedure to cover all the outcomes per the probable range of uncertain variables, and methods based on a combination of these two approaches. The studies presented in [5-7] can be mentioned as one of the most recent efforts lying in the first group. In [5-6] an Artificial Neural Network (ANN) forecast technique is employed and followed by risk analysis based on the error in the forecast data. Then, the so called pre-processed data is directly taken as the input to the optimization process. Relying on the forecast tools, such methods suffer from high inaccuracy or ex-ante underestimation of the available power which increases the scheduled generation and reserve costs. Anyway, this approach is useful as it accounts for the temporal correlation between the random variables representative of each time step of the scheduling period, in terms of time-series models. On the other hand, in [8-9] which belong to the second group, the focus is on the stochastic scenario analysis rather than the forecasting methods. The usage of this approach also has its own advantages, as it tries to model the likely range of values for the random variables. However, the efficiency of this approach largely depends on the accuracy and reliability of their probabilistic analysis; based on which the potential scenarios are built.

The most effective approach is associated with the third group, which applies the advantages of both forecast techniques and scenario-based optimization approach. Reference [10] presents a computational framework for integrating a numerical weather prediction (NWP) model in stochastic unit commitment/economic dispatch formulations that describes the wind power uncertainty. In [11], the importance of stochastic optimization tools from the viewpoint of the profit maximization of power generation companies is investigated. The exposed financial losses regarding the wind speed forecast errors are discussed. A stochastic model is also presented in [12]which uses a heuristic optimization method for the reduction of random wind power scenarios. The wind speed data is assumed to follow the normal PDF. A similar approach is introduced in [13] whereas the wind speed error distribution is considered as a constant percentage of the forecasted data. In [14], the Auto-Regressive Moving Average (ARMA) time series model was chosen to estimate the wind speed volatility. Based on the model, the temporal correlation of wind speed at a time step with respect to the prior time steps is well analyzed.

In this chapter, the authors present a framework for stochastic modeling of random processes including wind speed and solar irradiation which are involved in the power generation scheduling optimization problems. Based on a thorough statistical analysis of the accessible historical observations of the random variables, a set of scenarios representing the available level of wind and solar power for each time step of scheduling are estimated. To this aim, the Kernel Density Estimation (KDE) method is proposed to improve the accuracy in modeling the Probability Distribution Function (PDF) of wind and solar random variables. In addition, the concept of aggregation of multi-area wind/solar farms is analyzed using Copula method. Taking the advantage of this method, we can reflect the interdependency and spatial correlation of the power generated by several wind farms or PV farms that are spread over different locations in the power system. A final framework is developed to perform the stochastic analysis of the random variables to be input into the stochastic optimization process, as discussed in the following sections.

2. Methodology of data processing

2.1. Probability distribution function and data sampling

In order to generate sample data for random variables, the random behavior should be simulated somehow that the model follows the historical data pattern with the most homology to real data. In order to specify the pattern of a random variable, the PDF should be obtained. There are two classes of methods to determine the PDF of a random variable including parametric and non-parametric methods [15]. In parametric methods, the data samples are fitted to one of the well-known standard PDFs (such as Normal, Beta, Weibull, etc.) so that the most possible adaptation between the PDF and the existing data is achieved. The values associated with the PDF parameters are evaluated using Goodness of Fit (GoF) methods such as Kolmogorov-Smirnov test [16]. On the other hand, the nonparametric methods do not employ specific well-known PDF models.

The use of parametric methods in some studies in which simulation of probabilistic models for wind and solar data is included have been reported, as in [6, 12]. Similarly, authors in [9] employ a fixed experimental equation to represent the PDF of wind data. However, this approach to PDF estimation can bring about some defects as follows:

1. The parametric methods may show significant deviation to the actual distribution of data, mainly because the actual distribution does not characterize the underlying symmetry in the standard PDFs. As an example, Figure1 shows the distribution function for yearly solar irradiation sample data at 11 AM in a region. As seen in the figure, the parametric distribution fittings are not capable of modeling the right side skewness in the actual distribution, which will reveal considerable error in the outcoming samples.

2. Some random variables in general and particularly solar irradiation and wind speed are very time-dependent in behavior. In other words, their patterns change with different time periods, months and seasons. Hence, the nonparametric approach is advantageous in terms of time period adaptation, because it does not consider a specific type of distribution. However, the parametric approach tries to nominate a certain type of PDF to each random phenomenon in all circumstances. For instance, it is common to associate a Weibull pattern to wind speed data, which may not be the most appropriate option to be generalized to all time periods.

Based on the aforementioned facts, in this study, it is desired to obtain the most accurate distribution model taking the advantage of Kernel Density Estimation (KDE), categorized as a non-parametric method.

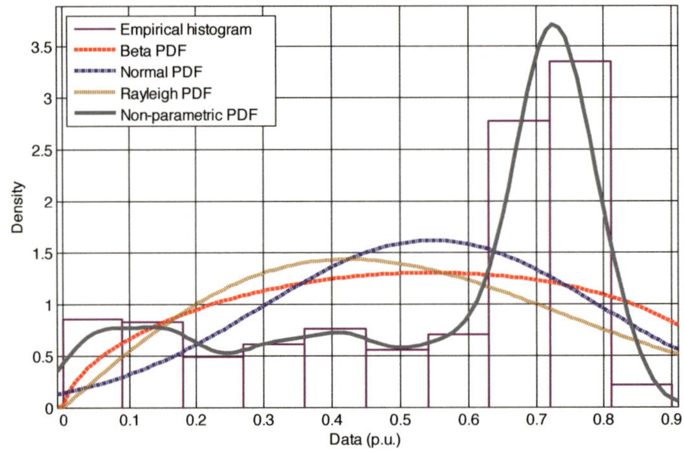

Figure 1. Solar irradiation data histogram for samples at 11 AM during a season and the parametric and nonparametric PDF fits to the data

2.2. Kernel Density Estimation (KDE)

The simplest and most frequently used nonparametric method is to use a histogram of historical samples. As a brief description of the method, the distance covering the range of samples is divided into equal sections called "bin". For each bin, a sample value is considered as the kernel of that bin. A number of rectangular blocks equal to the number of samples in each bin, each with unity area, are located on each bin. In this way, a discrete curve is obtained that somewhat describes the probability distribution of samples. However, the overall curve is largely dependent on the size of bins and their marginal points, because with the alterations in the bin size, the number of samples in each bin will be changed [17]. Besides, the obtained curve suffers from high raggedness. Hence, KDE method was introduced to solve the mentioned drawbacks. In this method, considering the samples as the kernels of each bin, the blocks are with a unity width and a height equal to the inverse of the number of samples for each sample value ($\frac{1}{\mathrm{n}}$) totally gaining a block with unity area. The accumulation of these blocks builds the PDF curve. In order to smooth the curve and eliminate the dependency to the block width, continuous kernels such as Gaussian or Cosine along with kernel smoother methods are used [18]. Figure 2 demonstrates a KDE with Gaussian kernel. The overall PDF will be obtained by:

$$\hat{f}(x)=\frac{1}{N}\sum_{i=1}^{N}\frac{1}{h_i}.K\left(\frac{x-X_i}{h}\right) \tag{1}$$

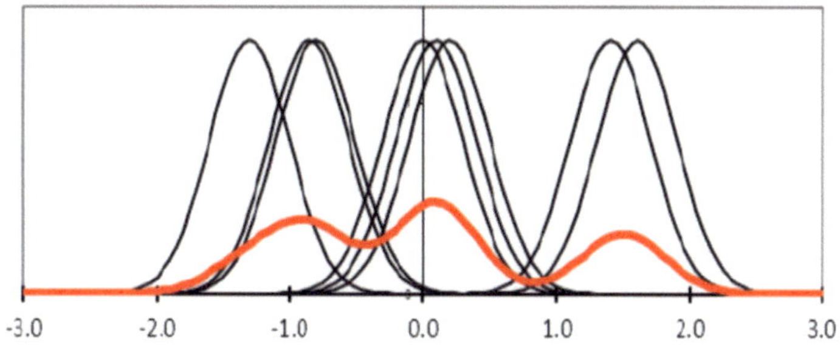

Figure 2. Demonstration of a KDE with Gaussian kernel

where h is the smoothing factor. The kernel function in Gaussian case is given by:

$$K(x) = \frac{1}{N}\sum_{i=1}^{N}\frac{1}{\sqrt{2\pi}}.e^{-\frac{(x-X_i)}{2h^2}} \tag{2}$$

In the present study, the KDE method is used to obtain the PDF of the seasonally wind speed and solar irradiation data for each hour in a day. The method is implemented using the *dfit tool* function in MATLAB.

2.3. Correlation of random variables

Sample generation from a random variable is possible simply using a Monte Carlo simulation of its corresponding PDF. However, this is more cumbersome for a group of random variables which may have underlying dependence or correlation. Neglecting the correlation will result in the inaccurate multivariate PDF and then to irrelevant and deviated samples.

There are several correlation coefficients to quantify the correlation among a number of random variables, among which the most famous one is the Pearson coefficient:

$$\rho = \frac{\mathrm{cov}(i,j)}{\sqrt{\mathrm{cov}(i,i).\mathrm{cov}(j,j)}} \tag{3}$$

where *cov* is the covariance function. This analysis reflects only the *linear* correlation among the random variables. Nevertheless, in many cases, random variables reveal nonlinear correlation and more complicated relationships among themselves, especially when the PDF of the variables are not of similar patterns [19]. In a case with a large number of random variables, neglecting the nonlinear relation will result in more significant deviation of output samples from what they actually should be. For instance, the solar irradiation behavior in two different regions in power system may not establish a linear relation between each other, though they are not completely independent. The nonlinear correlation concept is presented in Figure 3.

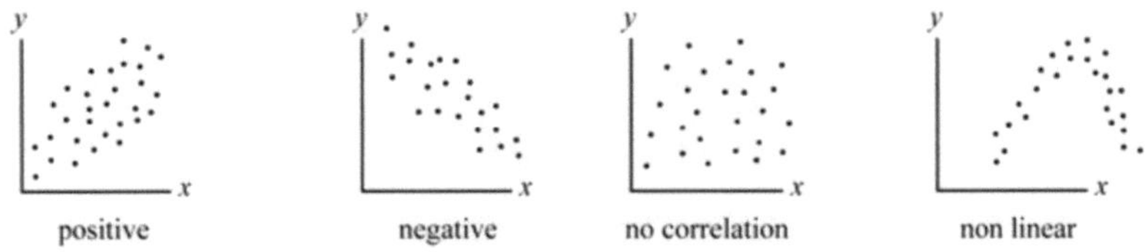

Figure 3. A representation of linear vs. nonlinear correlation between two random variables.

In the problem under study, i.e. power system scheduling in the presence of uncertain renewables, we consider the presence of multiple wind farms and solar farms throughout the power system. The solar power and wind power as well as load demand are three distinct stochastic processes. They can be discriminated into 24 random variables representing 24 hours of the day.

The random variables within each random process have their own temporal relation which can be modeled by time series prediction methods [14]. However, there may also be spatial correlation among the random variables from different processes, although it might seem unlikely. Here, we are going to deal with the tangible nonlinear correlation between the hourly random variables for a wind farm and another farm located in a close region, as well as for a PV farm and another one located in a close region. The interested reader may examine other possible dependence structures between random variables / processes. Obviously, taking into account these relations results in more accuracy and enhancement of the models and solutions. Figure 4 presents how neglecting the correlations and directly using single variable PDFs to generate samples for a multivariate process may lead to model malfunction. For two random variables with similar Log-normal distribution and Pearson correlation of 0.7 (with diagonal covariance matrix), 1000 samples have been generated considering total independence (Figure 4 (a)) and linear dependence (Figure 4 (b)). It is observed in Figure 4 (b) that X1 values tend to be closer to X2 values especially in the upper range, in comparison with Figure 4 (a).

In order to describe the correlations between random variables including nonlinear correlations, a method named Copula can be employed which is described in the following section.

2.3.1. Copula method

The correlation between random variables or samples is measured by the Copula concept. Embrechts & McNeil introduced Copula functions for application in financial risk and portfolio assessment problems [20]. Recently, much attention is being paid to this method in statistical modeling and simulation problems.

Copulas provide a way to generate distribution functions that model the correlated multivariate processes and describe the dependence structure between the components. The cumulative distribution function of a vector of random variables can be expressed in terms of marginal distribution functions of each component and a copula function.

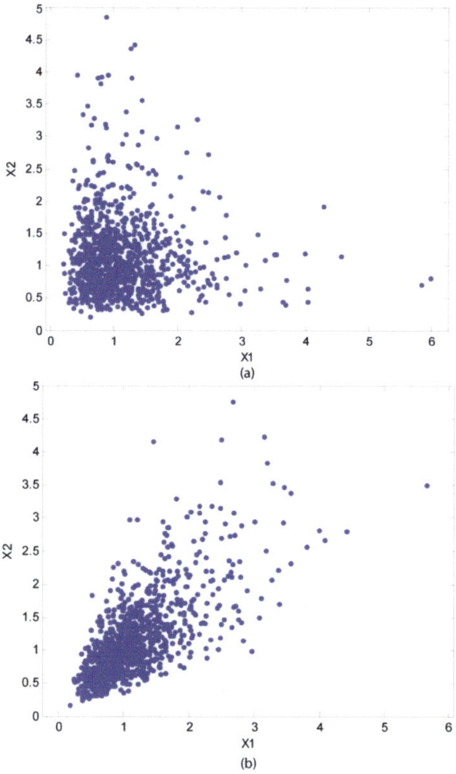

Figure 4. The effect of correlation on generation of random samples: (a) the samples are generated assuming complete independence and (b) the samples are generated considering the correlation.

The basic idea behind the copula method is described as the Sklar's theorem [21]. It shows that a multivariate cumulative distribution function (CDF) can be expressed in terms of a multivariate uniform distribution function with marginal density functions U(0,1). In fact, if we have n random variables with an n-variable CDF, F, with margins $F_1, F_2, \ldots, F_n$, there is an n-variable distribution function, C, given by:

$$F(x_1, x_2, \ldots, x_n) = C(F_1(x_1), \ldots, F_n(x_n)) \tag{4}$$

This equation can be rewritten to extract the Copula of the joint distribution function of the random variables, as follows:

$$C(u_1, \ldots, u_n) = F(F_1^{-1}(u_1), \ldots, F_n^{-1}(u_n)) \tag{5}$$

where $F_i^{-1}(u_i)$ is the inverse CDF. If F is a continuous multivariate PDF with continuous single-variable PDFs, the implicit copula distribution function is obtained as follows:

$$c(u) = \frac{f(F_1^{-1}(u_1),\ldots,F_n^{-1}(u_n))}{f(F_1^{-1}(u_1)).f(F_2^{-1}(u_2))\ldots f(F_n^{-1}(u_n))} \tag{6}$$

This equation can be restated as:

$$c(F_1(x_1),\ldots,F_n(x_n)) = \frac{f(x_1,\ldots,x_n)}{f_1(x_1)\ldots.f_n(x_n)} \tag{7}$$

where c is the PDF corresponding to C. Therefore, a multivariate PDF can be written in terms of the product of its single-variable marginal distributions and its underlying copula (c):

$$f(x_1,\ldots,x_n) = c(F_1(x_1),\ldots,F_n(x_n)).f_1(x_1)\ldots f_n(x_n) \tag{8}$$

Various copula functions are introduced by present. They are generally classified into explicit and implicit types. The implicit copulas are inspired by standard distribution functions and have complicated equations, whereas the explicit ones are simpler and do not follow the specific functions. Among the most widely used implicit copulas, Gaussian copula and t-Student copula and among the explicit ones, Clayton copula and Gumbel copula can be mentioned. The selection of the most appropriate copula is a complicated problem itself. Here, the t-Student copula is employed because of its simplicity and flexibility. The t-Student copula is formulated as [21]:

$$C_{\rho,\upsilon}(u,v) = \int_{-\infty}^{t_\upsilon^{-1}(u)} \int_{-\infty}^{t_\upsilon^{-1}(v)} \frac{1}{2\pi(1-\rho^2)^{1/2}} \left\{1 + \frac{s^2 - 2\rho st + t^2}{\upsilon(1-\rho^2)}\right\}^{-(\upsilon+2)/2} dsdt \tag{9}$$

where (ρ,υ) are the copula parameters, $t_\upsilon^{-1}(.)$ is the T distribution function with υ degrees of freedom, mean of zero and variance of $\upsilon/(\upsilon - 2)$. The best values for these parameters can be estimated using Inference Functions for Margins (IFM) or Canonical Maximum Likelihood (CML) methods. In both methods, at first the parameters of the single-variable marginal distribution functions are computationally or experimentally determined. Then, by substitution of these functions into Copula likelihood functions, the Copula functions are calculated so that the Copula likelihood functions are maximized. Further discussion on the mathematical background and the calculation methods can be found in [21-23] which is suggested to be pursued by the interested reader.

In the current study, the authors employed the two-dimensional Copula method to present the correlation of the wind speed patterns between two wind farms and the correlation of the solar irradiation patterns between two PV farms, for every hour of the day. The available data for three years are initially normalized:

$$s = \frac{x - \mu_x}{\sigma_x} \tag{10}$$

where μ_x, σ_x are the mean and standard deviation of data x, respectively. The simulation results for wind speed data distribution in farm 1 (x axis) and farm 2 (y axis) for 7 and 11 AM in fall season are shown in Figure 5. Figure 6 shows the samples for solar irradiation data distribution on two farms. Also, the values of linear correlation coefficients, the Kendall's tau correlation and the t-Student copula parameters for wind speed are presented in Table 1. The results state that the correlation of wind speed (and similarly solar irradiation) between two farms may not be negligible since they have similarities in their climatic conditions.

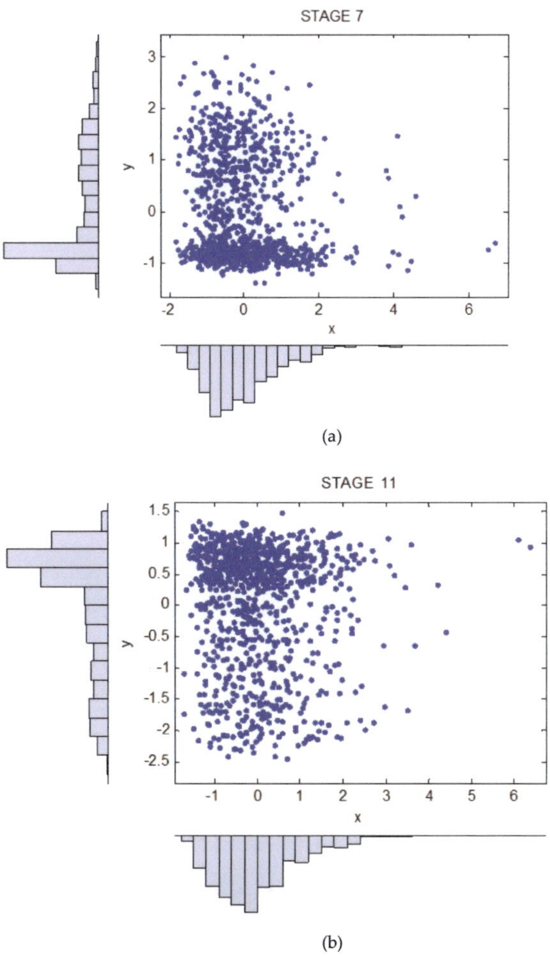

Figure 5. Correlated samples for wind speed of two farms for 7 and 11 AM in fall season.

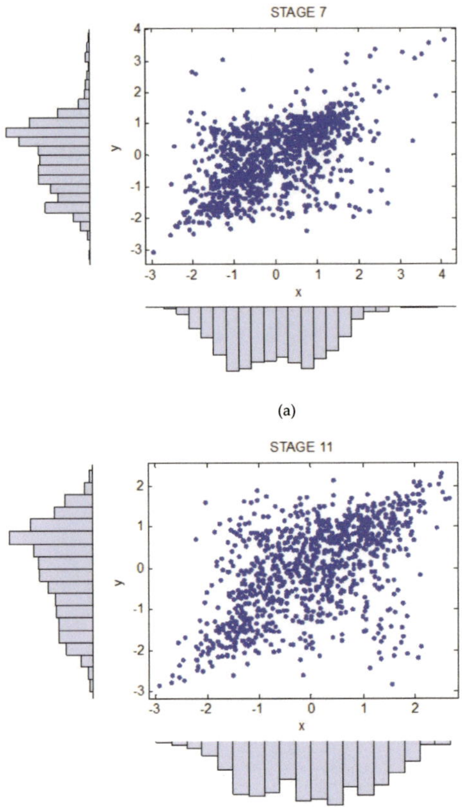

(a)

(b)

Figure 6. Correlated samples for solar irradiation of two farms for 7 and 11 AM in fall season

Kendall's tau correlation	Freedom degree (υ)	Linear correlation (ρ)	Hour
-0.14	3.7	-0.21	7
-0.12	15.58	-0.19	10
-0.08	7.39	-0.13	11
-0.12	24.31	-0.18	16

Table 1. The parameters of the two-variable t-Student copula distribution for wind speed of two farms at different hours of the day in fall season

2.4. Time-series prediction of wind speed and solar irradiation

As mentioned earlier, besides the spatial correlation among different farms, the wind speed and solar irradiation random variables assigned to the scheduling time steps exhibit temporal correlation, i.e., they are dependent on the condition of random variables at previous time steps (hours). In order to take into account the temporal correlation, time-series prediction models can be used. Here, For the purpose of day-ahead scheduling of power system, an initial prediction of random variables should be performed using ANN. Other forecast tools such as ARMA model are reported [6, 24], but ANN is preferred due to its capability of reflecting nonlinear relations among the time-series samples and better performance for long-term applications. Afterwards, the distribution of forecast errors is analyzed to determine the confidence interval around the forecasted values for the upcoming potential wind speed and solar irradiation data on the scheduling day.

The forecast process is performed using two Multi-Layer Perceptron (MLP) neural networks [25] for wind speed and solar irradiation. Each network is configured with three layers including one hidden layer. The input is a 24 hour structure in which a vector of 90 data samples forms its arrays (representing each hour of the day for three month of a season). The available data is divided into three groups proportional to 70%, 15% and 15% for training, validation and test steps, respectively. The hourly data of wind speed and solar irradiation for the first farm are presented in Figure 7 and Figure 8, respectively. The plots of forecast results along with the actual data for one week are followed in Figure 9 and Figure 10.

2.5. Estimation of the confidence interval and risk analysis for wind speed and solar irradiation scenarios

From the power system planning viewpoint, the important aim is to reduce as far as possible the uncertainty and risk associated with generation and power supply. The risk is more crucial when the ex-ante planned generation is less than the ex-post actual generation. The error in the forecast data can be estimated with a level of confidence (LC), in order to determine a reliable level of generation to be considered in the planning stage. Here, the confidence interval method [26], known in risk assessment problems is proposed as a constraint to specify a lower and upper band for the wind speed and solar irradiation scenarios. For example, *LC=90%* means that the probability of forecast error ($Power_{risk}$) being less than a definite value obtained from the distribution of forecast error

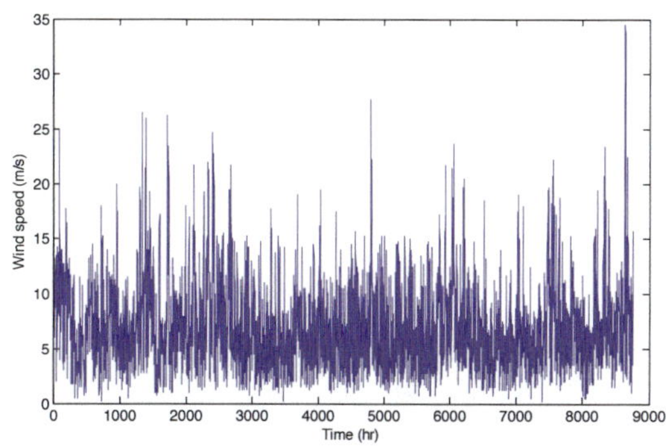

Figure 7. Hourly wind speed data in farm 1

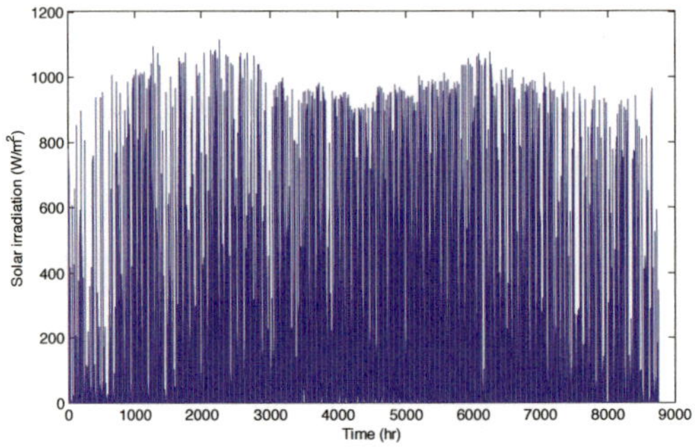

Figure 8. Hourly solar irradiation data in farm 1

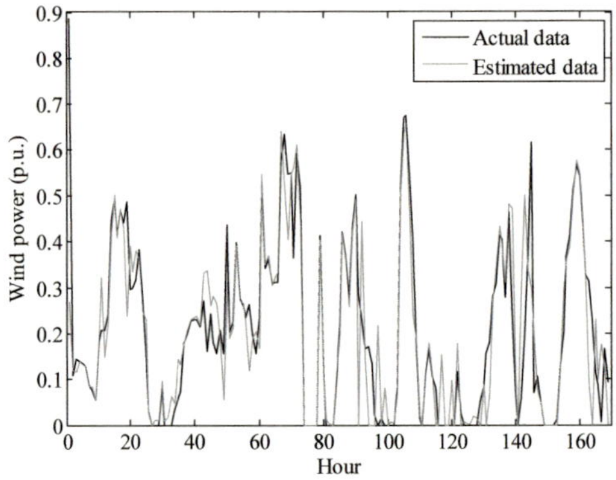

Figure 9. forecast results of wind power along with the actual data for one week

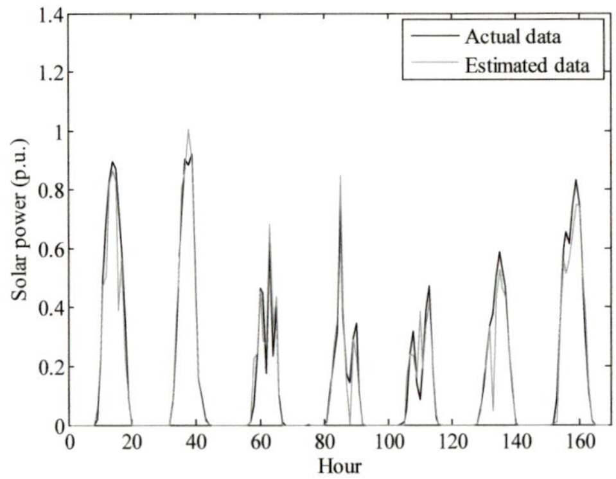

Figure 10. forecast results of wind power along with the actual data for one week

($Power_{risk,\max}$) is more than 90%. The value of $Power_{risk,\max}$ is referred to as the "Value at Risk" (VAR).

$$P\left[Power_{risk} \leq Power_{risk,max}\right] > LC \tag{11}$$

In order to calculate the VAR of forecast based on the defined LC, the PDF of forecast error should be analyzed. In fact, the maximum value of $Power_{risk,\max}$ that satisfies equation (11) is chosen as the VAR. Given the forecast error PDF to be a normal distribution with mean μ_e and standard deviation σ_e, the VAR is calculated by (Figure 11):

$$\alpha = 100 - LC \tag{12}$$

$$P[e \geq \mu_e + z_\alpha \sigma_e] < \frac{\alpha}{100} \tag{13}$$

$$Power_{risk,max} = \tilde{e} = \mu_e + z_\alpha \sigma_e \tag{14}$$

where $\tilde{e}$ is the maximum forecast error at which the condition is satisfied. The error is expressed in terms of μ_e and σ_e where z_α is the coefficient of σ_e. Then, this value is subtracted from and added to the forecasted data to give the lower band and upper band of the confidence interval, respectively. This band limits the scenarios generated for wind speed and solar irradiation and provides a reliable range for generation of scenarios that will be considered in the stochastic optimization process. Figure 12 and Figure 13 present an example of the distribution of forecast error of wind speed and solar irradiation at 11 AM, respectively. A representation of the confidence interval method applied to the wind speed data is depicted in Figure 14.

2.6. Scenario generation and reduction

The final step of data processing before performing the stochastic optimization procedure is the scenario generation and reduction. This makes the main distinction between the deterministic programming and the stochastic programming approach. The deterministic programming deals with determined inputs and pre-defined parameters of the system model, whereas the stochastic programming combines the process of assignment of optimum values to the control variables with the stochastic models of the existing random variables. The variables with stochastic behavior are indeed represented by a bunch of scenarios reflecting the most probable situations that is likely to occur for them. Here, in the scenario generation step, using the multivariate distribution function obtained, a large set of random vector samples will be generated using the Monte Carlo simulation. Then, since all of the generated scenarios may not be useful and some of them may exhibit similarities and correlation with other scenarios, the scenario reduction techniques are applied to eliminate low-probability scenarios and merge the similar ones to extract a limited number of scenarios keeping the whole probable region of the variables covered. Furthermore, the scenario reduction technique increases the computational efficiency of the optimization process. The most well-known methods for scenario reduction are fast backward, fast forward/backward and the fast forward method [27-28]. The first step in scenario reduction is clustering. By clustering, the scenarios which are close to each other are put in one cluster. In the following, we present a description on the fast backward method.

The backward method is initialized by selection of scenarios as the candidates to be eliminated. The selection criterion is based on the minimum value for the product of each sample's probability by the probabilistic distance of that sample to others. The probabilistic distance of each sample to others is considered as the minimum distance of that sample to each of the other samples in the same set. In the next step, the same analysis is performed on the remaining scenarios, however, in this step the product of the eliminated scenarios probabilities in the previous step by the distance of the current sample to other samples are also included. This process continues until the probabilistic distance obtained in each step (iteration) would be less than a predefined value as the convergence criterion. Mathematically speaking, the algorithm can be summarized as follows:

$$
\begin{aligned}
&\textbf{STEP [0]:}\\
&\quad J^{[0]} := 0\\
&\quad Z^{i} = p_i \min_{i \neq j} \left\| \xi^{i} - \xi^{j} \right\|^{r} \; ; \; i = 1,\ldots,N\\
&\quad J^{[1]} = Min\, Z^{i} \quad ; \; i = 1,\ldots,N\\
&\textbf{STEP [i]:}\\
&\quad j_i \in \arg \underset{J^{[i-1]}}{Min} \sum_{J^{[i-1]} \cup i} p_k \underset{J^{[i-1]} \setminus i}{Min} \left\| \xi^{k} - \xi^{l} \right\|^{r}\\
&\quad if\; \underset{J^{[i-1]}}{Min} \sum_{J^{[i-1]} \cup i} p_k \underset{J^{[i-1]} \setminus i}{Min} \left\| \xi^{k} - \xi^{l} \right\|^{r} \leq \varepsilon\\
&\qquad J^{[i]} = J^{[i-1]} \cup j_i\\
&\quad Else \quad J = J^{[i-1]}\\
&\textbf{END}
\end{aligned}
\tag{15}
$$

where N is the total number of scenarios, n is the number of remaining scenarios, and $|J| = N - n$. ξ^{i} is scenario i with probability p_i. Here, 1000 initial scenarios are generated based on the obtained Copula multi-variable distribution function within the calculated confidence interval. Then, the scenarios are reduced to 20 uncorrelated scenarios using SCENRED function in GAMS software. Figure 15 and Figure 16 demonstrate the final set of scenarios for a day of scheduling.

The power generation scheduling problem including units with uncertain and volatile characteristic is commonly treated as a stochastic optimization problem. In this approach, the objective function calculation is repeated per all final scenarios in each iteration of the optimization process, where the summation of these objective values commonly defines the overall objective function value to be optimized. As a comparison, the mean square error (MSE) of the calculated scenarios with respect to the real data is calculated in two cases. In the first case, the scenarios are obtained from the Copula distribution and within the calculated confidence interval, based on the proposed framework. In the second case, the scenarios are generated from the single-variable distributions without modeling the underlying temporal and spatial correlations. Table 2 exhibits that the error in the first case is less than that in the second case in three days under study.

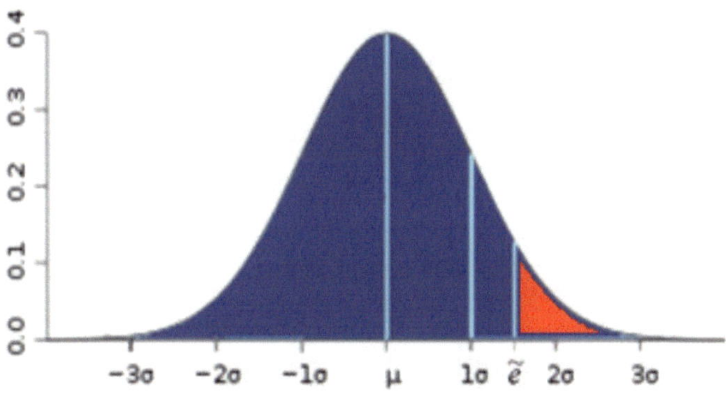

Figure 11. Value at risk calculation based on the distribution of forecast error

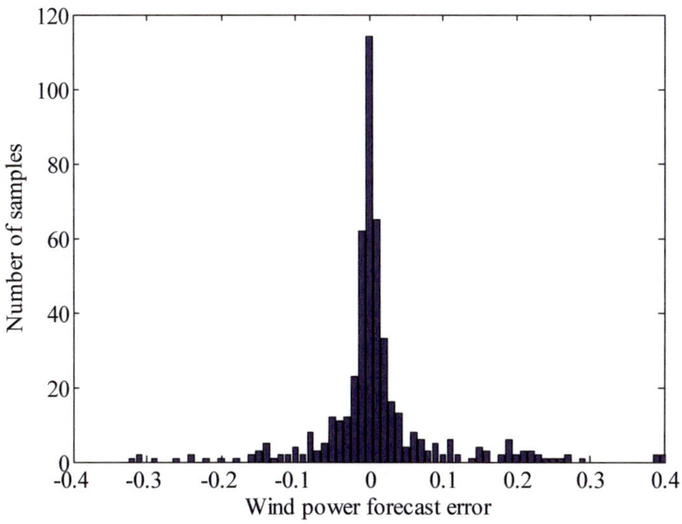

Figure 12. Distribution of error for wind speed forecast at 11 AM

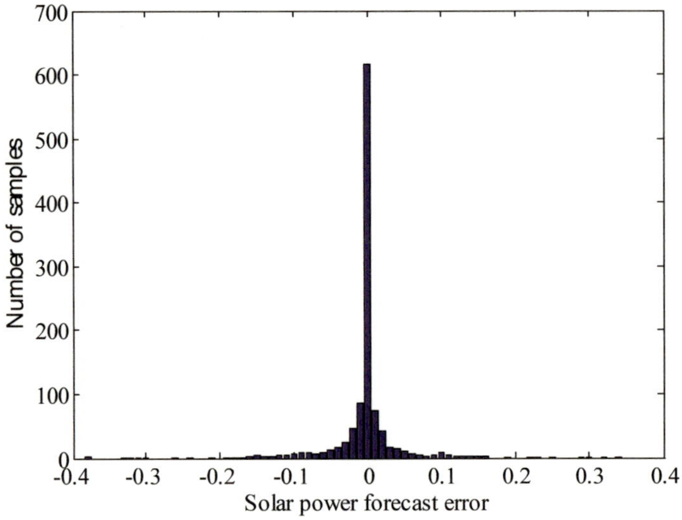

Figure 13. Distribution of error for solar irradiation forecast at 11 AM

MSE in Case 2	MSE in Case 1	
1.97e-3	1.4e-3	Day 1
5.99e-3	4.05e-3	Day 2
4.11e-3	3.31e-3	Day 3

Table 2. MSE in Case 1 and 2 for the total power of two wind farms

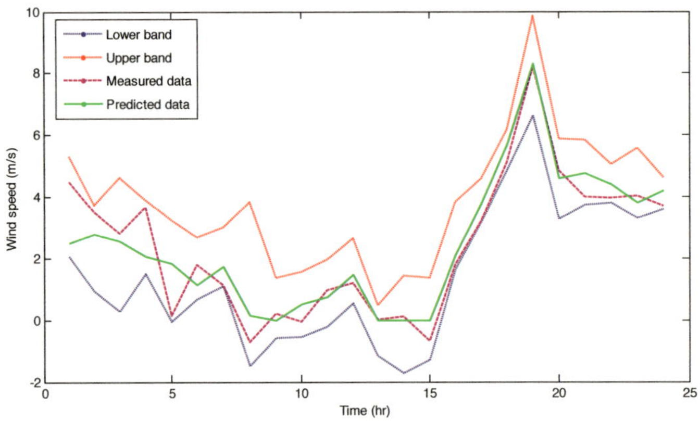

Figure 14. Upper and lower band of confidence interval for wind speed data on one day

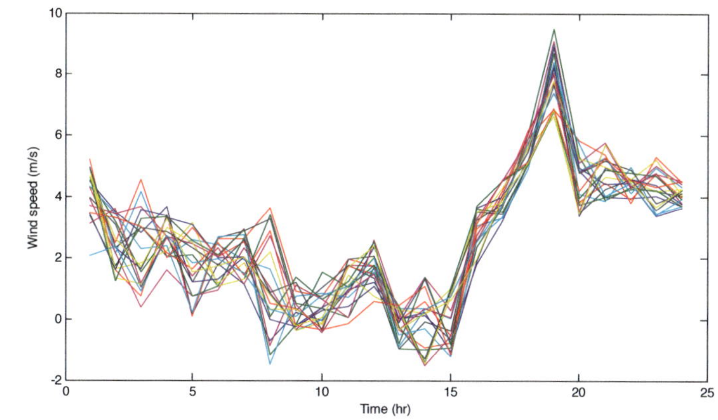

Figure 15. Final scenarios of wind speed within its confidence interval

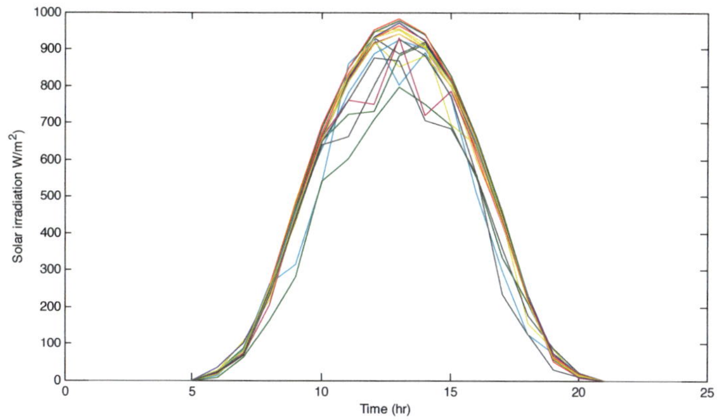

Figure 16. Final scenarios of solar irradiation within its confidence interval

3. Conclusion

Natural characteristics of wind and solar energy impose uncertainty in their design and operation. Hence, considering various possible scenarios in the model of these resources can lead to more realistic decisions. The uncertain parameters are expressed by probability distributions, showing the range of values that a random variable could take, and also accounting for the probability of the occurrence of each value in the considered range. Therefore, the way the random processes are modeled in terms of their PDF is a significant problem. The possible spatial correlations have been addressed and shown effective using the Copula method. Similarly, the possible temporal correlations have been taken into account using a time-series analysis. In summary, the overall framework can be listed as follows:

1. Take historical data of random variables;
2. Data normalization;
3. Calculate the PDF for each random variable using KDE;
4. Calculate multivariate Copula PDF;
5. Forecast the data time-series over the scheduling horizon;
6. Calculate the confidence interval of potential scenarios for the vector of random variables based on error analysis of forecast data;
7. Generate initial set of scenarios within the confidence interval obtained from the previous step, based on the PDF of step 4;
8. Perform scenario reduction;
9. Perform stochastic optimization based on the final set of scenarios for each random process.

Author details

Sajjad Abedi[1], Gholam Hossein Riahy[1], Seyed Hossein Hosseinian[1] and Mehdi Farhadkhani[2]

1 Electrical Engineering Department, Amirkabir University of Technology, Tehran, Iran

2 Electricity Markets Research Group, Niroo Research Institute, Tehran, Iran

References

[1] 20% "Wind Energy By 2030: Increasing Wind Energy's Contribution to U.S. Electricity Supply,," Office of Energy Efficiency and Renewable Energy DOE/GO-102008-2567, 2009.

[2] REN21 Steering Committee, Renewables Global Status Report 2009 Update [Online]. Available: http://sitedev.cci63.net/europe/actu/2009/0609/doc/energie_renouv2009_Update.pdf

[3] S. Abedi, *et al.*, "Risk-Constrained Unit Commitment of Power System Incorporating PV and Wind Farms," *ISRN Renewable Energy*, vol. 2011, 2011.

[4] S. Abedi, *et al.*, "A comprehensive method for optimal power management and design of hybrid RES-based autonomous energy systems," *Renewable and Sustainable Energy Reviews*, vol. 16, pp. 1577-1587, 2012.

[5] S. Abedi, *et al.*, "Risk-Constrained Unit Commitment of Power System Incorporating PV and Wind Farms," *ISRN Renewable Energy,* vol. 2011, p. 8, 2011.

[6] K. Methaprayoon, *et al.*, "An integration of ANN wind power estimation into unit commitment considering the forecasting uncertainty," *Industry Applications, IEEE-Transactions on,* vol. 43, pp. 1441-1448, 2007.

[7] B. C. Ummels, *et al.*, "Impacts of wind power on thermal generation unit commitment and dispatch," *Energy Conversion, IEEE Transactions on,* vol. 22, pp. 44-51, 2007.

[8] C. L. Chen, "Optimal wind–thermalgenerating unit commitment," *Energy Conversion, IEEE Transactions on,* vol. 23, pp. 273-280, 2008.

[9] M. A. Ortega-Vazquez and D. S. Kirschen, "Estimating the spinning reserve requirements in systems with significant wind power generation penetration," *Power Systems, IEEE Transactions on,* vol. 24, pp. 114-124, 2009.

[10] E. M. Constantinescu, *et al.*, "A computational framework for uncertainty quantification and stochastic optimization in unit commitment with wind power generation," *Power Systems, IEEE Transactions on,* pp. 1-1, 2011.

[11] [L. V. L. Abreu, *et al.*, "Risk-Constrained Coordination of Cascaded Hydro Units With Variable Wind Power Generation," *Sustainable Energy, IEEE Transactions on,* vol. 3, pp. 359-368, 2012.

[12] V. S. Pappala, *et al.*, "A stochastic model for the optimal operation of a wind-thermal power system," *Power Systems, IEEE Transactions on,* vol. 24, pp. 940-950, 2009.

[13] H. Siahkali and M. Vakilian, "Stochastic unit commitment of wind farms integrated in power system," *Electric Power Systems Research,* vol. 80, pp. 1006-1017, 2010.

[14] R. Billinton, *et al.*, "Unit commitment risk analysis of wind integrated power systems," *Power Systems, IEEE Transactions on,* vol. 24, pp. 930-939, 2009.

[15] J. W. GOETHE, "Non-Parametric Statistics," 2009.

[16] G. Marsaglia, *et al.*, "Evaluating Kolmogorov's distribution," *Journal of Statistical Software,* vol. 8, pp. 1-4, 2003.

[17] A. R. Mugdadi and E. Munthali, "Relative efficiency in kernel estimation of the distribution function," *J. Statist. Res,* vol. 15, pp. 579–605, 2003.

[18] R. D. Deveaux, "Applied Smoothing Techniques for Data Analysis," *Technometrics,* vol. 41, pp. 263-263, 1999.

[19] K. Sutiene and H. Pranevicius, "Copula Effect on Scenario Tree," *International journal of applied mathematics,* vol. 37, 2007.

[20] P. Embrechts, *et al.*, "Correlation and dependence in risk management: properties and pitfalls," *Risk management: value at risk and beyond,* pp. 176-223, 2002.

[21] U. Cherubini, *et al.*, *Copula methods in finance* vol. 269: Wiley, 2004.

[22] D. Berg, "Copula goodness-of-fit testing: an overview and power comparison," *The European Journal of Finance,* vol. 15, pp. 675-701, 2009.

[23] M. Dorey and P. Joubert, "Modelling copulas: an overview," *The Staple Inn Actuarial Society,* 2005.

[24] G. Riahy and M. Abedi, "Short term wind speed forecasting for wind turbine applications using linear prediction method," *Renewable Energy,* vol. 33, pp. 35-41, 2008.

[25] M. T. Hagan, *et al., Neural network design*: PWS Boston, MA, 1996.

[26] D. Y. Hsu, et al., Spatial Error Analysis: A Unified, Application-oriented Treatment: IEEE Press, 1999.

[27] *GAMS/SCENRED Documentation [Online].* Available: http://www.gams.com/docs/document.htm

[28] N. G.-K. J. Dupacová, and W. Römisch, "Scenario Reduction in Stochastic Programming: An Approach Using Probability Metrics," *Math. Program,* pp. 493–511, 2000.

Chapter 6

Modeling of Photovoltaic Cell Using Free Software Application for Training and Design Circuit in Photovoltaic Solar Energy

Miguel Pareja Aparicio, José Pelegrí-Sebastiá,
Tomás Sogorb and Vicente Llario

Additional information is available at the end of the chapter

http://dx.doi.org/10.5772/51925

1. Introduction

There are numerous studies that develop the mathematical modeling of photovoltaic cells and verified by software, for example [1] or [2]. The model presented is based on an equivalent circuit implemented in free software. Free software used is Quite Universal Circuit Simulator (*QUCS*), [3]. *QUCS* uses a generic diode for adjust the current and voltage curve (IV curve) at photovoltaic cell. Additionally, you can use equations to define the model of photovoltaic cell and represent the characteristic curves on the same page, [4]. QUCS is a multiplatform application that runs on Windows and Linux, this software is available in Linux distributions for electronics. [5].

This model can be used for training in photovoltaic solar energy, using: subcircuits, curves, tables and equations. Further, can be used an attractive presentation to the student with a real representation of PV cell. Also, can be used to test circuit with photovoltaic solar cell as power supply, in applications such as: micropower systems for harvesting energy, stand alone PV system for control battery charge.

The model of PV cell can be used to simulate a PV module, because PV module is an association of cells in series and parallel. The model PV module can use to study mismatch effects due to different electrical characteristics of PV cells and the use of pass diode to reduce loss due to partial shadows. Then, can be use PV module study PV grid connection and energy production prediction.

2. Equivalent circuit of photovoltaic cell

The equivalent circuit of an ideal cell is formed by a current source in parallel with a diode (figure 1a). There are several circuits that include resistors for real effects of a photovoltaic cell, for example, figure 1b includes a resistor in series, [2], figure 1c includes parallel and series resistance, [1] and [6]. Other models include two diodes as in figure 1d, [7] and [8].

The circuit of figure 1c is the more commonly used, although in several simulations simplifies the parallel resistance value with a high value, using the series resistance to include effect of fill factor, gets a similar circuit of figure 1d and used *Rp* to avoid problem with simulation. Then, this circuit has a simple and accurate model to simulate a photovoltaic cell.

The problem is the parameter values of circuit components. Therefore, in Section 4 are calculated parameters using data from the photovoltaic cell indicated in datasheets, for equivalent circuit on*figure 1c*.

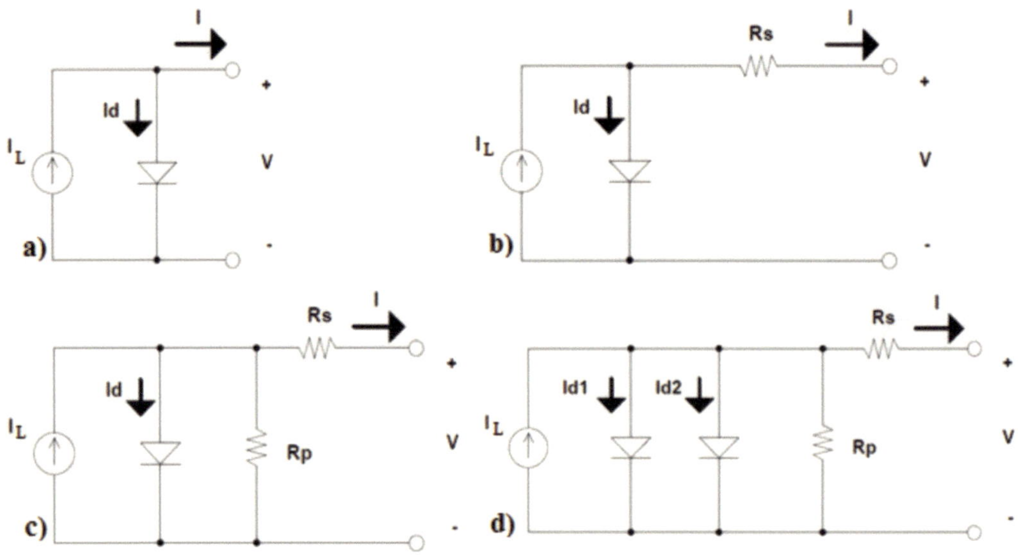

Figure 1. Equivalents circuits of photovoltaic cell

3. Current-Voltage (I-V) Curve

Calculate equivalent circuit parameters need to know the *I-V curve*. In the *I-V curve* (*figure 2*) can extract the electrical characteristics of the photovoltaic cell in standard conditions of measurement (*SCM*): I_{SC} (short circuit current) is maximum intensity that can generate a photovoltaic cell or module when measuring the current if performing a short circuit (output voltage of 0 volts), V_{OC} (open circuit voltage) is maximum voltage that can generate a

http://dx.doi.org/10.5772/51925

photovoltaic cell or module when measuring the voltage if not flowing current (current of 0 amps), P_{MAX} (Maximum power) is maximum power that can generate a photovoltaic cell or module and it's the product of maximum voltage and current, V_{MAX} (Maximum Voltage) is the voltage at maximum power (around 80% of open circuit voltage) and I_{MAX} (Maximum Current) is the current at maximum power.

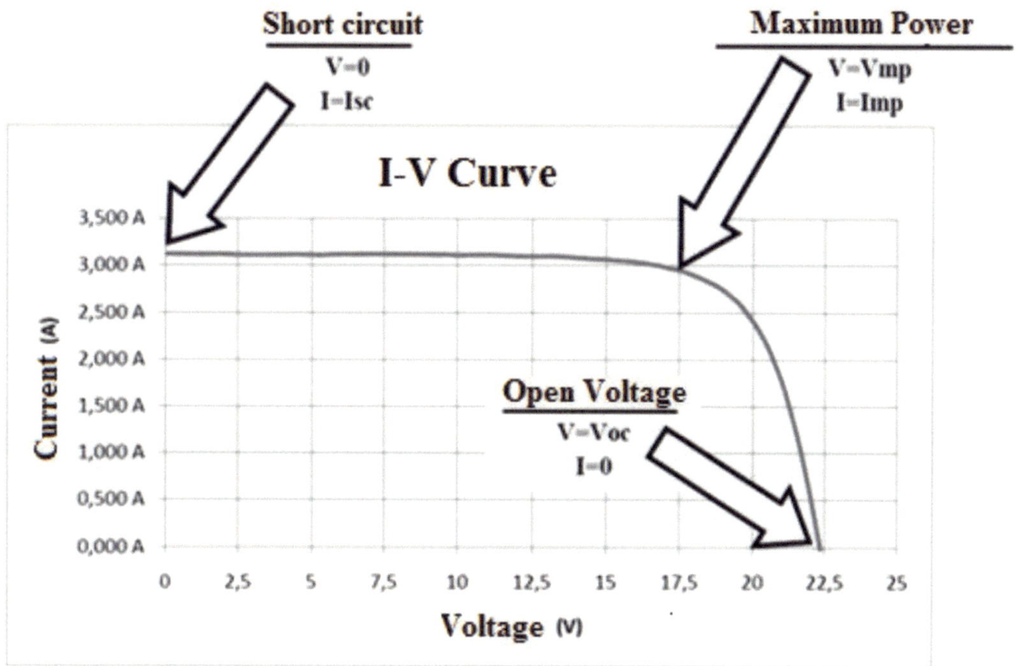

Figure 2. IV Curve of photovoltaic module

Compare the figure 2 with equivalent circuit (figure 1b), value of I_{SC} related with I_L and value of V_{OC} related with voltage and current diode (*Id*). Then diode includes effects of exponential of I-V curve. Finally *Rs* adjusts value of P_{MAX}. All parameter information is extract to datasheets, and *SCM* are: *25ºC* for cell temperature (T_C), *1000 W/m²* for irradiance (*G*), *1,5* for spectrum solar and *3m/s* for wind speed.

4. Modeling equations of photovoltaic cell

Reference to the circuit of figure 1c, then show all equations needed to obtain all the parameters that define the model in standard conditions of measurement (*SCM*). In equation 1 shows the intensity value generated by the photovoltaic cell, [9]: *I* is output current of photovoltaic cell, *V* is output voltage of photovoltaic cell, I_L is the photogenerated current, I_0 is the saturation current of diode, R_S is series resistance due to the junction between the semiconductor and the metal contacts (interconnects), R_P is parallel resistance due to no linearity of union *PN*, *m* is ideal factor of diode and *Vt* is thermovoltage shown in equation 2 (where: *k* is the Boltzmann constant, *q* is the electron charge and *T* is temperature in degree Kelvin).

$$I = I_L - I_0\left[e^{\left(\frac{V+I\cdot R_S}{m\cdot Vt}\right)} - 1\right] - \left[\frac{V + I\cdot R_S}{R_P}\right] \tag{1}$$

$$Vt = \frac{k\cdot T}{q} \tag{2}$$

Equation 1 can simplify the last term with a high value of R_P (for example 100kΩ). Furthermore, I_L is considered equal to the short circuit current in *SCM* (I_{SC_SCM}), [1]. Then we obtain the equation 3 at *SCM*.

$$I = I_{SC\_SCM} - I_{0\_SCM}\left[e^{\left(\frac{V+I\cdot R_S}{m\cdot Vt}\right)} - 1\right] \tag{3}$$

The value of I_0 is obtained for *SMC* (I_{0_SMC}) using equation 4, based on [10], considering open voltage circuit in *SCM* (V_{OC_SCM}) and cell temperature in *SCM* (T_{C_SCM}).

$$I_{0\_SCM} = \frac{I_{SC\_SCM}}{\left(e^{\left(q\cdot V_{OC\_SCM} / m\cdot k\cdot T_{C\_SCM}\right)} - 1\right)} \tag{4}$$

There is an empirical relationship between the value of V_{OC_SCM} and I_{SC_SCM} with R_S, [9]. Then, needs calculate fill factor of ideal device (FF_0) at equation 5, using parameter *voc* of equation 6, and calculate R_S using equation 8. The fill factor (*FF*) of photovoltaic cell shows at equation 7. This approach only use when R_P is high, therefore fill factor depends of R_S value.

$$FF_0 = \frac{voc - \ln(voc + 0{,}72)}{voc + 1} \tag{5}$$

$$voc = \frac{V_{OC\_SCM}}{Vt} \tag{6}$$

$$FF = \frac{V_{MAX\_SCM}\cdot I_{MAX\_SCM}}{I_{SC\_SCM}\cdot V_{OC\_SCM}} \tag{7}$$

$$R_S = \left(1 - \frac{FF}{FF_0}\right) \cdot \left(\frac{V_{OC\_SMC}}{I_{SC\_SMC}}\right) \quad (8)$$

Changes in temperature affect the values of $I_{SC,}$ *voc* and P_{MAX}, when cell temperature increase the V_{OC} decrease, same with the P_{MAX}, and when irradiation increase the I_{SC} also increase. The datasheet used parameters of *table 1*, establishing the relationship between units (voltage, current and power) and temperature. Temperature can be expressed on degree Celsius or Kelvin, depends of manufacturer.

Parameters	Affect to...	Units
β	V_{OC}	mV/ºC or %/ºC
α	I_{SC}	mA/ºC or %/ºC
δ	P_{MAX}	mW/ºC or %/ºC

Table 1. Parameters that include temperature variations on photovoltaic cell

Also, changes in irradiance (*G*) affect the value of I_{SC} and V_{OC}. Therefore, using approximation of Luque-Sala and Duffie & Beckman, [10], we obtain the equation 9 for I_L. Besides empirical tests simulation to study the effects of temperature on I_0, used for get equation 10, based on the approximation Duffie&Beckman.

$$I_L = I_{SC\_SCM} \cdot \frac{G}{G_{C\_SCM}} \cdot \left(1 + \alpha \cdot (T_C - T_{C\_SCM}\right) \quad (9)$$

$$I_0 = \frac{I_{SC\_SCM} \cdot \frac{G}{G_{SCM}} \cdot \left(1 + \alpha \cdot (T_C - T_{C\_SCM}\right)}{\left(e^{\left(q \cdot \left(V_{OC\_SCM} + \beta \cdot (Tc - T_{C\_SCM})\right) \big/ m \cdot k \cdot T_{C\_SCM} \right)} - 1 \right)} \quad (10)$$

The relationship between temperature ambient (T_A) and cell (T_C), can used equations 11 *and 12*, [11]. It is based on normal operating temperature cell (T_{NOC}), is defined as the average temperature of equilibrium within a photovoltaic cell to an irradiance of *800W/m²* and an outside temperature of *20ºC*. Use this approximation is interesting because there are statistics of temperature ambient on geographic situation but temperature cell depends to PV cell and module.

$$T_C - T_A = C \cdot G \Rightarrow Tc = C \cdot G + T_A \quad (11)$$

$$C = \frac{T_{NOC} - 20}{800\ W / m^2} \tag{12}$$

Figure 3 shows the simulation window with all the necessary equations and the visualization of results using *QUCS*. Photovoltaic cell simulated in figure 3 is *C3ISF200SB* of Isofoton, [12]. QUCS allows represented on the same page a circuit and results of simulation, for example in figure 3 included: *IV Curve*, output power curve, output current curve, output voltage curve and a table with numerical results. For functions used in *QUCS* see on [13], in figure 3 used following equations: eqn1 for parameters of photovoltaic cell, eqn2 for change temperature ambient to cell temperature, eqn3 for parameters adjust to equivalent circuit and eqn4 to calculate variables to represented results on graphical depends to output measurement of equivalent circuit.

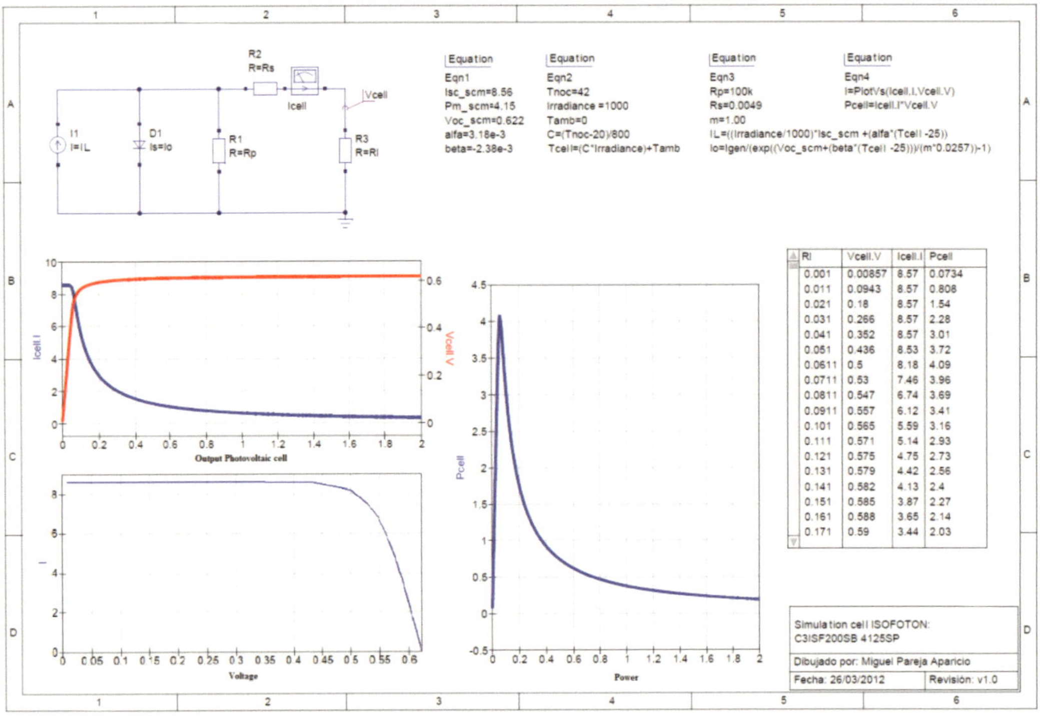

Figure 3. Final result of the simulation Qucs photovoltaic cell

The equivalent circuit is formed by following components: current source (*dc current source* on *source* library), diode (*diode* on *non linear components* library) and resistors (*resistor* on *lumped components* library). The value of current source is calculate on variable *IL* (current generate) on equ3 based on equation 9, the value of current saturation on diode is calculate on variable *I0* on equ3 based on equation 10, the value of resistor series and parallel its calculate manually and indicate on eqn3. The model of diode simulation includes ohmic series resistence (*Rs*) and zero bias junction capacitance (*Cjo*), to adjust to PV cell model must changue to *0Ω* and *0F* respectively.

CurrentmeasurementofPVcellmodelusedcurrentprobe(*Icell*)onprobelibrary,forvoltagemeasurementusedawirelabel(*Vcell*)togetvoltageonnode.Powergenerate(*Pcell*)ofPVcelliscalculates using equ4 as the product of current (*Icell.I*) and voltage (*Vcell.V*) measurement. To represent a *IV curve* isneedstogetavariable*I*usingfunction*PlotVS()*onequ4.Theresultsonsimulationshowon graphical(*Cartesian*on*diagrams*library)andtable(*Tabular*on*diagrams*library).

In *QUCS* is a used component of library simulations for configuring simulation, for example to get *IV curve* need components: *dc simulation* and *parameter sweep* (figure 4). The configuration of simulation that show on figure 4, changes value of variable *Rl* from *0,01Ω* to *10Ω*, variable *Rl* is used to change value of resistor *R3* (figure 3).

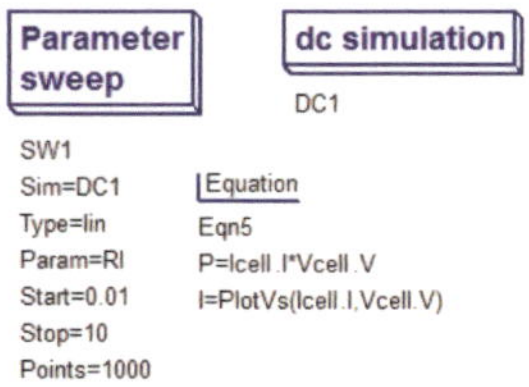

Figure 4. Simulation configuration

Also, if changes value of *Irradiance* variable on eqn2 (figure 3), changes the solar condition and current generate of PV cell. For example on figure 5 shows *IV curve* to different values of solar irradiance at the same value of cell temperature (*25ºC*): *1000W/m²* (*G_1000*), *750W/m²* (*G_750*) and *500W/m²* (*G_500*).

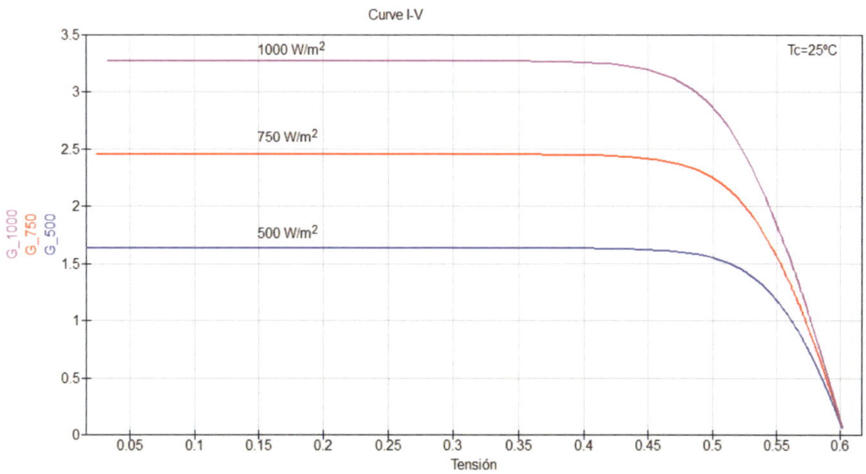

Figure 5. IV Curve for different values of irradiation

Further, if changes value *Tamb* variable on eqn2 (figure3), change ambient temperature and therefore the cell temperature condition based on equations 10 *and* 11. For example on figure 6 shows *IV curve* to different values of cell temperature at the same value of irradiance ($1000W/m^2$): *50ºC (T_50)*, *25ºC (T_25)* and *0ºC (T_0)*. Then, combining the two variables can adjust weather conditions.

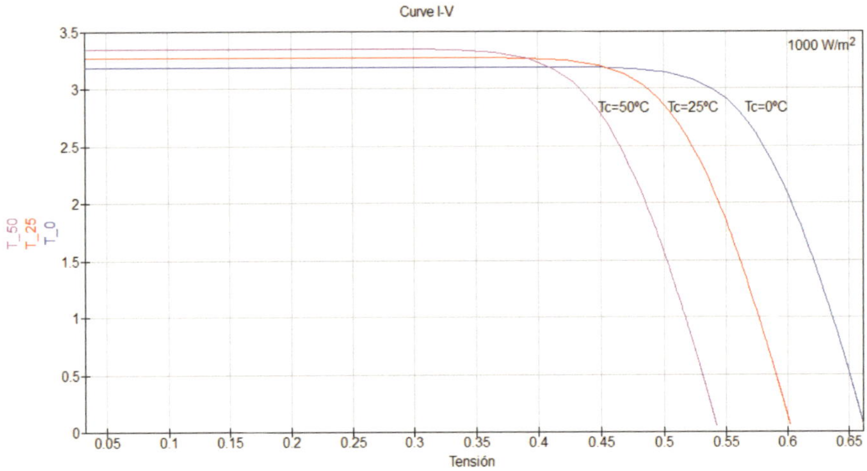

Figure 6. IV Curve for different values of cell temperature

5. Educational application

The model shown in Section 4 can be used for the formation of PV system. In particular, using subcircuit for an attractive presentation for the student, [10]. In addition to evaluating the effects of: association series and parallel, potential losses, weather conditions, non-ideality of photovoltaic cells and effect of partial shadow. Figure 7 shows steps for modelling by subcircuits of a PV cell and module, first represented a equivalent circuit and include parameters gets of datasheet, second create a symbol to represented a PV cell, third the subcircuit used to external variables for irradiance (G) and cell temperature (*Tc*), fourth associate cells to build a PV module, and finally create a symbol to represented a PV module.

To create a subcircuit needs connection for output PV cell, after to select all component of equivalent circuit (figure 3) the output connection of PV cell (in series with *Rs* for positive connection) connects used insert port (figure 8). Once finish equivalent circuit can be edit representation of subcircuit pressing *F9*, for edit representation can be used painting library, [13].

Model for PV module can be create using PV cell subcircuit and connection in series and parallel, for example a PV module for 12V nominal voltage can be formed by 36 PV cells connects in series, in figure 9 shows connections of PV module with 2 pass diodes and external ports connections (positive for *P1* and negative for *P2*). Representation used for PV cell and module shows on figure 7, after can be used subcircuit of PV cell or module on different practices.

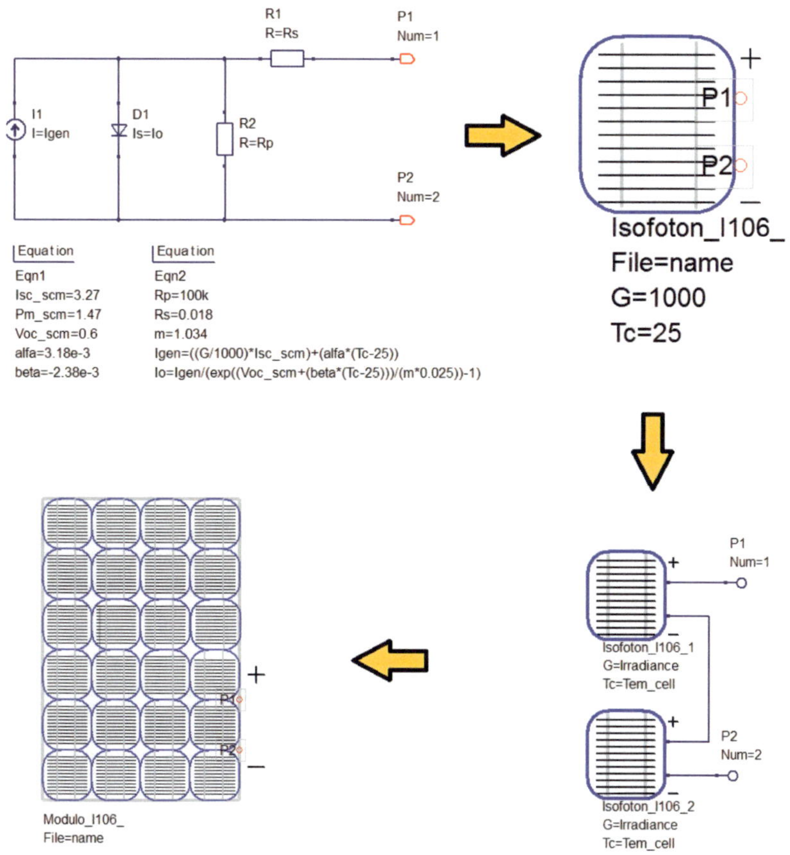

Figure 7. Steps modeling subcircuit

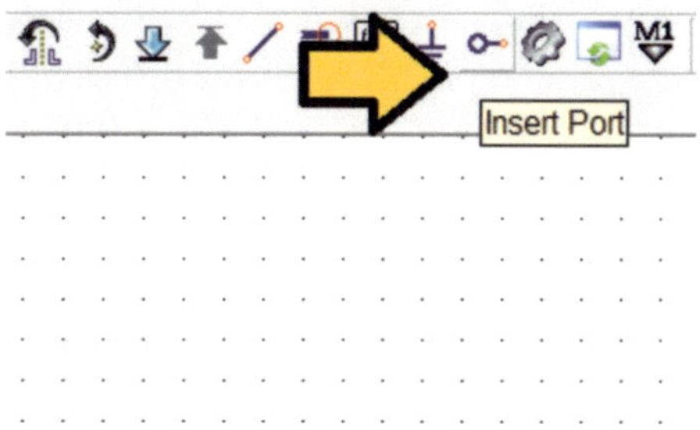

Figure 8. Insert connection to subcircuit

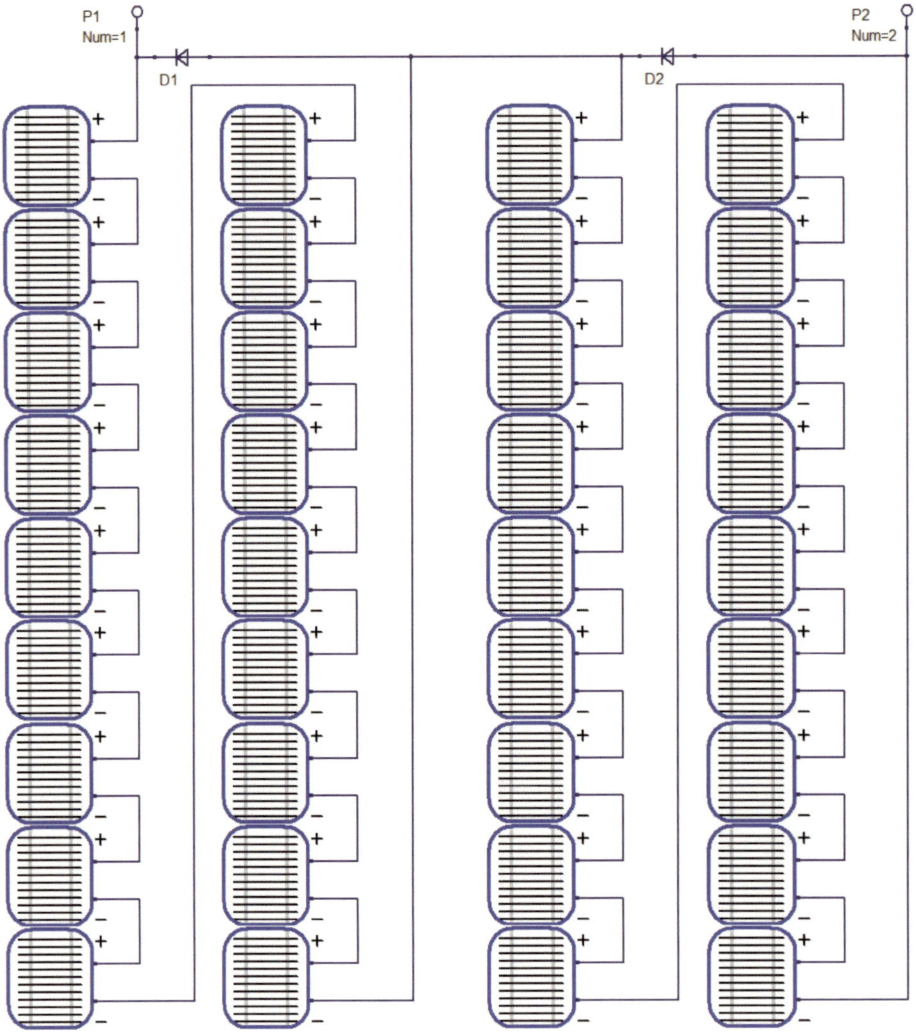

Figure 9. Model of PV module based on subcircuit PV cell

Then, student works directly with subcicuits independently of equivalent circuit model, because student can be used the same representation for different PV cells. The material used will consist of a compressed file, which contains a project *QUCS* (for example *practices_solar_prj.rar*), this file contains a number of files to be need for practical work (files used on *QUCS* use extension *sch* and a project contains all files on *QUCS* directory). The practices developed show in table 2, for study for photovoltaic training on stage generation, after can be complete with model of all components on photovoltaic system (battery, regulator or inverter) and can be used on renewable energy training. Practices on table 2 can be complete with a previous work: select values for equivalent circuit and introduction of *QUCS*.

http://dx.doi.org/10.5772/51925

Practiques	Descriptions
Obtaining IV Curve	This practice consists in the practical realization of obtaining an IV curve, connecting a resistor to photovoltaic cell, how it is done in laboratory, the student modify the values of resistance load and take measurements of voltage and current generate by photovoltaic cell.
Connecting PV cells	In this practice the student performed tests: vacuum, short circuit and load. Also, performed measurements with various connections: serial, parallel and mixed. This practice can be extent compare with a real measurement.
Effect of irradiance and temperature on PV cell	This practice is to see the effects on curve IV by changing irradiance and temperature. Simulation can be used by scanning parameter to be directly represents the curve IV and only change the values of irradiance and cell temperature. Can use the same practice of PV module.
Partial and total shadowing effect	This practice associated series and parallel solar cells and see the effects for partial shadowing about currentand voltage generate by PV cell. After can repeat practice for total shadowing effect. After, the student can extract conclusion about the shadow problem on PV cell.
Using bypass diodes and losses from partial shading on PV module	This practice is to see the utility about bypass diodes in the manufacturing on PV module. Must create two models about PV module, one with diode and other without diode. After the student compare results.
Connection blocking diodes	This practice is to see the effect and necessity of blocking diodes. Must connect a circuit PV cell and battery between blocking diode, after remove blocking diode and compare results.
Mismatch losses in PV Module	This practice is to see effects to use different PV cells to manufacturing PV module, must create two models about PV module, one with ideal PV cells and other with different values of short current. After the student compare results.
Voltage drop in PV cells connection	This practice is to see the losses associated with wiring to connect photovoltaic devices. Can show the effect of using one or another section, and the length between the module and circuit regulator. It creates a new subcircuit corresponding to the calculation of the resistance of the wiring with copper conductors.

Table 2. Examples of practical using modeling and GPL application

Figure 10 shows several examples for use subcircuits model for educational, and are directly applicable to any practices described. As parameters such as temperature (variable Tc in ºC) and irradiance (variable G in W/m^2) cell, see the effects of partial shading on the association of cells and the use of bypass diodes in the manufacture of photovoltaic modules, and the measurement of short circuit current.

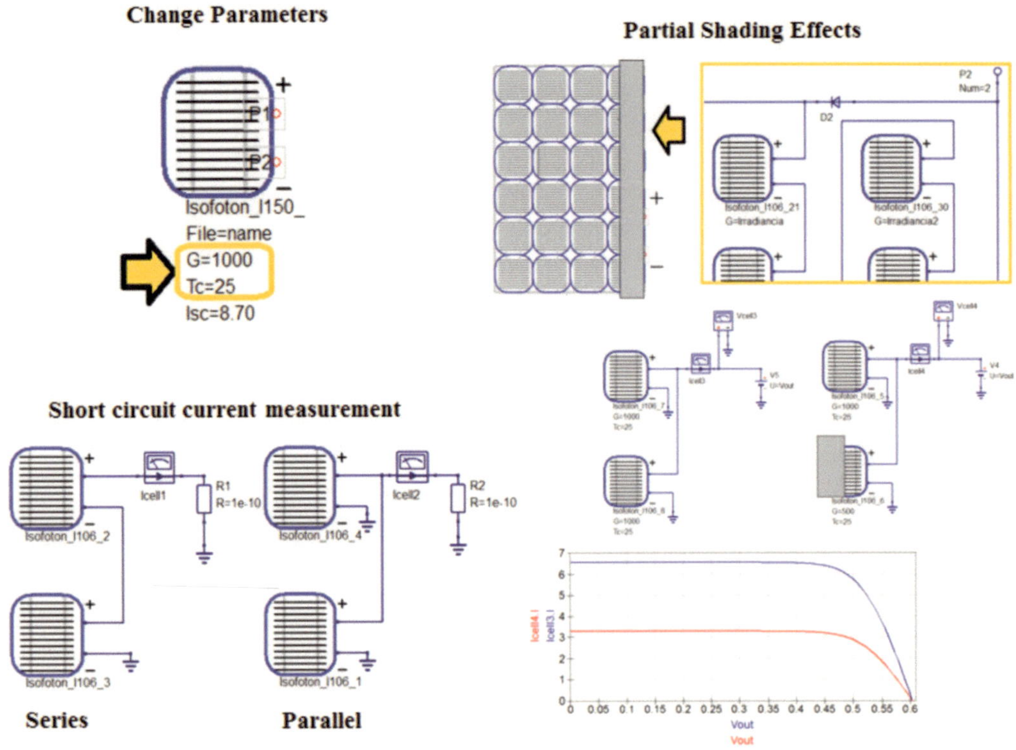

Figure 10. Examples for subcircuit model

6. Design circuit application

The availability of the model described in section 4, allows design circuits power management when energy source are PV cells or modules. For example, in power supply of sensor nodes, [14], to design efficient harvesting energy control, because *QUCS* dispose different electronic components for simulation by means of changing parameters to adjust to real component. *QUCS* has generic devices electronic *on non lineal* library and adjust parameter on properties menu to adjust real component, or use component library (selected on *Tools* menu or *Ctrl+5*) for used component with properties adjusted to real device (figure 11).

Also, working in selection PV cells to power supplies performing a comparative commercial device to adapted for applications at climatologic conditions of system localization, for example to study PV cell to weather station [15] or irrigation actuator [16]. Further, to test regulator circuit using in stand-alone PV system and control circuit to obtain maximum power point tracker, [17].

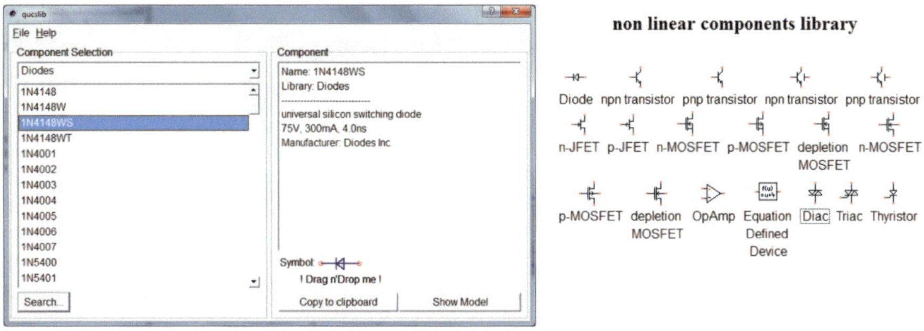

Figure 11. Component library for QUCS

Other application, its predict power generation from PV cells and modules, using a variable parameter for irradiance and ambient temperature. In addition, can be include partial shadow for same object (for example: tree or building) in PV grid connection or stand-alone.

For example, in figure 12a shows curve of irradiation variation of irradiance (G) around day. Value used can be extracted to Photovoltaic Geographical Information Systems (*PVGIS*) [18] mean values per moths. After connection a constant resistor to photovoltaic cell and get outputs: power, voltage and current (figure 12b). After can complete modifying load connected. Also, can included ambient temperature around day using variable T_C. The PV cell characteristics used on figure 12 is: *3,27A* for I_{SC} and *0,6V* for V_{OC} on *SCM*.

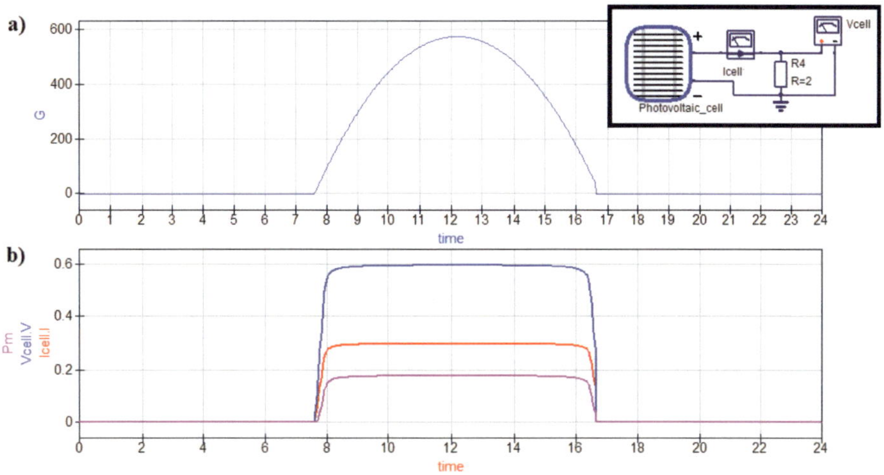

Figure 12. Variation Irradiance (G) around day with a constant load

Other example shows on figure 13 to use PV cell model for simulate a circuit control to obtain the maximum power point to PV cell. In figure 13a show a subcircuit connect to dc converter, and figure 13b show output voltage and control signal, the output is controlled modify duty cycle, then the duty cycle decrease when the output voltage arrive to maximum power voltage and increase when output voltage away from maximum power voltage. The PV cell characteristics used on figure 13 is: *150mA* for I_{SC} and *0,62V* for V_{OC} on *SCM*.

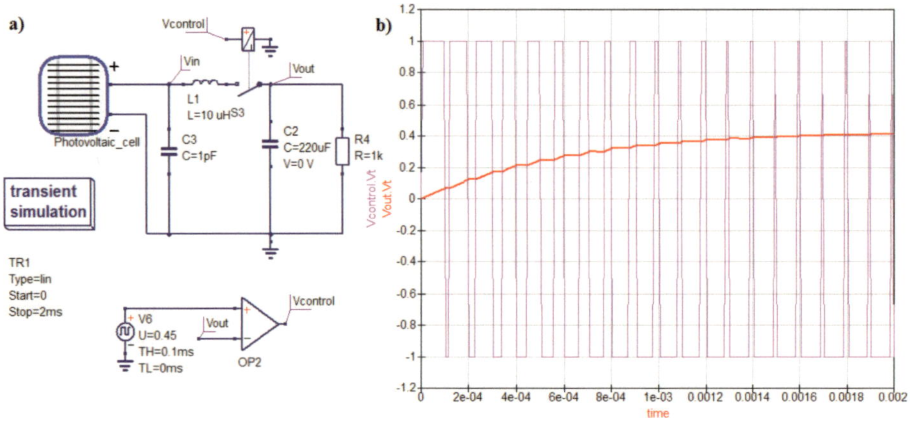

Figure 13. Used model to study circuit control of PV cell

Also, model PV cell can be used to study number of PV cells need to supply energy to the system, for example figure 14 show *16 PV cells* for simple circuit supply of *5V* source formed by: block diode (*D1*) and a regulator circuit (*LM140*).Then, change number of PV cells and configure climatic conditions (Irradiance and temperature) can see if has enough or need more PV cells, too if has more PV cells that is need. On figure 14 show: output voltage (*V_load.V*), output current (*I_load.I*) and voltage cells (*V_cells.V*), to different values of irradiation, therefore study irradiance needs to obtain *5V* on load resistor. Conclusions on figure 14 is than needs *16 PV cells* on series and irradiance value around *500-550 W/m²* or higher to obtain *5V* and *0,5A* output (on load). The PV cell characteristics used on figure 14 is: *1A* for I_{SC} and *0,6V* for V_{OC} on *SCM*.

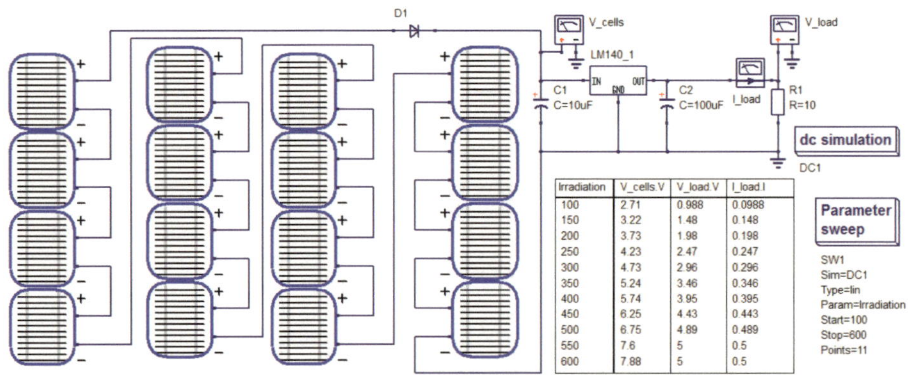

Irradiation	V_cells.V	V_load.V	I_load.I
100	2.71	0.988	0.0988
150	3.22	1.48	0.148
200	3.73	1.98	0.198
250	4.23	2.47	0.247
300	4.73	2.96	0.296
350	5.24	3.46	0.346
400	5.74	3.95	0.395
450	6.25	4.43	0.443
500	6.75	4.89	0.489
550	7.6	5	0.5
600	7.88	5	0.5

Figure 14. Study irradiation for simple regulator circuit

http://dx.doi.org/10.5772/51925

7. PV cells of module

Some case has available information of PV module on *SCM* to emulate PV module (figure 15), then needs calculate values for PV cells: division between module open voltage (V_{OC_MODULE}) and number PV cells series (N_{CELLS_SERIES}) to obtein open voltage of cell (V_{OC_CELL}), division between module current short (I_{SC_MODULE}) and number of strings cells connection ($N_{CELLS\_STRINGS}$) to obtain current short of cell (I_{SC_CELL}), and repeat by module maximum values of voltage (V_{MAX_MODULE}) and current (V_{MAX_MODULE}) to obtain maximum values of voltage (V_{MAX_CELL}) and current (V_{MAX_CELL}) of cell; equations 13 to *16* respectively.

$$V_{OC\_CELL} = \frac{V_{OC\_MODULE}}{N_{CELLS\_SERIES}} \tag{13}$$

$$I_{SC\_CELL} = \frac{I_{SC\_MODULE}}{N_{CELLS\_STRINGS}} \tag{14}$$

$$V_{MAX\_CELL} = \frac{V_{MAX\_MODULE}}{N_{CELLS\_SERIES}} \tag{15}$$

$$I_{MAX\_CELL} = \frac{I_{MAX\_MODULE}}{N_{CELLS\_STRINGS}} \tag{16}$$

Values obtained on equations 13 *and* 14 used to obtain: *voc*, R_S, I_L, and I_0, on *equations*: *6*, *8*, *9* and *10*. The values obtained on equations 13 to *16* used to obtain *FF* on equation 7. This approximation is based on a PV module is union of PV cells connected in series and parallel strings. Model of PV module included connections cells loss inside module.

On figure 15 shows an example for PV module based on subcircuit on PV cells, characteristic of PV module on *SCM* is: *150W* to P_{MAX}, *22,6V* to V_{OC}, *8,7A* to I_{SC}, *18,5V* to V_{MAX} and *8,12A* for I_{MAX}. The module used in figure 15 used 36 cells connected in series on a string.

Advantage to use model PV cells on model PV module is that change parameter of irradiation and temperature by cell (figure 7), and so study effects: partial shading, number pass diode, different connections of pass diode, hot cells, etc. Also can be study effect on mismatch on module used PV cell with different electrical characteristics. For example on figure 16 show effects of partial shading on module using 2 diodes pass (figure 9): figure 16a without shadow, figure 16b shadow affect to same number of cell connects on parallel by diode and figure 16c shadow affect only to cells connects on parallel with 1 diode pass.

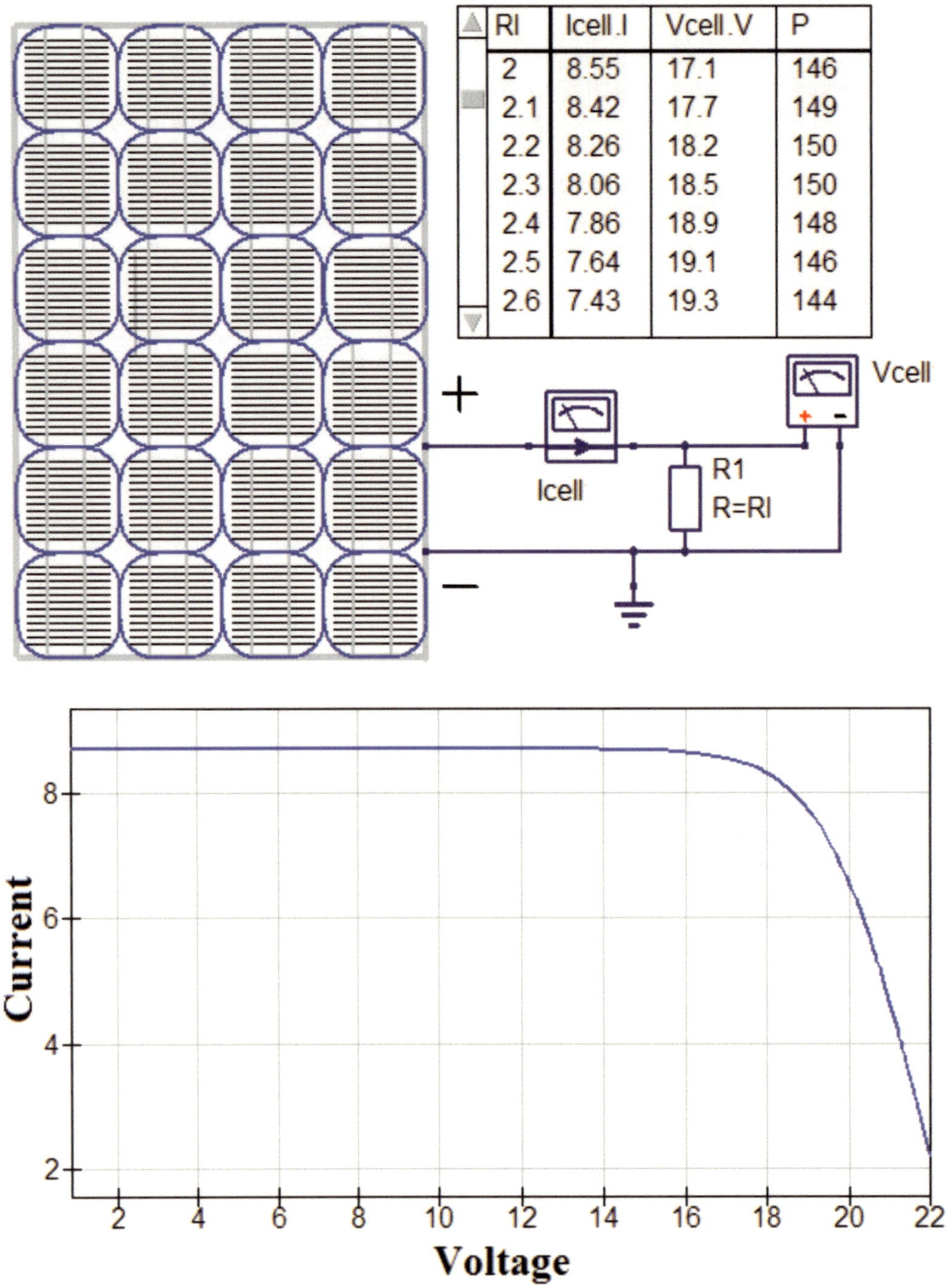

Rl	Icell.I	Vcell.V	P
2	8.55	17.1	146
2.1	8.42	17.7	149
2.2	8.26	18.2	150
2.3	8.06	18.5	150
2.4	7.86	18.9	148
2.5	7.64	19.1	146
2.6	7.43	19.3	144

Figure 15. Simulate PV module based on PV cell model

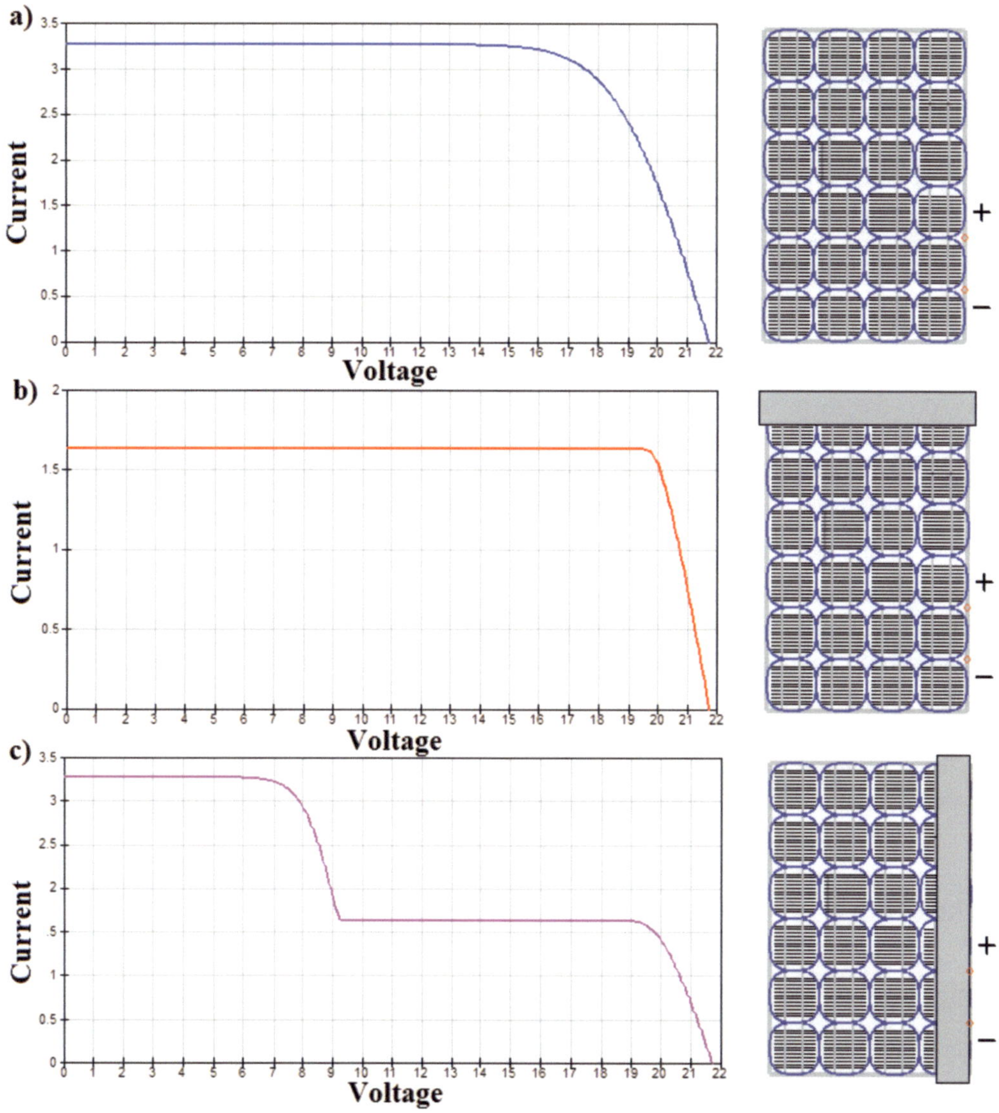

Figure 16. Partial shading on PV module use model PV cell.

8. Conclusion

In this chapter show a equivalent circuit for simulate PV Cell, then show equation to obtain all parameters to circuit based on PV cell datasheet. To check model is implemented on free software and compare results of output PV Cell model represented on IV Curve and output measurements with datasheet information. The software application selected is QUCS because can be include equations to calculate circuit parameters and a real representation of subcircuit.

The PV cell model can modelled any PV cell using datasheet information, also the model include variations of temperature and irradiance for output PV cell. Therefore used PV cell model connected in series and parallel can modelled a PV module. In section 4 shows all necessary equations to obtain parameters of circuit to PV cell model and section 7 shows process and equations to obtain PV module cell.

Using *QUCS* to model a PV cell allows subcircuit and a real representation to a attractive presentation for teaching. In section 5 show examples of practices used on formation, further can be used on: courses of photovoltaic, online formation or distance learning, because only need download *QUCS* application, and is a good complement to a previous works on laboratory or concepts review for theory. Advantage to used *QUCS* is that allow several PV cells with a few mouse click, also does not needs buy additional PV cells to used on laboratory because can be modelled the PV cell available on laboratory. Further, is not a problem the availability material on laboratory, because the material of PV system can be expensive, then is best provide a good photovoltaic devices that a devices for all student in a class.

Other application for model of PV cell or module is used to design of supply circuits, for example in a network sensors node, to design power control. In section 6 shows examples to used electronic devices available on *QUCS* library to control output PV cell, with the advantage to change weather condition to study operation circuit and is surer for designer and electronic devices. Also, can emulate various conditions to which the PV cell work (irradiance and temperature), which can reduce design time, detecting errors. Further, the availability of weather conditions that need for tested circuit.

Author details

Miguel Pareja Aparicio, José Pelegrí-Sebastiá, Tomás Sogorb and Vicente Llario

Universidad Politécnica de Valencia, Spain

References

[1] Castañer, L., & Silvestre, S. Modeling of photovoltaic cell using free software application for training and design circuit in photovoltaic solar energy. John Wiley and sons LTD; (2002).

[2] Altas I.H. and Sharaf A.M.th European Photovoltaic Solar Energy Conference: Array simulation model for Matlab-Simulink GUI Environment, Glasgow, UK; (2000).

[3] Scordilis T., Brinson M., Kraut G., Jahn S. and Pitcher C. Workbook of QUCS; 2007. http://qucs.sourceforge.net/docs.hml.

[4] QUCSProject: http://qucs.sourceforge.net/index.html.

[5] Fedora electroniclab: http://spins.fedoraproject.org/fel/#publications

[6] González, F. M. o congreso Iberoamericano de estudiantes de ingeniería eléctrica, electrónica y computación: Model of photovoltaic module in matlabTM, Puerto la Cruz, Venezuela, II CIBELEC; (2006).

[7] Ishaque K. and Salam Z., An Accurate MATLAB Simulink PV System Simulator Based on the Two-diode Model. Journal of Power Electronics 2011, 11(2) 179-187.

[8] Ishaque, K., Salam, Z., Simple, Fast., Accurate-Diode, Two., Model, for., & Photovoltaic, Modules. (2011). *Solar Energy Materials and Solar Cells*, 95(2), 586-594.

[9] Green M.A. Solar cell: principles, technology and system applications.Prentice Hall; (1981).

[10] Pareja M., Modeling and simulation of cell / PV module with GPL software: application to teaching practice.Master thesis. Universidad Internacional de Andalucía (UNIA). Spain; (2012).

[11] E. Skoplaki, J.A. Palyvos, Operating temperature of photovoltaic modules: A survey of pertinent correlations, Renewable Energy, Volume 34, Issue 1, January 2009, Pages 23-29, ISSN 0960-1481, 10.1016/j.renene.2008.04.009. (http://www.sciencedirect.com/science/article/pii/S0960148108001353)

[12] Isofotonmanufacture. http://www.isofoton.com/ingles/solar_cells.html

[13] Brinson M., A tutorial: Component, compact device and circuit modeling using symbolic equations; 2007. http://qucs.sourceforge.net/docs.hml.

[14] Pelegrí, J., Lajara, R., Alberola, J., Solar, power., source, for., & autonomous, sensors. Renewable Energy, T J Hammons (Ed.), 978-9-53761-952-7InTech, Available from: http://www.intechopen.com/books/renewable-energy/Modeling of photovoltaic cell using free software application for training and design circuit in photovoltaic solar energy.

[15] Alberola, J., Pelegri, J., Sogorb, T., & Vicente, Llario. J. Ultra Low Power Wireless Weather Station. Sensor Technologies and Applications, (2007). SensorComm 2007.

[16] Lajara, R.; Alberola, J.; Pelegrí-Sebastiá, J. A Solar Energy Powered Autonomous Wireless Actuator Node for Irrigation Systems. Sensors 2011, 11, 329-340.

[17] Dunlop, J.P.: Batteries and Charge Control in Stand-Alone Photovoltaic Systems Fundamentals and Application; Florida Solar Energy Center, prepared for Sandia National Laboratories, Photovoltaic Systems Applications Dept, January 1997.

[18] PhotovoltaicGeographical Information System. PVGIS. Geographical Assessment of Solar Resource and Performance of Photovoltaic Technology. http://re.jrc.ec.europa.eu/pvgis/.

Chapter 7

Steady State Modeling of Three Phase Self–Excited Induction Generator Under Unbalanced/Balanced Conditions

A. Alsalloum and A. I. Alolah

Additional information is available at the end of the chapter

http://dx.doi.org/10.5772/53914

1. Introduction

Due to the increased emphasis on the energy issues and problems, concentration has been focused upon developing autonomous electric power supplies to be operated in remote and rural areas where electric services is unavailable from existing or nearby grids. These types of power sources can be used even in regions supplied by network grids in the event of power interruptions. Among such types that have received a notable attention and importance is the three-phase self-excited induction generator due to its numerous advantages such as simple design, robustness, and low installation and maintenance costs [1-4]. Experimental works and computer simulations have been extensively performed in order to model and analyze both steady state and transient performance of the SEIG under balanced operating conditions. However, the unbalanced operation of the SEIG has been given little attention despite its practical needs. There are two main methods to predict the steady state performance of the SEIG under balanced operating conditions. The first method is based on the generalized machine theory [5]. The second method is based on the analysis of the generalized per-phase equivalent circuit of the induction machine by applying either the loop impedance or the nodal admittance concept [6,7]. Furthermore, other studies have concentrated only on the single-phase self-excited induction generator and its voltage regulation improvement [8,9]. The influence of the terminal capacitance has been investigated in [10,12]. The previous studies have centralized mainly on modeling and analyzing the performance of SEIG under only balanced operating conditions. The major contribution of this chapter is to model the SEIG together with its excitation and load. In our steady state study, the performance of the SEIG was determined for No-load, balanced and unbalanced load and/or excitation for different SEIG and load connections. The operating conditions were found by solving the proposed model iteratively. An experimental setup has been built to verify the results obtained from the theoretical model. The model is generalized to cover more connection types of SEIG and/or load. It is clear that the theoretical results are in good agreement with those reported experimentally. The effect of the machine core losses is considered by representing the core resistance as a second order polynomial in terms of X_m. Furthermore, the magnetizing reactance has been included in the negative sequence equivalent circuit as a variable. The positive and negative sequence equivalent circuits are used to model the SEIG. The final characteristic equation is reached by equating both the positive-sequence and negative-sequence voltages across the SEIG and the load.

2. Star connected generator–star connected load without a neutral connection

The equivalent load impedance shown in Figure 1 may be described as follows;

$$Z_a = Z_{L_a} // -jX_{C_a} \tag{1}$$

$$Z_b = Z_{L_b} // -jX_{C_b} \tag{2}$$

$$Z_c = Z_{L_c} // -jX_{C_c} \tag{3}$$

where,

$Z_{L_{a,b,c}}$ = Load impedance at base frequency connected across phase *a*, *b* and *c*, respectively.

$X_{C_{a,b,c}}$ = Excitation capacitor reactance at base frequency connected across phase *a*, *b* and *c*, respectively.

$Z_{a,b,c}$ = Equivalent impedance of load and excitation capacitor at base frequency connected across phase *a*, *b* and *c* respectively.

At the load side, the phase voltages are:

$$\begin{bmatrix} V_{ak} \\ V_{bk} \\ V_{ck} \end{bmatrix} = \begin{bmatrix} Z_a & 0 & 0 \\ 0 & Z_b & 0 \\ 0 & 0 & Z_c \end{bmatrix} \begin{bmatrix} I_a \\ I_b \\ I_c \end{bmatrix} \tag{4}$$

Since the load and/or the excitation capacitors are expected to be unbalanced, it is more appropriate to describe the different quantities involved in Eq. (4) in terms of their symmetrical components. Using the symmetrical components technique, the following is found:

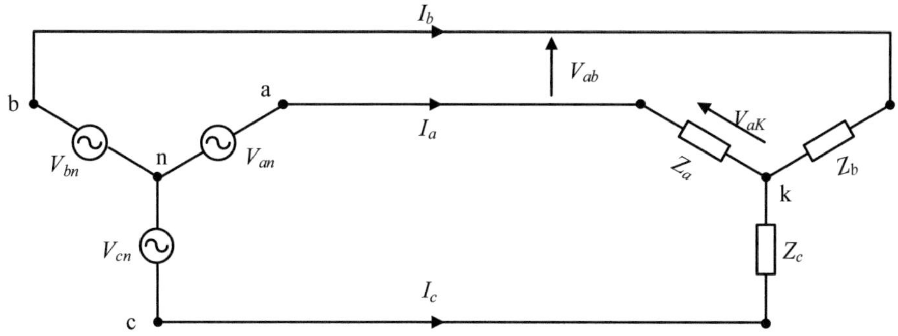

Figure 1. Star connected generator - star connected load without neutral connection.

$$\begin{bmatrix} V_{ak0} \\ V_{ak1} \\ V_{ak2} \end{bmatrix} = \begin{bmatrix} Z_0 & Z_2 & Z_1 \\ Z_1 & Z_0 & Z_2 \\ Z_2 & Z_1 & Z_0 \end{bmatrix} \begin{bmatrix} I_{a0} \\ I_{a1} \\ I_{a2} \end{bmatrix} \tag{5}$$

Where the subscripts 0, 1, and 2 stands for zero, positive and negative sequence components, respectively. The symmetrical components of the load phase voltages may be found from the three-phase values as follows:

$$\begin{bmatrix} V_{ak0} \\ V_{ak1} \\ V_{ak2} \end{bmatrix} = \begin{bmatrix} 1 & 1 & 1 \\ 1 & \gamma & \gamma^2 \\ 1 & \gamma^2 & \gamma \end{bmatrix} \begin{bmatrix} V_{ak} \\ V_{bk} \\ V_{ck} \end{bmatrix} \tag{6}$$

where $\gamma=1\angle 120$

On the other hand, the three phase voltages may be found in terms of their symmetrical components by using the following transformation:

$$\begin{bmatrix} V_{ak} \\ V_{bk} \\ V_{ck} \end{bmatrix} = \begin{bmatrix} 1 & 1 & 1 \\ 1 & \gamma^2 & \gamma \\ 1 & \gamma & \gamma^2 \end{bmatrix} \begin{bmatrix} V_{ak0} \\ V_{ak1} \\ V_{ak2} \end{bmatrix} \tag{7}$$

The transformation matrix shown in Eq. (6) can be used to find the symmetrical components of currents, namely, I_{a0}, I_{a1}, and I_{a2}. The symmetrical components of the three phase impedances Z_a, Z_b, and Z_c are as follows;

$$\begin{bmatrix} Z_0 \\ Z_1 \\ Z_2 \end{bmatrix} = \frac{1}{3}\begin{bmatrix} 1 & 1 & 1 \\ 1 & \gamma & \gamma^2 \\ 1 & \gamma^2 & \gamma \end{bmatrix}\begin{bmatrix} Z_a \\ Z_b \\ Z_c \end{bmatrix} \tag{8}$$

Since in an isolated-neutral star-connected load, the zero sequence component of line current (phase current) equals to zero, substituting I_{a0}=0 in Eq. (5) and expanding yields:

$$V_{ak1} = Z_0 I_{a1} + Z_2 I_{a2} \tag{9}$$

$$V_{ak2} = Z_1 I_{a1} + Z_0 I_{a2} \tag{10}$$

It can be shown using the symmetrical components technique that the relation between the positive and negative sequence components of both the line and the phase voltages are as follows:

$$V_{LL1} = \left(1 - \gamma^2\right) V_{LN1} \tag{11}$$

$$V_{LL2} = \left(1 - \gamma\right) V_{LN1} \tag{12}$$

Hence,

$$V_{ab1} = \left(1 - \gamma^2\right) V_{ak1} \tag{13}$$

$$V_{ab2} = \left(1 - \gamma\right) V_{ak2} \tag{14}$$

Now looking at the generator side, the following positive and negative sequence circuits of Figure 2 can be used to model the generator. As can be seen in Figure 2, the core loss resistance is taken into consideration in the positive-sequence equivalent circuit of the SEIG.

As the core loss is variable according to saturation, the core loss resistance is expressed as a function of the magnetizing reactance (X_m) as shown in Eq. (15). Figure 3 shows the variation of the core resistance as a function of magnetizing reactance. Although this will increase the complexity of the model but the model will be closer to the actual case.

$$\begin{aligned} & R_c(X_m) = a_2 X_m^2 + a_1 X_m + a_0 \quad p.u. \\ & where, \\ & a_2 = -17.159,\ a_1 = 33.372,\ a_0 = 35.699 \end{aligned} \tag{15}$$

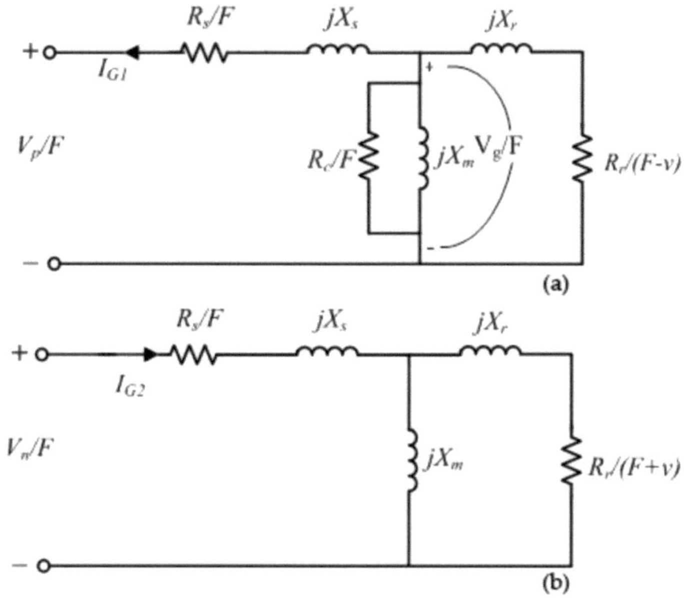

Figure 2. Induction generator equivalent circuits: (a) +ve seq. (b) –ve seq.

Furthermore, the air gap voltage may be approximated over the saturated region as a function of X_m by the following sixth order polynomial. Figure 4 shows this variation.

$$V_g / F = b_6 X_m^6 + b_5 X_m^5 + b_4 X_m^4 + b_3 X_m^3 + b_2 X_m^2 + b_1 X_m + b_0 \quad p.u.$$
$$\text{where,} \qquad (16)$$
$$b_6 = 0.95,\ b_5 = -8.28,\ b_4 = 28.91,\ b_3 = -51.78,\ b_2 = 50.18,\ b_1 = -25.07,\ b_0 = 6.21$$

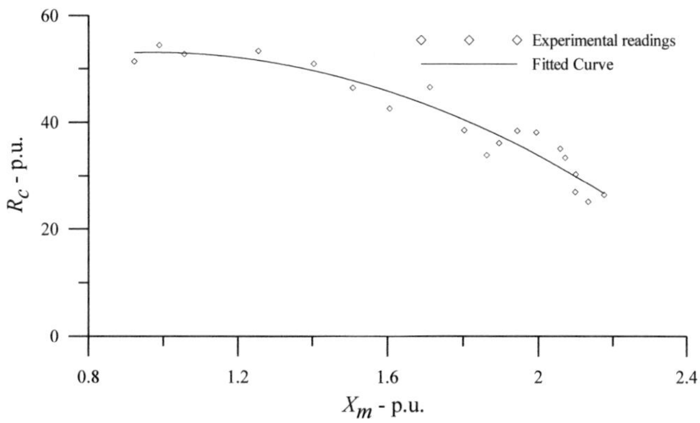

Figure 3. Variation of core resistance as a function of magnetizing reactance.

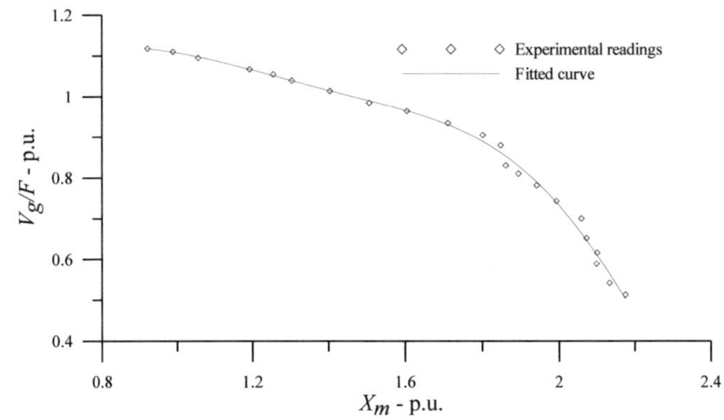

Figure 4. Variation of air gap voltage as a function of magnetizing reactance.

The terminal voltage of the positive and negative sequence equivalent circuits is given by;

$$V_{an1} = -I_{G1} Z_{G1} \tag{17}$$

$$V_{an2} = I_{G2} Z_{G2} \tag{18}$$

Eq. (11), Eq. (12), Eq. (17), and Eq. (18) yield the following,

$$V_{ab1} = -\left(1-\gamma^2\right) I_{G1} Z_{G1} \tag{19}$$

$$V_{ab2} = \left(1-\gamma\right) I_{G2} Z_{G2} \tag{20}$$

where

Z_{G1}, Z_{G2}Input impedance of positive and negative sequence equivalent circuits, respectively.

Equating symmetrical components of line-to-line voltages yields:

$$-I_{G1} Z_{G1} = Z_0 I_{a1} + Z_2 I_{a2} \tag{21}$$

$$I_{G2} Z_{G2} = Z_1 I_{a1} + Z_0 I_{a2} \tag{22}$$

Since the phase current in a star connected generator is the same as the line current, hence,

$$I_{a1} = I_{G1} \tag{23}$$

$$I_{a2} = I_{G2} \tag{24}$$

Substituting Eq. (23) and Eq. (24) into Eq. (21) and into Eq. (22) and rearranging, yields the following:

$$(Z_{G1} + Z_0)\, I_{a1} + Z_2 I_{a2} = 0 \tag{25}$$

$$Z_1 I_{a1} + (Z_0 - Z_{G2}) I_{a2} = 0 \tag{26}$$

Solving these two equations simultaneously yields,

$$(Z_0 + Z_{G1})(Z_0 - Z_{G2}) - Z_1 Z_2 = 0 \tag{27}$$

This is the characteristic equation of an isolated-neutral star connected induction generator. It consists of two parts, namely, the real part and the imaginary part. I_{a1} and I_{a2} does not equal zero because self-excitation is assumed to occur, hence, the real and imaginary parts of Eq. (27) must equal to zero. By substituting the machine parameters, speed, excitation capacitor values, a nonlinear equation with constant coefficients in F and X_m can be found. Solving iteratively to find the real roots of the equation that satisfies the constraints, the values of F and X_m are found; hence, the performance of the generator under these conditions can be determined. This procedure is carried out using MATHCAD® software. The flow chart describing the performance evaluation is shown in Figure 5.

3. Star connected generator–star connected load with a neutral connection

The connection for this case is shown in Figure 6. In this type of connection the zero sequence component of line currents is present (i.e. $I_{a0} \neq 0$) while the zero sequence component of phase voltages ($V_{ak0} = 0$) equals zero.

Expanding Eq. (5) yields:

$$V_{ak0} = Z_0\, I_{a0} + Z_2\, I_{a1} + Z_1\, I_{a2} \tag{28}$$

$$V_{ak1} = Z_1\, I_{a0} + Z_0\, I_{a1} + Z_2\, I_{a2} \tag{29}$$

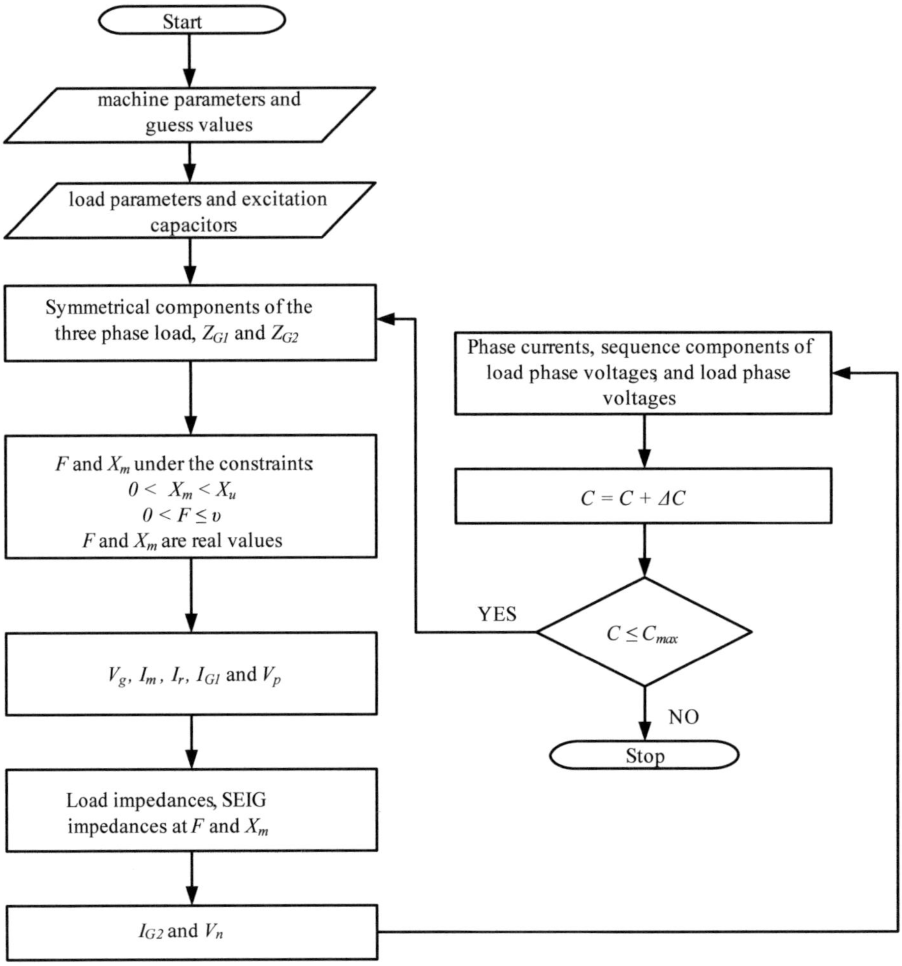

Figure 5. Flow chart describing performance evaluation.

$$V_{ak2} = Z_2 I_{a0} + Z_1 I_{a1} + Z_0 I_{a2} \tag{30}$$

Substituting $V_{ak0} = 0$ in Eq. (28) and rearranging, yield:

$$I_{a0} = -(Z_2 I_{a1} + Z_1 I_{a2}) / Z_0 \tag{31}$$

Substituting this result in Eqs. (29) and (30), yields:

$$V_{ak1} = \left(Z_0 - \frac{Z_1 Z_2}{Z_0} \right) I_{a1} + \left(Z_2 - \frac{Z_1^{\,2}}{Z_0} \right) I_{a2} \tag{32}$$

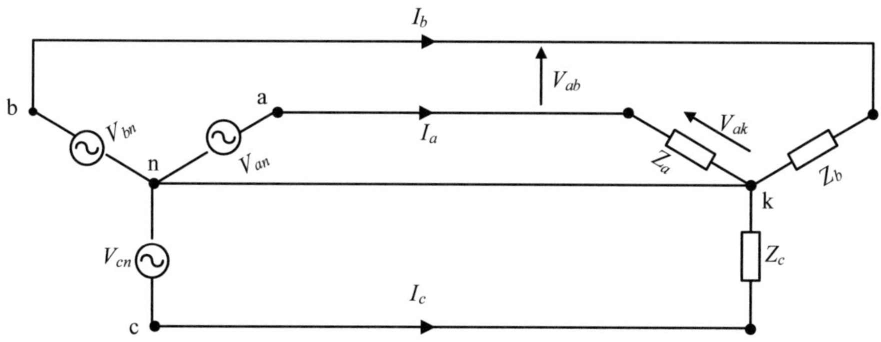

Figure 6. Star connected generator - star connected load with a neutral connection.

$$V_{ak2} = \left(Z_1 - \frac{Z_2^{\,2}}{Z_0} \right) I_{a1} + \left(Z_0 - \frac{Z_1 Z_2}{Z_0} \right) I_{a2} \tag{33}$$

Since the phase voltage of both the generator and the load are equal, hence,

$$V_{ak1} = V_{an1} = - I_{G1} Z_{G1} \tag{34}$$

$$V_{ak2} = V_{an2} = I_{G2} Z_{G2} \tag{35}$$

Substituting Eqs. (34) and (35) into Eqs. (32) and (33), and taking into consideration that $I_{G1} = I_{a1}$ and $I_{G2} = I_{a2}$, yield,

$$-I_{a1} Z_{G1} = \left(Z_0 - \frac{Z_1 Z_2}{Z_0} \right) I_{a1} + \left(Z_2 - \frac{Z_1^{\,2}}{Z_0} \right) I_{a2} \tag{36}$$

$$I_{a2} Z_{G2} = \left(Z_1 - \frac{Z_2^{\,2}}{Z_0} \right) I_{a1} + \left(Z_0 - \frac{Z_1 Z_2}{Z_0} \right) I_{a2} \tag{37}$$

Since sequence currents does not equal to zero, hence, the characteristics equation of this system equals to zero;

$$\left(Z_0 - \frac{Z_1 Z_2}{Z_0} + Z_{G1} \right)\left(Z_0 - \frac{Z_1 Z_2}{Z_0} - Z_{G2} \right) - \left(Z_1 - \frac{Z_2^{\,2}}{Z_0} \right)\left(Z_2 - \frac{Z_1^{\,2}}{Z_0} \right) = 0 \tag{38}$$

4. Delta connected generator–delta connected load

A delta-connected generator feeding a delta-connected load is shown in Figure 7, where the elements of the delta-connected load may be defined as follows,

$$Z_{ab} = Z_{L_{ab}} // -jX_{C_{ab}} \tag{39}$$

$$Z_{bc} = Z_{L_{bc}} // -jX_{C_{bc}} \tag{40}$$

$$Z_{ca} = Z_{L_{ca}} // -jX_{C_{ca}} \tag{41}$$

The symmetrical components for this type of load connection are as follows:

$$\begin{bmatrix} Z_0 \\ Z_1 \\ Z_2 \end{bmatrix} = \frac{1}{3} \begin{bmatrix} 1 & 1 & 1 \\ 1 & \gamma & \gamma^2 \\ 1 & \gamma^2 & \gamma \end{bmatrix} \begin{bmatrix} Z_{ab} \\ Z_{bc} \\ Z_{ca} \end{bmatrix} \tag{42}$$

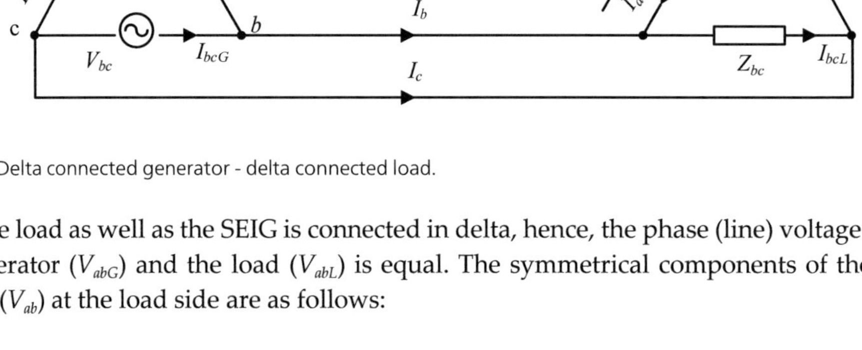

Figure 7. Delta connected generator - delta connected load.

Since the load as well as the SEIG is connected in delta, hence, the phase (line) voltage of both the generator (V_{abG}) and the load (V_{abL}) is equal. The symmetrical components of the phase voltage (V_{ab}) at the load side are as follows:

$$\begin{bmatrix} V_{abL0} \\ V_{abL1} \\ V_{abL2} \end{bmatrix} = \begin{bmatrix} Z_0 & Z_2 & Z_1 \\ Z_1 & Z_0 & Z_2 \\ Z_2 & Z_1 & Z_0 \end{bmatrix} \begin{bmatrix} I_{abL0} \\ I_{abL1} \\ I_{abL2} \end{bmatrix} \tag{43}$$

It is known that for a Delta connected load, $V_{abL0} = 0$, hence, from Eq. (43)

$$V_{abL0} = 0 = Z_0 I_{abL0} + Z_2 I_{abL1} + Z_1 I_{abL2} \tag{44}$$

This equation yields:

$$I_{abL0} = -(Z_2 I_{abL1} + Z_1 I_{abL2}) / Z_0 \tag{45}$$

From Eq. (43)

$$V_{abL1} = Z_1 I_{abL0} + Z_0 I_{abL1} + Z_2 I_{abL2} \tag{46}$$

$$V_{abL2} = Z_2 I_{abL0} + Z_1 I_{abL1} + Z_0 I_{abL2} \tag{47}$$

Substituting Eq. (45) in Eqs. (46) and (47), yields

$$V_{abL1} = \left(Z_0 - \frac{Z_1 Z_2}{Z_0} \right) I_{abL1} + \left(Z_2 - \frac{Z_1^{\ 2}}{Z_0} \right) I_{abL2} \tag{48}$$

$$V_{abL2} = \left(Z_1 - \frac{Z_2^{\ 2}}{Z_0} \right) I_{abL1} + \left(Z_0 - \frac{Z_1 Z_2}{Z_0} \right) I_{abL2} \tag{49}$$

Since both the generator and the load are Delta connected, hence:

$$V_{abG1} = V_{abL\ 1} \text{and } V_{abG2} = V_{abL\ 2}$$

However,

$$V_{abG1} = -I_{abG1} Z_{G1} \quad \text{and } V_{abG2} = I_{abG2} Z_{G2}$$

Substituting in Eqs. (48) and (49), yields:

$$-I_{abG1} Z_{G1} = \left(Z_0 - \frac{Z_1 Z_2}{Z_0} \right) I_{abL1} + \left(Z_2 - \frac{Z_1^{\ 2}}{Z_0} \right) I_{abL2} \tag{50}$$

$$I_{abG2} Z_{G2} = \left(Z_1 - \frac{Z_2^{\ 2}}{Z_0} \right) I_{abL1} + \left(Z_0 - \frac{Z_1 Z_2}{Z_0} \right) I_{abL2} \tag{51}$$

It can be shown using symmetrical components technique that the sequence components of phase and line currents are related as follows:

$$I_{a1} = (1-\gamma)\, I_{ab1} \tag{52}$$

$$I_{a2} = (1-\gamma^2)\, I_{ab2} \tag{53}$$

Substituting Eqs. (52) and (53) into Eqs. (50) and (51)), yields

$$-\frac{I_{a1}}{(1-\gamma)} Z_{G1} = \left(Z_0 - \frac{Z_1 Z_2}{Z_0}\right)\frac{I_{a1}}{(1-\gamma)} + \left(Z_2 - \frac{Z_1^{\,2}}{Z_0}\right)\frac{I_{a2}}{(1-\gamma^2)} \tag{54}$$

$$\frac{I_{a2}}{(1-\gamma^2)} Z_{G2} = \left(Z_1 - \frac{Z_2^{\,2}}{Z_0}\right)\frac{I_{a1}}{(1-\gamma)} + \left(Z_0 - \frac{Z_1 Z_2}{Z_0}\right)\frac{I_{a2}}{(1-\gamma^2)} \tag{55}$$

Since excitation is assumed to occur and rearranging yields,

$$\left(\frac{Z_1 Z_2}{Z_0} - Z_0 - Z_{G1}\right)\left(Z_0 - \frac{Z_1 Z_2}{Z_0} - Z_{G2}\right) - \left(Z_2 - \frac{Z_1^{\,2}}{Z_0}\right)\left(\frac{Z_2^{\,2}}{Z_0} - Z_1\right) = 0 \tag{56}$$

5. Delta connected generator–star connected load

The connection for this case is shown in Figure 8. It is known that $I_{a0} = 0$ for a star connected load, substituting in Eq. (5) and expanding V_{ak1}, V_{ak2};

$$V_{ak1} = Z_0\, I_{a1} + Z_2 I_{a2} \tag{57}$$

$$V_{ak2} = Z_1\, I_{a1} + Z_0\, I_{a2} \tag{58}$$

The load sequence components of line-to-line voltage may be expressed in terms of the sequence components of line to neutral voltage as;

$$V_{abL1} = (1-\gamma^2)\, V_{ak1} \tag{59}$$

$$V_{abL2} = (1-\gamma)\, V_{ak2} \tag{60}$$

The positive and negative sequence components of generator voltage in terms of input impedances Z_{G1} and Z_{G2} are;

$$V_{abG1} = -\, I_{abG1}\, Z_{G1} \tag{61}$$

$$V_{abG2} = I_{abG2}\, Z_{G2} \tag{62}$$

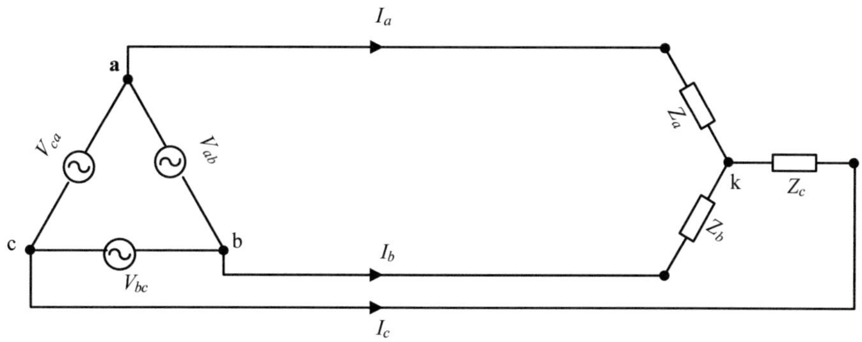

Figure 8. Delta connected generator - star connected load

These voltages equal to the load line voltages, as;

$$V_{abG1} = V_{abL1} \tag{63}$$

$$V_{abG2} = V_{abL2} \tag{64}$$

Substituting Eq. (59), Eq. (60), Eq. (61), and Eq. (62) into Eq. (63) and Eq. (64), yields:

$$(1-\gamma^2)\, V_{ak1} = -\, I_{abG1}\, Z_{G1} \tag{65}$$

$$(1-\gamma)\, V_{ak2} = I_{abG2}\, Z_{G2} \tag{66}$$

Substituting Eq. (57) and Eq.(58) into Eq. (65) and Eq. (66), yields:

$$-\, I_{abG1}\, Z_{G1} = (1-\gamma^2)\,(Z_0\, I_{a1} + Z_2\, I_{a2}) \tag{67}$$

$$I_{abG2}\,Z_{G2} = (1-\gamma)(Z_1\,I_{a1} + Z_0\,I_{a2}) \tag{68}$$

The symmetrical components of line current are related to the symmetrical components of phase current in a delta connected generator as follows;

$$I_{a1} = (1-\gamma)\,I_{abG1} \tag{69}$$

$$I_{a2} = (1-\gamma^2)\,I_{abG2} \tag{70}$$

Since excitation is assumed to occur, by substituting Eq. (69) and Eq. (70) in Eqs. (67) and (68), and rearranging, yields:

$$(3\,Z_0 + Z_{G1})(3\,Z_0 - Z_{G2}) - 9\,Z_1\,Z_2 = 0 \tag{71}$$

6. Star connected generator–delta connected load

At the load side of Figure 9, the symmetrical components of the load are as follows:

$$\begin{bmatrix} Z_0 \\ Z_1 \\ Z_2 \end{bmatrix} = \frac{1}{3}\begin{bmatrix} 1 & 1 & 1 \\ 1 & \gamma & \gamma^2 \\ 1 & \gamma^2 & \gamma \end{bmatrix}\begin{bmatrix} Z_{ab} \\ Z_{bc} \\ Z_{ca} \end{bmatrix} \tag{72}$$

It is known that for a delta connected load, $V_{abLO} = 0$. Hence, from Eq. (43);

$$V_{abLO} = 0 = Z_0\,I_{abLO} + Z_2\,I_{abL1} + Z_1\,I_{abL2} \tag{73}$$

from this equation, it can be shown that;

$$I_{abLO} = -(Z_2\,I_{abL1} + Z_1 I_{abL2})\,/\,Z_0 \tag{74}$$

From Eq. (43);

$$V_{abL1} = Z_1\,I_{abLO} + Z_0\,I_{abL1} + Z_2\,I_{abL2} \tag{75}$$

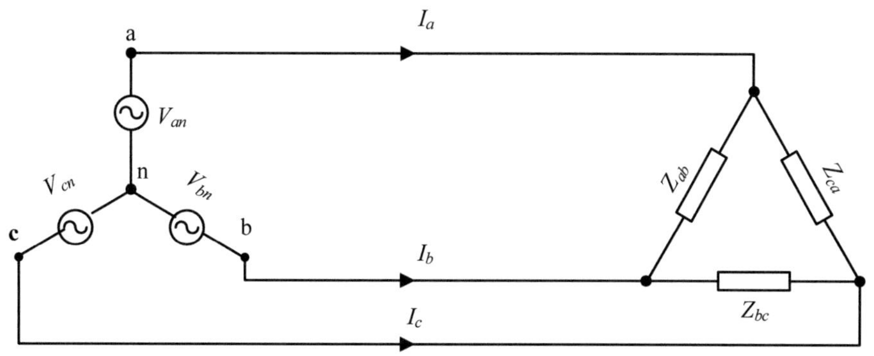

Figure 9. Star connected generator - delta connected load

$$V_{abL2} = Z_2\, I_{abL0} + Z_1\, I_{abL1} + Z_0\, I_{abL2} \tag{76}$$

Substituting Eq. (74) in Eqs. (75) and (76), yields:

$$V_{abL1} = \left(Z_0 - \frac{Z_1 Z_2}{Z_0} \right) I_{abL1} + \left(Z_2 - \frac{Z_1^{\ 2}}{Z_0} \right) I_{abL2} \tag{77}$$

$$V_{abL2} = \left(Z_1 - \frac{Z_2^{\ 2}}{Z_0} \right) I_{abL1} + \left(Z_0 - \frac{Z_1 Z_2}{Z_0} \right) I_{abL2} \tag{78}$$

at the generator side,

$$V_{abG1} = (1\text{-}\gamma^2)\, V_{an1} \tag{79}$$

$$V_{abG2} = (1\text{-}\gamma)\, V_{an2} \tag{80}$$

It is known that

$$V_{an1} = -I_{G1}\, Z_{G1} \tag{81}$$

$$V_{an2} = I_{G2}\, Z_{G2} \tag{82}$$

Substituting Eq. (81) and Eq. (82) into Eq.(79) and Eq. (80), gives,

$$V_{abG1} = -(1-\gamma^2)\, I_{G1} Z_{G1} \tag{83}$$

$$V_{abG2} = (1-\gamma\ )\, I_{G2} Z_{G2} \tag{84}$$

Since the line voltages at the generator and the load side are equal, hence,

$$-(1-\gamma^2) I_{G1} Z_{G1} = \left(Z_0 - \frac{Z_1 Z_2}{Z_0} \right) I_{abL1} + \left(Z_2 - \frac{Z_1^{\ 2}}{Z_0} \right) I_{abL2} \tag{85}$$

$$(1-\gamma) I_{G2} Z_{G2} = \left(Z_1 - \frac{Z_2^{\ 2}}{Z_0} \right) I_{abL1} + \left(Z_0 - \frac{Z_1 Z_2}{Z_0} \right) I_{abL2} \tag{86}$$

It is known that,

$$I_{G1} = I_{a1} \tag{87}$$

$$I_{G2} = I_{a2} \tag{88}$$

$$I_{a1} = (1-\gamma) I_{abL1} \tag{89}$$

$$I_{a2} = (1-\gamma^2)\, I_{abL2} \tag{90}$$

Since excitation is assumed to occur, by substituting Eqs. (87), (88), (89), and (90) into Eqs. (85) and (86), and rearranging yield,

$$\left(Z_0 + 3Z_{G1} - \frac{Z_1 Z_2}{Z_0} \right)\left(Z_0 - 3Z_{G2} - \frac{Z_1 Z_2}{Z_0} \right) - \left(Z_1 - \frac{Z_2^{\ 2}}{Z_0} \right)\left(Z_2 - \frac{Z_1^{\ 2}}{Z_0} \right) = 0 \tag{91}$$

7. Generalization of steady state model

The characteristics equations derived in the previous sections can be represented by one single general equation of the following form:

http://dx.doi.org/10.5772/53914

$$\begin{aligned}&(\beta_1 Z_0 Z_{G1} + \beta_2 Z_0^2 + \beta_3 Z_1 Z_2)(\beta_4 Z_0 Z_{G2} + \beta_5 Z_0^2 + \beta_6 Z_1 Z_2)\\&-(\beta_7 Z_0 Z_1 + \beta_8 Z_2^2)(\beta_9 Z_0 Z_2 + \beta_{10} Z_1^2) = 0\end{aligned} \tag{92}$$

where for each connection the appropriate value of β parameters are given in table 1 below.

Connection SEIG-Load	β1	β2	β3	β4	β5	β6	β7	β8	β9	β10
Δ - Δ	-1	-1	1	-1	1	-1	-1	1	1	-1
Δ - Y	1	3	0	-1	3	0	3	0	3	0
Y - Δ	3	1	-1	-3	1	-1	1	-1	1	-1
Y - Y	1	1	0	-1	1	0	1	0	1	0
Y - Y with neutral	-1	-1	1	-1	1	-1	-1	1	1	-1

Table 1. Values of Parameter β of characteristics equation

8. Conclusions

A mathematical model based on the sequence equivalent circuits of the SEIG and the sequence components of the three-phase load was developed to study the performance of the SEIG in the steady state condition. Core loss resistance is included in the model as a function of X_m.

Furthermore, the magnetizing reactance X_m is taken as a variable in the negative sequence equivalent circuit. The performance of the SEIG was determined for no-load, balanced and unbalanced load and/or excitation for different SEIG and load connections. The operating conditions were found by solving the proposed model iteratively using MATHCAD® software as described in section 1. Self-excited induction generator uses excitation capacitors at the terminals. For a given speed, and load situation, there is a specific value of the excitation capacitor that ensures voltage build up. Residual magnetism is a must in SEIG to initiate excitation. In balanced mode of operation, per phase equivalent circuit is solved to find F and X_m. Unbalanced operation of SEIG can be analyzed through the method of symmetrical components and the phase voltages may differ from each other notably.

An experimental setup has been built to verify the results obtained from the theoretical model. It is found that the theoretical results are in good agreement with those recorded experimentally. The model is generalized to cover all connection types of SEIG and/or load. The characteristic equation of each type may be found by substituting the appropriate parameters, i.e., β_1 up to β_{10} from table 1, in the general model.

$$\begin{aligned}&(\beta_1 Z_0 Z_p + \beta_2 Z_0^2 + \beta_3 Z_1 Z_2)(\beta_4 Z_0 Z_n + \beta_5 Z_0^2 + \beta_6 Z_1 Z_2)\\&-(\beta_7 Z_0 Z_1 + \beta_8 Z_2^2)(\beta_9 Z_0 Z_2 + \beta_{10} Z_1^2) = 0\end{aligned}$$

Author details

A. Alsalloum and A. I. Alolah*

*Address all correspondence to: alolah@ksu.edu.sa

EE Depatrment, College of English, King Saud University, Riyadh, Saudi Arabia

References

[1] Murthy, S, Malik, O, & Tandon, A. Analysis of Self-Excited Induction Generators", IEE Proc., pt. B, (129), (1982). (6), 260-265.

[2] Elder, J, Boys, J, & Woodward, J. The Process of Self Excitation in Induction Generators", ibid, (130), pt. B., (1983). (2), 103-107.

[3] Quazene, I, & Mcpherson, G. Analysis of Self-Excited Induction Generators", IEEE Trans., (1983). , PAS-102, 2793-2797.

[4] Raina, G, & Malik, O. Wind Energy Conversion Using a Self-Excited Induction Generator", ibid., Power System Apparatus, (1983). , PAS-102, 3933-3936.

[5] Grantham, C, Sutanto, D, & Mismail, B. Steady State and Transient Analysis of Self-Excited Induction Generators", IEE Proc., pt. B, (136), (1989). (2), 61-68.

[6] Malik, N, & Mazi, A. Capacitance Requirements for Isolated Self-Excited Induction Generators", IEEE Trans., EC-2(1), (1987). , 62-69.

[7] Al-jabri, A, & Alolah, A. Capacitance Requirements for Isolated Self-Excited Induction Generators", IEE Proc., (137), pt. B, (1990). (3), 154-159.

[8] Singh, B. and Shilpakar, "Steady State Analysis of Single Phase Self-Excited Induction Generator", ibid., (146), (1999). (5), 421-427.

[9] Ojo, O. Performance of Self-Excited Single Phase Induction Generators with Shunt, Short Shunt and Long Shunt Excitation Connections", IEEE Trans., EC-11(3), (1996). , 477-482.

[10] Wagner, C. F, & Evans, R. D. Symmetrical Components, McGraw-Hill, Book, (1933).

[11] Malik, N, & Al-bahrani, A. Influence of the Terminal Capacitor on the Performance Characteristics of a Self Excited Induction Generator", IEE Proc., pt. C, (1990). , 137(2), 168-173.

[12] Al-bahrani, A, & Malik, N. Steady State Analysis and Performance Characteristics of a Three-Phase Induction Generator Self Excited with a Single Capacitor", IEEE Trans., EC-4(4), (1990). , 725-732.

Chapter 8

Maximum Power Extraction from Utility-Interfaced Wind Turbines

Ali M. Eltamaly, A. I. Alolah and Hassan M. Farh

Additional information is available at the end of the chapter

http://dx.doi.org/10.5772/54675

1. Introduction

Wind energy is one of the most promising renewable energy resources for producing electricity due to its cost competitiveness compared to other conventional types of energy resources. It takes a particular place to be the most suitable renewable energy resources for electricity production. It isn't harmful to the environment and it is an abundant resource available in nature. Hence, wind power could be utilized by mechanically converting it to electrical power using wind turbine, WT. Various WT concepts have a quick development of wind power technologies and significant growth of wind power capacity during last two decades. Variable speed operation and direct drive WTs have been the modern developments in the technology of wind energy conversion system, WECS. Variable-speed operation has many advantages over fixed-speed generation such as increased energy capture, operation at MPPT over a wide range of wind speeds, high power quality, reduced mechanical stresses, aerodynamic noise improved system reliability, and it can provide (10-15) % higher output power and has less mechanical stresses in comparison with the operation at a fixed speed [1, 2]. WTs can be classified according to the type of drive train into direct drive (DD) and gear drive (GD). The GD type uses a gear box, squirrel cage induction generator (SCIG) and classified as stall, active stall and pitch control WT and work in constant speed applications. The variable speed WT uses doubly-fed induction generator, (DFIG) especially in high power WTs. The gearless DD WTs have been used with small and medium size WTs employing permanent-magnet synchronous generator (PMSG) with higher numbers of poles to eliminate the need for gearbox which can be translated to higher efficiency. PMSG appears more and more attractive, because the advantages of permanent magnet, (PM) machines over electrically excited machines such as its higher efficiency, higher energy yield, no additional power supply for the magnet field excitation and higher reliability due to the absence of mechanical components such as slip rings. In addition, the performance of PM materials is improving, and the cost is decreasing in recent years. Therefore, these advantages make direct-drive PM wind turbine systems more attractive in application of small and medium-scale wind turbines [1, 3-4].

Robust controller has been developed in many literatures [5-15] to track the maximum power available in the wind. They include tip speed ratio (TSR) [5, 13], power signal feedback (PSF) [8, 14], and the hill-climb searching (HCS) [11-12] methods. The TSR control method regulates the rotational speed of the generator to maintain an optimal TSR at which power extracted is maximum [13]. For TSR calculation, both the wind speed and turbine speed need to be measured, and the optimal TSR must be given to the controller. The first barrier to implement TSR control is the wind speed measurement, which adds to system cost and presents difficulties in practical implementations. The second barrier is the need to obtain the optimal value of TSR, this value is different from one system to another. This depends on the turbine-generator characteristics results in custom-designed control software tailored for individual wind turbines [14]. In PSF control [8, 14], it is required to have the knowledge of the wind turbine's maximum power curve, and track this curve through its control mechanisms. The power curves need to be obtained via simulations or off-line experiment on individual wind turbines or from the datasheet of WT which makes it difficult to implement with accuracy in practical applications [7-8, 15]. The HCS technique does not require the data of wind, generator speeds and the turbine characteristics. But, this method works well only for very small wind turbine inertia. For large inertia wind turbines, the system output power is interlaced with the turbine mechanical power and rate of change in the mechanically stored energy, which often renders the HCS method ineffective [11-12]. On the other hand, different algorithms have been used for maximum power extraction from WT in addition to the three method mentioned above. For example, Reference [1] presents an algorithm for maximum power extraction and reactive power control of an inverter through the power angle, δ of the inverter terminal voltage and the modulation index, m_a based variable-speed WT without wind speed sensor. Reference [16] presents an algorithm for MPPT via controlling the generator torque through q-axis current and hence controlling the generator speed with variation of the wind speed. These techniques are used for a decoupled control of the active and reactive power from the WT through q-axis and d-axis current respectively. Also, reference [17] presents a decoupled control of the active and reactive power from the WT, independently through q-axis and d-axis current but maximum power point operation of turbine system has been produced through regulating the input dc current of the dc/dc boost converter to follow the optimized current reference [17]. Reference [18] presents an algorithm for MPPT through directly adjusting duty ratio of the dc/dc boost converter and modulation index of the PWM- VSC. Reference [19] presents MPPT control algorithm based on measuring the dc-link voltage and current of the uncontrolled rectifier to attain the maximum available power from wind. Finally, references [20-22] present MPPT control based on a fuzzy logic control (FLC). The function of FLC is to track the generator speed with the reference speed for maximum power extraction at variable speeds. The MPPT algorithms can be divided into two categories, the first one is MPPT algorithms for WT with wind speed sensor and the second one is MPPT algorithms without wind speed sensor (sensorless MPPT controller). Wind speed sensor normally used in conventional wind energy conversion systems, WECS [10, 23] for implementing MPPT control algorithm. This algorithm

http://dx.doi.org/10.5772/54675

increases cost and reduces the reliability of the WECS in addition to inaccuracies in measuring the wind speed. Therefore, some MPPT control methods estimate the wind speed; however, many of them require the knowledge of air density and mechanical parameters of the WECS [88-92]. Such methods require turbine generator characteristics result in custom-design software tailored for individual wind turbines. Air density, on the other hand, depends upon climatic conditions and may vary considerably over various seasons. Therefore, this technique is not favorite in modern design of WT and a lot of research efforts are focused on developing wind speed sensorless MPPT controller which does not require the knowledge of air density and turbine mechanical parameters [1, 9-11, 22-25]. Therefore, the cost and maintenance of the power control system is decreased and implementation of the power control system is not difficult compared to the sensored MPPT controller.

2. Wind Energy Conversion Systems (WECS)

Figure 1 shows the schematic diagram of the variable-speed wind energy conversion system based on a synchronous generator. This system is directly connected to the grid through power conversion system. There are two common types of the power conversion systems, the first configuration is a back-to-back PWM-VSC connected to the grid. This configuration has a lot of switches, which cause more losses and voltage stress in addition to the presence of Electromagnetic Interference (EMI). The presence of a dc-link capacitor in PWM-VSC system provides a decoupling between the two converters, it separates the control between these two converters, allowing compensation of asymmetry of both on the generator side and on the grid side, independently [16]. The second configuration consists of a diode-bridge rectifier, a boost converter and a PWM-VSC connected to the grid. This configuration is, simple, less expensive, robust, and rigid and needs simple control system. But, with this configuration the control of the generator power factor is not possible, which in turn, affects generator efficiency. Also, high harmonic distortion currents are obtained in the generator that reduce efficiency and produce torque oscillations [22].

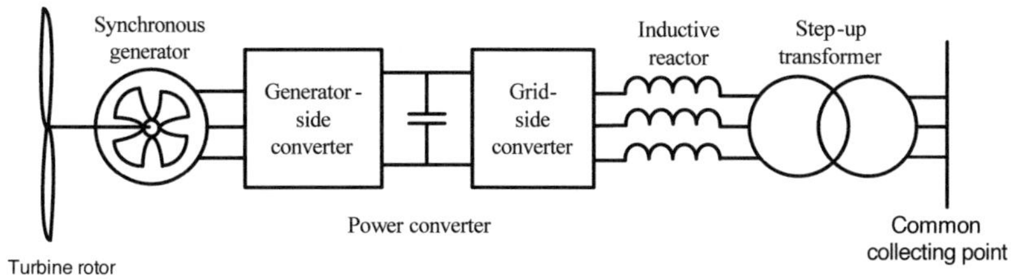

Figure 1. Wind energy conversion system based on a synchronous generator [26].

Wind turbine converts the wind power to a mechanical power, which in turn, runs a generator to generate electrical power. The mechanical power generated by wind turbine can be expressed as [15]:

$$P_m = \frac{1}{2} C_P(\lambda, \beta) \rho A u^3 \tag{1}$$

where

C_p: Turbine power coefficient.

ρ: Air density (kg/m^3).

A: Turbine sweeping area (m^2).

u: wind speed (m/s).

λ: tip speed ratio of the wind turbine which is given by the following equation [1];

$$\lambda = \frac{r_m \omega_r}{u} \tag{2}$$

Where r_m is the turbine rotor radius, $\omega_r$is the angular velocity of turbine (rad/s).

The turbine power coefficient, C_p, describes the power extraction efficiency of the wind turbine. It is a nonlinear function of both tip speed ratio, λ and the blade pitch angle, β. While its maximum theoretical value is approximately 0.59, it is practically between 0.4 and 0.45 [15]. There are many different versions of fitted equations for C_p made in the literatures. A generic equation has been used to model $C_p(\lambda, \beta)$ and based on the modeling turbine characteristics as shown in the following equation [27]:

$$C_P(\lambda, \beta) = 0.5176\left(116 * \frac{1}{\lambda_i} - 0.4\beta - 5\right) e^{-\frac{21}{\lambda_i}} + 0.0068\lambda \tag{3}$$

With

$$\frac{1}{\lambda_i} = \frac{1}{\lambda + 0.08\beta} - \frac{0.035}{1+\beta^3} \tag{4}$$

The C_p-λ characteristics, for different values of the pitch angle β, are illustrated in Figure 2. The maximum value of C_p is achieved for $\beta = 0$ degree and for λ_{opt}. The particular value of λ is defined as the optimal value (λ_{opt}). Continuous operation of wind turbine at this point guarantees the maximum available power which can be harvested from the available wind at any speed.

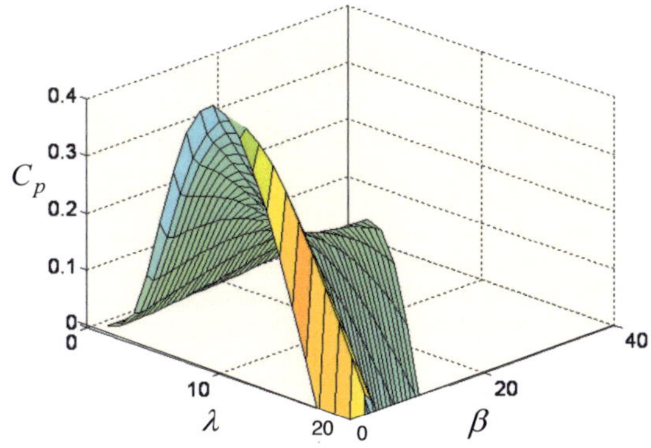

Figure 2. Aerodynamic power coefficient variation against λ and β.

2.1. Wind turbine arrangement with back-to-back PWM-VSCs

In this arrangement both the generator and the grid-side converters are PWM-VSCs as shown in Figure 3. The output voltage of the generator is converted into dc voltage through a PWM-VSC. As the previous model the dc-link voltage is converted to constant frequency voltage using grid side PWM-VSC. The dc-link voltage is controlled by the modulation index (m_a) and power angle (δ). Controlling the dc-link voltage and the bitch angle of the blades of WT will track the maximum power of WT in the case of variable pitch angle control. In the case of fixed pitch angle control the maximum power extraction is achieved by tracking the optimum shaft speed. The grid side PWM-VSC can be used to enhance the stability of the dc-link voltage and controls the active and reactive power from WT by controlling m_a and δ. The generator can be directly controlled by the generator side converter (controller-1) while the grid-side converter (controller-2) maintains the dc-link voltage at the desired value by exporting active power to the grid. Controller-2 also controls the reactive power exchange with the grid [26]. So, the main target of controller-1 is to track the maximum power available from the WTG and the function of controller-2 is to control the dc-link voltage and the reactive power injected to the electric utility.

2.2. Wind turbine arrangement with diode-based rectifier

Figure 4 shows the wind turbine with a diode-based rectifier as the generator-side converter. The diode bridge rectifier converts the generator output ac power to dc power and the PWM-VSC converts the dc power from the rectifier output to ac power. One method to control the operation of the wind turbine with this arrangement (assuming a PMSG) is illustrated in Figure 4. A dc–dc converter is employed to control the dc-link voltage (controller-1), the grid side converter controls the operation of the generator and the power flow to the grid (controller-2). With appropriate control, the generator and turbine speed can be adjusted as wind speed varies so that maximum energy is collected [26]. On the other hand, in most PMSG wind systems,

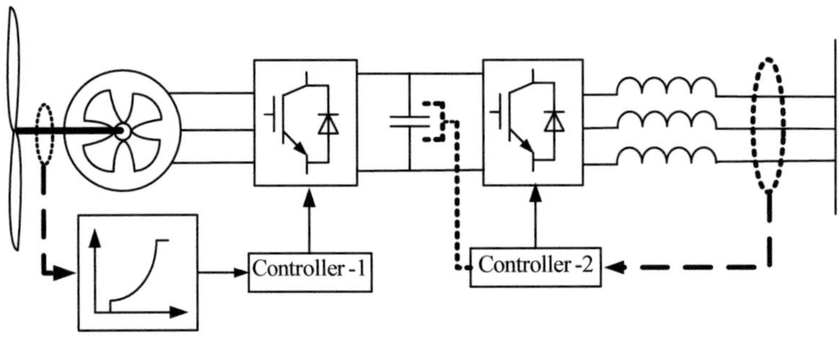

Figure 3. Wind turbine generator with back-to-back PWM-VSCs [26].

the output voltage of the generator is converted into dc voltage via a full-bridge diode rectifier and this dc voltage is adjusted to control the maximum power of turbine. The grid side converter is controlled by grid injected active and reactive power control method. The ac power output from PMSG is fed to a three-phase diode bridge forward by boost converter to effectively control the dc voltage level through the duty ratio of boost converter. The PWM-VSC is used to interface the WTG with the electrical utility also to track the maximum power available from PMSG. The modulation index of the PWM-VSC is controlled to enhance the stability of the dc link voltage as shown in Figure 4.

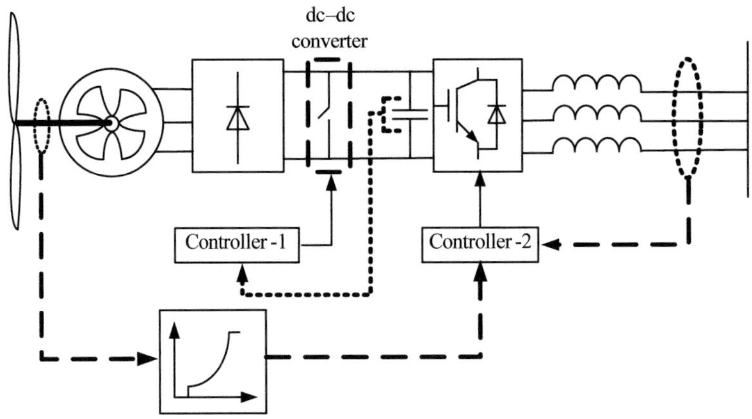

Figure 4. Wind turbine generator with a diode-based rectifier as the generator-side converter [26].

3. MPPT control strategies for the WECS

WECS has been attracting wide attention as a renewable energy source due to depleting fossil fuel reserves and environmental concerns as a direct consequence of using fossil fuel and nuclear energy sources. Wind energy varies continually as wind speed changes throughout the day, even though abundant. The Amount of power output from a WECS depends upon the accuracy of tracking the peak power points using the MPPT controller irrespective of the generator type used. The maximum power extraction algorithms can be classified into two categories. The two categories are MPPT algorithms with wind speed sensor and MPPT algorithms without wind speed sensor (sensor-less MPPT controller). These two algorithms have been discussed in the following sections.

3.1. MPPT algorithms for a WT with wind speed sensor

3.1.1. Tip Speed Ratio (TSR) technique

The TSR control method regulates the rotational speed of the generator to maintain an optimal TSR at which power extracted is maximum [13]. The target optimum power extracted from wind turbine can be written as [14]:

$$P_{\max} = K_{opt} * \omega_{opt}^{3} \tag{5}$$

Where $K_{opt} = 0.5 * \rho A * \left(\frac{r_m}{\lambda_{opt}}\right)^3 * C_{P-\max}$, and $\omega_{opt} = \frac{\lambda_{opt}}{r_m} * u$

The power for a certain wind speed is maximum at a certain value of rotational speed called optimum rotational speed, ω_{opt}. This optimum rotational speed corresponds to optimum tip speed ratio, λ_{opt}. In order to track maximum possible power, the turbine should always operate at λ_{opt}. This is achieved by controlling the rotational speed of the WT so that it always rotates at the optimum rotational speed. As shown in Figure 5, for TSR calculation, both the wind speed and turbine speed need to be measured, and the optimal TSR must be given to the controller. The first barrier to implement TSR control is the wind speed measurement, which adds to system cost and presents difficulties in practical implementations. The second barrier is the need to obtain the optimal value of TSR, this value is different from one system to another. This depends on the turbine-generator characteristics results in custom-designed control software tailored for individual wind turbines [14].

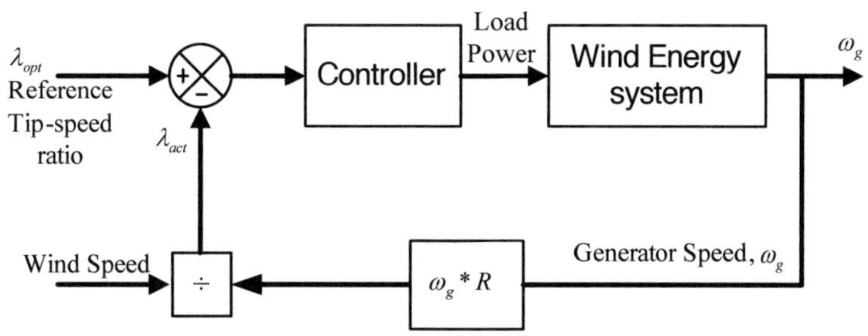

Figure 5. The block diagram of the tip speed ratio control of WECS [15].

3.1.2. Power Signal Feedback (PSF) control

In PSF control [14], it is required to have the knowledge of the wind turbine's maximum power curve, and track this curve through its control mechanisms. The maximum power curves need to be obtained via simulations or off-line experiment on individual wind turbines or from the datasheet of WT which makes it difficult to implement with accuracy in practical applications. In this method, reference power is generated using a maximum power data curve or using the mechanical power equation of the wind turbine where wind speed or the rotational speed is used as the input. Figure 6 shows the block diagram of a WECS with PSF controller for maximum power extraction. The PSF control block generates the optimal power command P_{opt} which is then applied to the grid side converter control system for maximum power extraction as follow [15]:

$$P_{opt} = K_{opt} * \omega_r^3 \tag{6}$$

The actual power output, P_t is compared to the optimal power, P_{opt} and any mismatch is used by the fuzzy logic controller to change the modulation index of the grid side converter, PWM-VSC as shown in Figure 6. The PWM-VSC is used to interface the WT with the electrical utility and will be controlled through the power angle, δ and modulation index, m_a to control the active and reactive power output from the WTG [15].

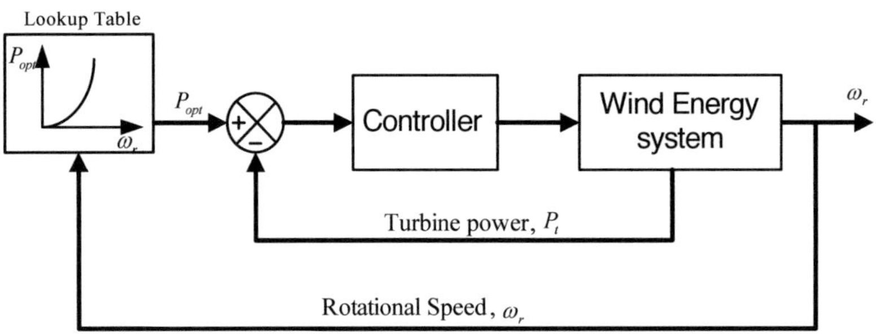

Figure 6. The block diagram of power signal feedback control [15].

3.1.3. Optimal torque control

The aim of the torque controller is to optimize the efficiency of wind energy capture in a wide range of wind velocities, keeping the power generated by the machine equal to the optimal defined value. It can be observed from the block diagram represented in Figure 7, that the idea of this method is to adjust the PMSG torque according to the optimal reference torque of the wind turbine at a given wind speed. A typical wind turbine characteristic with the optimal torque-speed curve plotted to intersect the $C_{P\text{-max}}$ points for each wind speed is illustrated in Figure 8. The curve T_{opt} defines the optimal torque of the device (i.e. maximum energy capture), and the control objective is to keep the turbine on this curve as the wind speed varies. For any wind speed, the MPPT device imposes a torque reference able to extract the maximum power. The curve T_{opt} is defined by [26]:

$$T_{opt} = K_{opt} * \omega^2{}_{opt} \tag{7}$$

Where

$$K_{opt} = 0.5 * \rho A * \left(\frac{r_m}{\lambda_{opt}} \right)^3 * C_{P-\max} \tag{8}$$

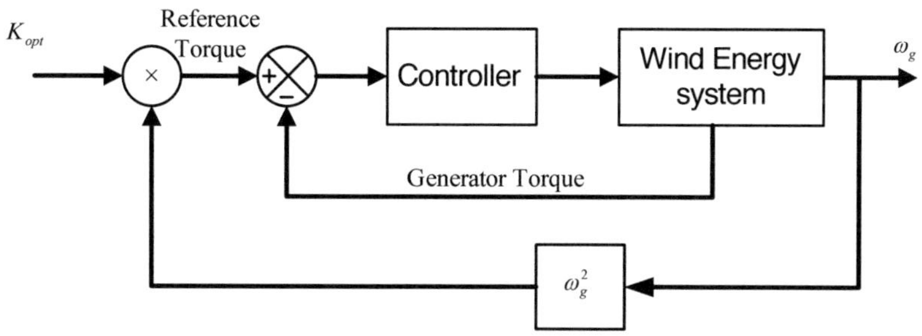

Figure 7. The block diagram of optimal torque control MPPT method.

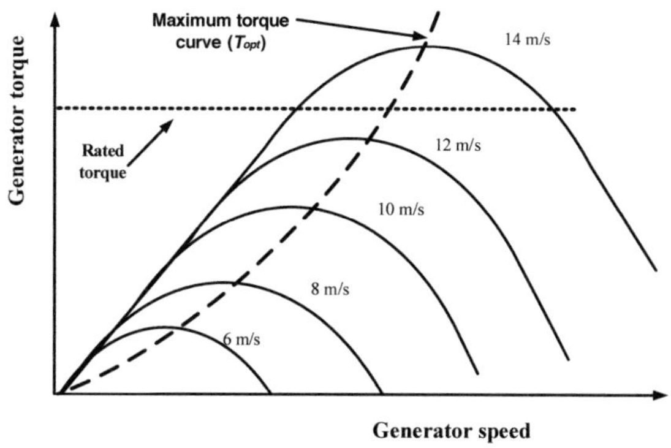

Figure 8. Wind turbine characteristic for maximum power extraction [26].

3.1.4. Load angle control

The load angle control can be explained by analyzing the transfer of active and reactive power between two sources connected by an inductive reactance as shown in Figure 9. The active power, P_S, and reactive power, Q_S, transferred from the sending-end to the receiving-end can be calculated from the following equation [26]:

$$P_s = \frac{V_s V_R}{X_{gen}} \sin\delta \tag{9}$$

$$Q_s = \frac{V_S^2}{X_{gen}} - \frac{V_S V_R}{X_{gen}} \tag{10}$$

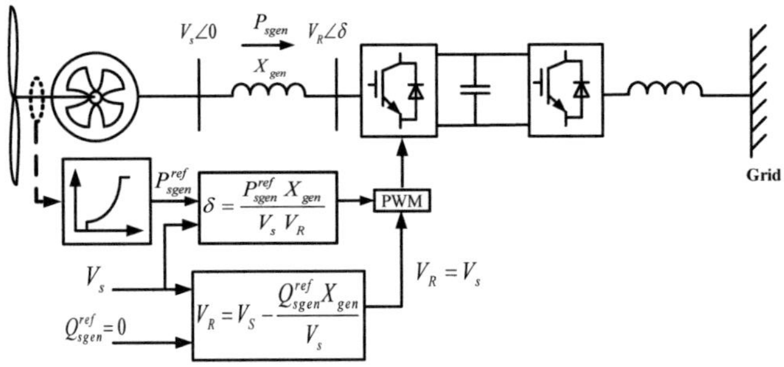

Figure 9. Load angle control of the generator-side converter [26].

3.1.4.1. Load angle control of the generator-side converter

The operation of the generator and the power transferred to the dc-link are controlled by adjusting the magnitude and angle of the voltage at the ac terminals of the generator-side converter. This can be achieved using the load angle control technique where the internal voltage of the generator is the sending source ($V_s\angle 0$), and the generator-side converter is the receiving source ($V_R\angle\delta$). The inductive reactance between these two sources is the synchronous reactance of the generator, X_{gen}, as shown in Figure 9 [26].

If it is assumed that the load angle δ is small, then $\sin\delta \approx \delta$ and $\cos\delta \approx 1$, Then the voltage magnitude, V_R, and angle magnitude, δ, required at the terminals of the generator-side converter are calculated using Equations (7) and (8) as shown [26]:

$$\delta = \frac{P_{Sgen}^{ref} X_{gen}}{V_S V_R} \tag{11}$$

$$V_R = V_S - \frac{Q_{Sgen}^{ref} X_{gen}}{V_S} \tag{12}$$

http://dx.doi.org/10.5772/54675

Where P_{Sgen}^{ref} is the reference value of the active power that needs to be transferred from the generator to the dc-link, and Q_{Sgen}^{ref} is the reference value for the reactive power. The reference value P_{Sgen}^{ref} is obtained from the characteristic curve of the machine for maximum power extraction for a given generator speed, ω_r. As the generator has permanent magnets, it does not require magnetizing current through the stator, thus the reactive power reference value can be set to zero, Q_{Sgen}^{ref}= 0 (i.e. V_S and V_R are equal in magnitude). The implementation of this load angle control scheme is illustrated in Figure 9. The major advantage of the load angle control is its simplicity. However, as the dynamics of the generator are not considered it may not be very effective in controlling the generator during transient operation [26].

3.1.4.2. Load angle control for the grid-side converter

The objective of the grid-side converter controller is to maintain the dc-link voltage at the reference value by exporting active power to the grid. In addition, the controller is designed to enable the exchange of reactive power between the converter and the grid as required by the application specifications. Also, the load angle control is a widely used grid side converter control method, where the grid-side converter is the sending source *($V_S\angle\delta$)*, and the grid is the receiving source *($V_R\angle 0$)*. As known, the grid voltage is selected as the reference; hence, the phase angle δ is positive. The reactance X_{grid} is the inductor coupling between these two sources [26]. The reference value for the active power,P_{Sgrid}^{ref} , that needs to be transmitted to the grid can be determined by examining the dc-link dynamics with the aid of Figure 10.

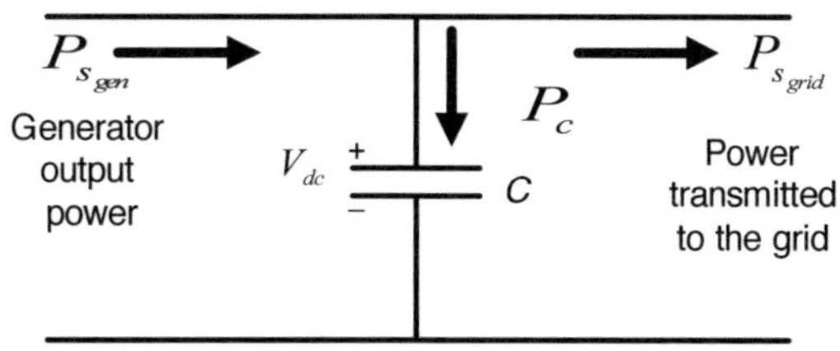

Figure 10. Power flow in the dc-link [26].

This figure illustrates the power balance at the dc-link [26] as shown in the following equation:

$$P_C = P_{Sgen} - P_{Sgrid} \tag{13}$$

where P_C is the power across the dc-link capacitor, C, P_{Sgen} is the active power output of the generator (and transmitted to the dc-link), and P_{Sgrid} is the active power transmitted from the dc-link to the grid.

The dc-link voltage V_{dc} can be expressed in terms of the generator output power, P_{Sgen}, and the power transmitted to the grid, P_{Sgrid}, as shown in the following [26]:

$$V_{dc} = \sqrt{\frac{2}{C}\int\left(P_{Sgen} - P_{Sgrid}\right)dt} \tag{14}$$

Equation (12) calculates the actual value of V_{dc}. The reference value of the active power, P_{Sgrid}^{ref}, to be transmitted to the grid is calculated by comparing the actual dc-link voltage, V_{dc}, with the desired dc-link voltage reference, $V_{dc\text{-}ref}$. The error between these two signals is processed by a PI-controller, whose output provides the reference active power P_{Sgrid}^{ref}, as shown in Figure 11. Figure 12 illustrates the implementation of the load angle control scheme for the grid-side converter with unity power factor [26].

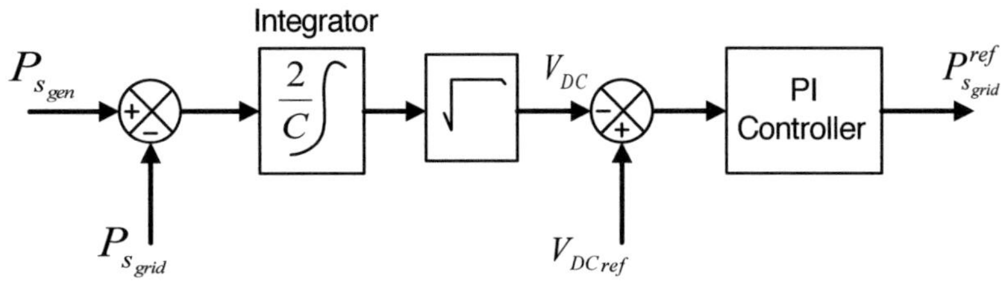

Figure 11. Calculation of active power reference, P_{Sgrid}^{ref}, (suitable for simulation purposes) [26].

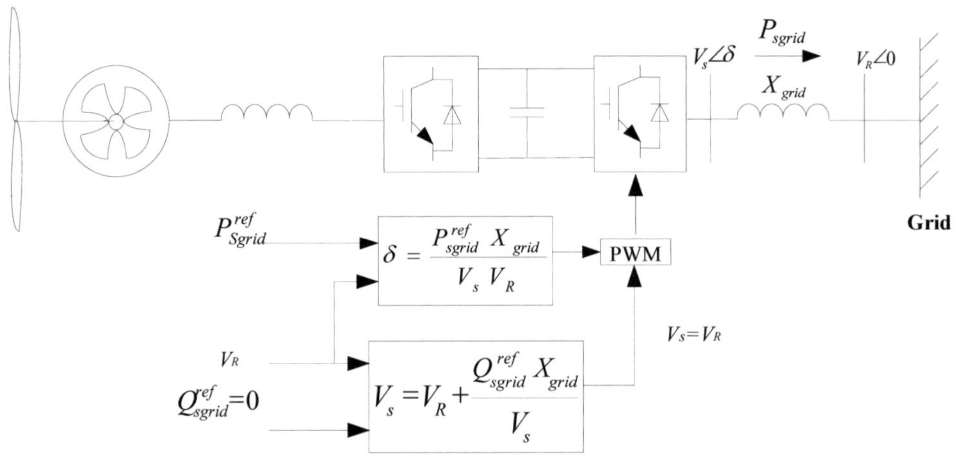

Figure 12. Load angle control of the grid-side converter [26].

3.2. MPPT algorithms for a WT without wind speed sensor

3.2.1. Hill-Climb Searching (HCS)

3.2.1.1. Principle of Hill-Climb Searching (HCS)

The HCS [11], control algorithm continuously searches for the peak power of the wind turbine. The maximum power can be extracted from WTG without requiring information about the wind and generator speeds (Hill-Climb Searching, HCS) [1, 11]. It can overcome some of the common problems normally associated with the other two methods, TSR and PSF. The tracking algorithm depends on the location of the operating point. According to the changes in power and speed the desired optimum signal has been computed in order to track the point of maximum power. Figure 13 shows the principle of HCS control where the operating point is moving toward or away from the maximum turbine power according to increasing (down-hill region) or decreasing the dc current, I_{dm} (up-hill region). The down-hill and up-hill regions are named according to the trend of the system output power with respect to the inverter dc-link voltage, V_{dc} for a wind energy system. If an increase of I_{dm} leads to an increase of the system output power, the HCS method considers the turbine running in the down-hill region, and I_{dm} should keep increasing toward the maximum power point; otherwise, the turbine is considered as running in the up-hill region, and I_{dm} decreasing will be the choice of the HCS method towrds the maximum power point [11].

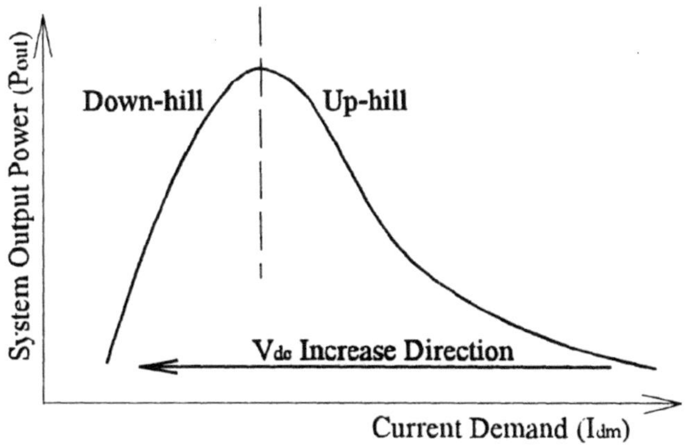

Figure 13. HCS control principle [11].

3.2.1.2. Advanced Hill-Climb Searching (HCS) method

Reference [11] introduces an advanced hill climb searching, AHCS which has been proposed to maximize P_m, through detecting the inverter output power and inverter dc-link voltage. The authors use a diode rectifier to convert the three-phase output ac voltage of a generator to V_{dc} as shown in Figure 14.

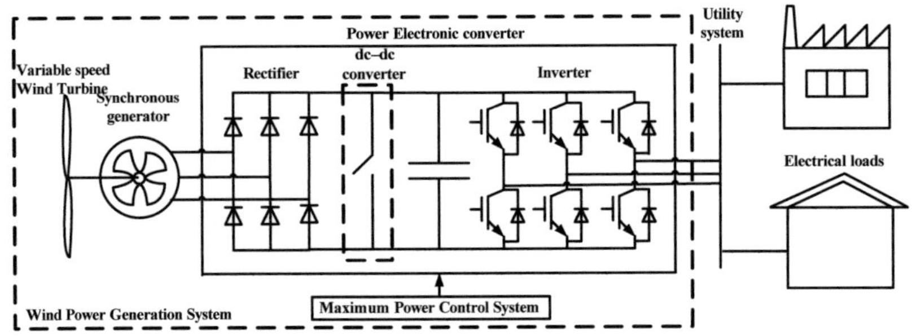

Figure 14. Typical wind power generation system connected to a utility grid [11].

V_{dc} is related to the generator angular rotational speed (ω_r) by a function of the generator field current (I_f) and the load current (I_g) as shown [11]:

$$V_{dc} = k(I_f, I_g) * \omega_r \tag{15}$$

The algorithm uses the relationship between the turbine mechanical power (P_m), and the electrical system output power (P_{out}) given by Equation (14). By differentiating Equation (14) to get a relationship for ΔP_m, Equation (15) is obtained:

$$P_m = P_{load} + T_f * \omega_r + \omega_r * J\frac{d\omega_r}{dt} = \frac{P_{out}}{\eta} + T_f * \omega_r + \omega_r * J\frac{d\omega_r}{dt} \tag{16}$$

$$\Delta P_m = \frac{\Delta P_{out}}{\eta} + T_f * \Delta\omega_r + \Delta(\omega_r * J\frac{d\omega_r}{dt}) \tag{17}$$

Authors noted that if the sampling period of the control system is adequately small then the term $k(I_f, I_g)$ can be considered as a constant value k during a sampling period. $T_f*\omega$ and η can also be considered as constant values in the same sampling period. Based on the above assumptions, Equation (15) leads to Equation (16) for digital control purposes.

$$\Delta P_m = \frac{\Delta P_{out}}{\eta} + JK^2 * \Delta(V_{dc} * \frac{dV_{dc}}{dt}) \tag{18}$$

In order to establish rules to adjust the system's operating point, this method evaluates the values of ΔP_{out} and $\Delta(V_{dc}*dV_{dc}/dt)$ (which represents $\Delta(\omega_r*d\omega_r/dt)$) based on Equation (14). Depending on the values of ΔP_{out} and $\Delta(V_{dc}*dV_{dc}/dt)$ the polarity of the inverter current demand control signal (I_{dm}) is decided according to Equation (16). There are three basic modes for this method, i) initial mode, ii) training mode, and iii) application mode as shown in Figure 15.

http://dx.doi.org/10.5772/54675

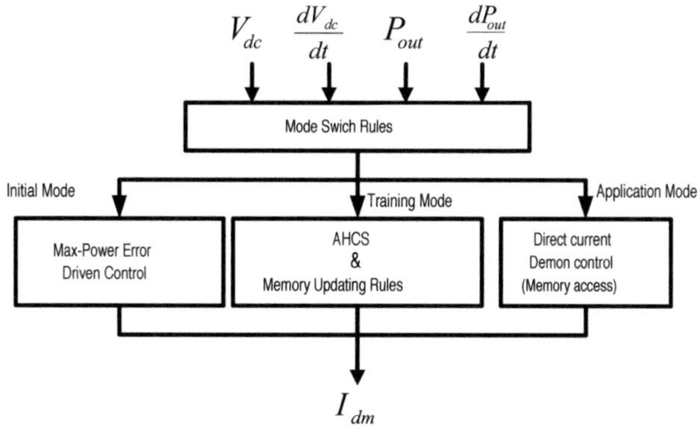

Figure 15. Structure of the intelligent maximum wind power extraction algorithm [11].

During its initial mode, before the algorithm has been trained, the magnitude of I_{dm} is determined by the maximum power error driven (MPED) control. MPED control is the implementation of the conventional HCS method in terms of wind energy system characteristics. During its training mode, the algorithm continually records and updates operating parameters into its programmable lookup table for its intelligent memory feature. Since this method is trainable with its intelligent memory, it allows itself to adapt to work with different WT. As a result, it is a solution to the customization problems of many algorithms. Another advantage of this algorithm is that it does not require mechanical sensors (like anemometers) which lowers its cost and eliminates its associated practical problems. However, it can be seen in [11] that the algorithm is relatively slow and complex as it has three different modes of operation. Another drawback is that the algorithm cannot take into account of the changes in air density, which affects the power characteristics quite significantly.

3.2.2. MPPT algorithm by directly adjusting the DC/DC converter duty cycle and modulation index of the PWM-VSC

MPPT Algorithm by Directly Adjusting the dc/dc Converter duty ratio, *D*, and Modulation Index of the PWM-VSC, m_a, is shown in Figure 16. In this direct drive converter, the mechanical power from the WT model is fed to the PMSG. The three-phase output voltages of the PMSG are fed to the three- phase diode bridge rectifier. There is no control on the output voltage of the diode bridge rectifier so it cannot be connected directly to the PWM because the PWM inverter needs constant dc voltage. So, a dc/dc converter should be used to control the dc-link voltage. Depending on the dc output voltage required from the dc/dc converter, boost or buck converter can be used. In this study, the dc output voltage, $V_{d,out}$ is required to be higher than input dc voltage, $V_{d,in,}$ so the boost converter is used. By controlling the dc voltage to be constant by controlling of D the boost converter and m_a the maximum available power from the wind can be extracted. The main drawback of this system is the diode bridge and boost converter are unidirectional power flow devices, so the PMSG has to work only in generator mode which may affect the stability of the system at abnormal conditions. A high capacitance of the dc link capacitor can remedy the effects of this drawback [18].

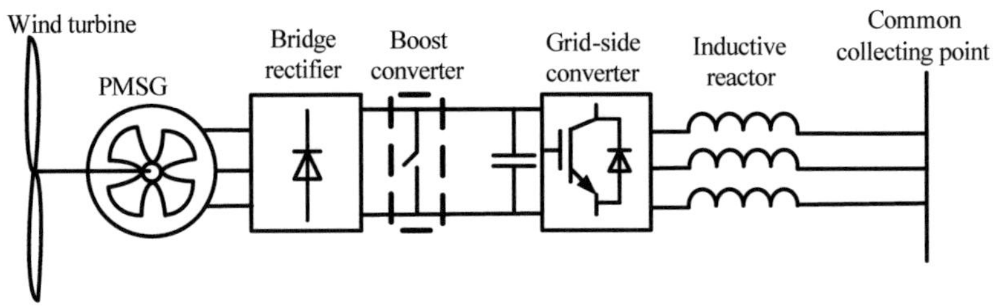

Figure 16. Modelling of wind turbine driving permanent magnet synchronous generator.

The active and reactive power can be obtained in terms of m_a and D of the boost converter as shown in Equation (17) and Equation (18), respectively [18].

$$P_{out} = \frac{\sqrt{3}\, m_a V_{d,in} V_{LLU} * \sin\delta}{2\sqrt{2}(1-D)X_s} \tag{19}$$

$$Q_{out} = \frac{\sqrt{3}\, m_a V_{d,in}}{2\sqrt{2}(1-D)X_s}\left(V_{LLU} * \cos\delta - \frac{\sqrt{3}\, m_a V_{d,in}}{2\sqrt{2}(1-D)} \right) \tag{20}$$

Where;

δ: torque angle at the electric utility side.

X_s: synchronous reactance of the electric utility.

It is clear from Equation (17) and Equation (18) that the active and reactive power can be controlled by controlling modulation index, m_a of the PWM inverter and duty ratio of the boost converter.

3.2.3. *Maximum power extraction and reactive power technique*

3.2.3.1. *Decoupled control of the active and reactive power, dependently*

This method presents an algorithm for maximum power extraction and reactive power control of an inverter based variable-speed wind-turbine generator without wind speed sensor. The algorithm does not require information about the wind and generator speeds or the inverter dc-link voltage and thus, is dependent of specifications of the wind turbine generation system [1].

Consider the wind-turbine generation system of Figure 17. The turbine mechanical power P_m and the generator output power P_g are related by

$$P_m = Pg + \omega_r * J\frac{d\omega_r}{dt} = \frac{P_{out}}{\eta} + \omega_r * J\frac{d\omega_r}{dt} \tag{21}$$

where P_{out} is the converter output power, ω_r is the generator speed, J is the combined turbine and generator moment of inertia and η is the overall system efficiency.

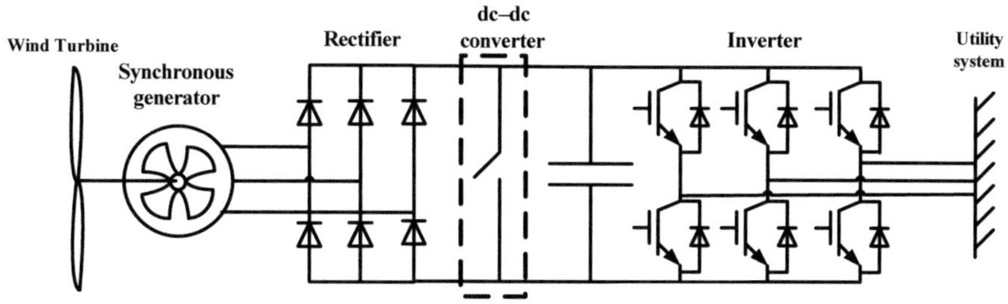

Figure 17. Grid-connected wind-turbine generation system [1].

A. Real power

This part shows how a maximum power can be extracted from a WTG without requiring information about the wind and generator speeds. In the system of Figure 17, the converter dc-link voltage is proportional to the generator speed, $V_{dc} = K\omega_r$, since the generator terminal voltage is proportional to the speed. Thus, Equation (19) can be expressed in terms of V_{dc} as [1]:

$$P_m = \frac{P_{out}}{\eta} + \frac{J}{K^2} V_{dc} \frac{dV_{dc}}{dt} \tag{22}$$

Taking the derivative of Equation (20), we deduce

$$\Delta P_m = \frac{\Delta P_{out}}{\eta} + \frac{J}{K^2} \Delta (V_{dc} \frac{dV_{dc}}{dt}) \tag{23}$$

V_{dc} is proportional to the inverter terminal voltage V_{inv} divided by the inverter amplitude modulation index, m_a. Thus,$\Delta V_{dc} = \Delta(V_{inv} / m_a)$.To extract maximum power from the wind, a small perturbation is applied to the angle of the inverter terminal voltage, δ. $\Delta P_{out}$and $\Delta(V_{inv} / m_a)$ are estimated and their signs determine whether the operating point is moving toward or away from the maximum turbine power, Figure 18. Depending on the operating point direction of movement, the decision is to increase or decrease the angle δ or to keep it constant. Table 1 describes the decision that is made based on the sign of the inverter output power variation $\Delta P_{out}$and that of the ratio of the inverter terminal voltage to the amplitude modulation index$\Delta(V_{inv} / m_a)$. With the proposed algorithm for maximum power extraction, only voltage and current at the inverter terminal need to be measured, and no information about the wind and generator speeds and the dc-link voltage is required [1].

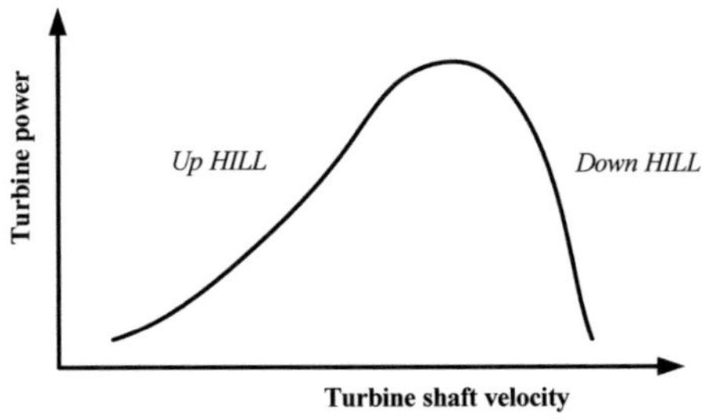

Figure 18. Wind turbine output power versus speed [1].

ΔP_{out}	$\Delta(\frac{V_{inv}}{m_a})$	Decision
> 0	> 0	Increase δ
< 0	< 0	Decrease δ
> 0	< 0	No change
< 0	> 0	No change

Table 1. Maximum power tracking algorithm [1].

B. Reactive power

The inverter should be able to regulate its output reactive power to provide the reactive power demand of the utility system, Figure 17. The inverter output reactive power must be controlled so as the maximum real power extraction is not violated. The real and reactive power components (P_{out}, Q_{out}) at the inverter output terminals are [1]:

$$P_{out} = \frac{V_{inv} V_{sys}}{X_T} \sin(\delta) \tag{24}$$

$$Q_{out} = \frac{V_{inv} V_{sys}}{X_T} \cos(\delta) - \frac{\left(V_{sys}\right)^2}{X_T} \tag{25}$$

V_{sys} is the utility system voltage and X_T is the reactance between the inverter and the utility system. It is seen from Equation (22) and Equation (23) that V_{inv} and δ can be controlled so as Q_{out} is regulated at a desired value while P_{out} is kept constant at its maximum corresponding to the wind speed. Substituting V_{inv} in Equation (22) by $\frac{\sqrt{3}}{2\sqrt{2}}$ m_a V_d, This equation can be deduced as follow:

$$P_{out} = \frac{\sqrt{3}\, m_a V_{dc} V_{sys}}{2\sqrt{2} X_T} \sin(\delta) \tag{26}$$

Assuming that V_{dc} does not change over the small sampling period T, P_{out} corresponding to the sampling time nT and $(n+1)T$ is

$$P_{out(n)} = \frac{\sqrt{3}\, m_{a(n)} V_{dc} V_{sys}}{2\sqrt{2}\, X_T} \sin(\delta_n) \tag{27}$$

$$P_{out(n+1)} = \frac{\sqrt{3}\, m_{a(n+1)} V_{dc} V_{sys}}{2\sqrt{2}\, X_T} \sin(\delta_{n+1}) \tag{28}$$

To keep the real power constant, i.e., $P_{out\,(n+1)} = P_{out\,(n)}$, while providing a desired reactive power, Equation (23), the inverter voltage angle must satisfy the following condition.

$$\delta_{n+1} = \sin^{-1}\left[\frac{m_{a(n)}}{m_{a(n+1)}} \sin(\delta_n) \right] \tag{29}$$

Figure 19 shows a flowchart of the proposed algorithm for maximum power extraction and reactive power control of a wind-turbine generator. The inputs are the three-phase voltages and currents at the inverter output terminals and the outputs are the required amplitude modulation index and the voltage angle of the inverter.

3.2.3.2 Decoupled control of the active and reactive power, independently

In this study [17], simple ac-dc-ac power conversion system and proposed modular control strategy for grid-connected wind power generation system have been implemented. Grid-side inverter maintains the dc-link voltage constant and the power factor of line side can be adjusted. Input current reference of dc/dc boost converter is decided for the maximum power point tracking of the turbine without any information of wind or generator speed. As the proposed control algorithm does not require any speed sensor for wind or generator speed, construction and installation are simple, cheap, and reliable. The main circuit and control block diagrams are shown in Figure 20. For wide range of variable speed operation, a dc-dc boost converter is utilized between 3-phase diode rectifier and PWM-VSC. The input dc current is regulated to follow the optimized current reference for maximum power point operation of turbine system. Grid PWM-VSC supply currents into the utility line by regulating the dc-link voltage. The active power is controlled by q-axis current through regulating the dc-link voltage whereas the reactive power can be controlled by d-axis current via adjusting the power factor

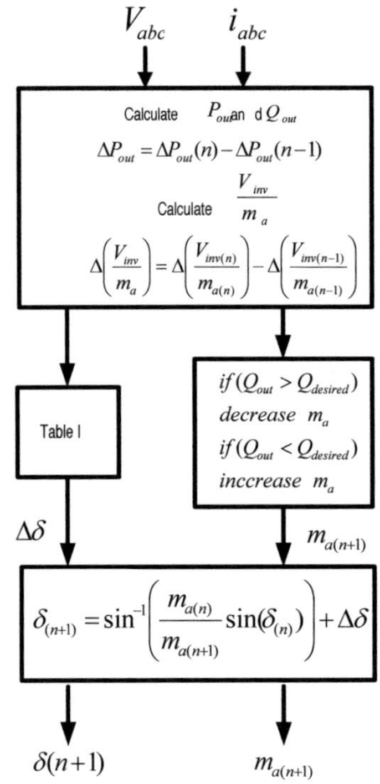

Figure 19. Flowchart of the proposed maximum active power and reactive power control [1].

of the grid side converter as shown in Figure 20. The phase angle of utility voltage is detected using Phased Locked Loop, *PLL*, in d-q synchronous reference frame [17].

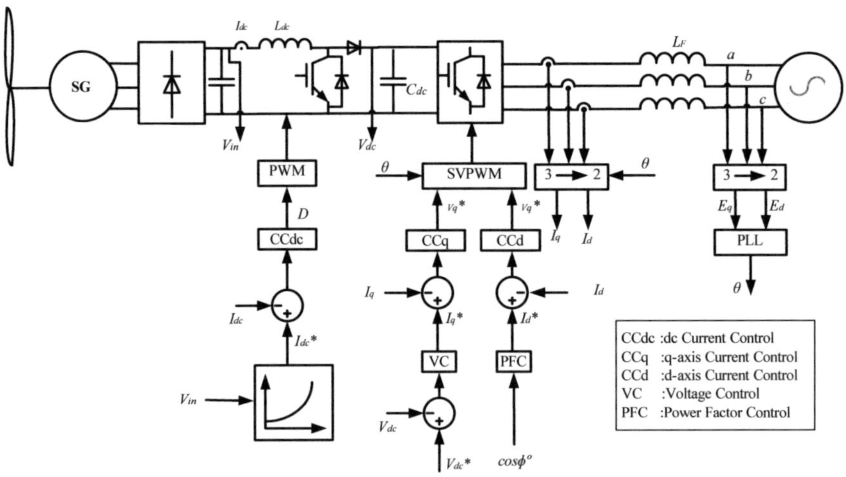

Figure 20. Block diagram of system control [17].

http://dx.doi.org/10.5772/54675

4. Co-simulation (PSIM/Matlab) program for interconnecting wind energy system with electric utility

In this study, the WECS is designed as PMSG connected to the grid via a back-to-back PWM-VSC as shown in Figure 21. MPPT control algorithm has been introduced using FLC to regulate the rotational speed to force the PMSG to work around its maximum power point in speeds below rated speeds and to produce the rated power in wind speed higher than the rated wind speed of the WT. Indirect vector-controlled PMSG system has been used for this purpose. The input to FLC is two real time measurements which are the change of output power and rotational speed between two consequent iterations (ΔP, and $\Delta \omega_m$). The output from FLC is the required change in the rotational speed $\Delta \omega_{m\text{-}new}{}^*$. The detailed logic behind the new proposed technique is explained in details in the following sections. Two effective computer simulation software packages (PSIM and Simulink) have been used to carry out the simulation effectively where PSIM contains the power circuit of the WECS and Matlab/Simulink contains the control circuit of the system. The idea behind using these two different software packages is the effective tools provided with PSIM for power circuit and the effective tools in Simulink for control circuit and FLC.

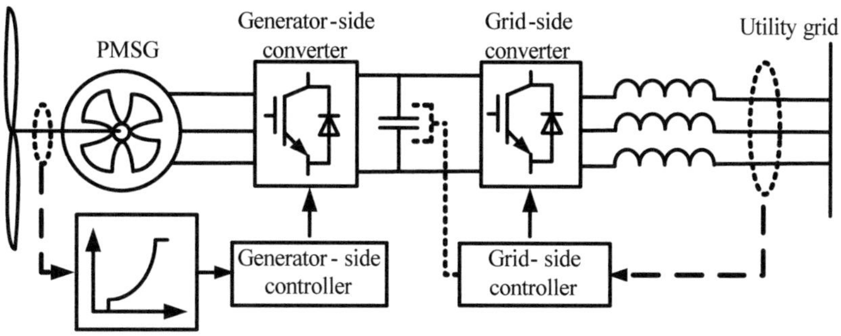

Figure 21. Schematic diagram of the overall system.

4.1. Wind energy conversion system description

Figure 22 shows a co-simulation (PSIM/Simulink) program for interconnecting WECS to electric utility. The PSIM program contains the power circuit of the WECS and Matlab/Simulink program contains the control of this system. The interconnection between PSIM and Matlab/Simulink has been done via the SimCoupler block. The basic topology of the power circuit which has PMSG driven wind turbine connected to the utility grid through the ac-dc-ac conversion system is shown in Figure 21. The PMSG is connected to the grid through back–to-back bidirectional PWM voltage source converters VSC. The generator side converter is used as a rectifier, while the grid side converter is used as an inverter. The generator side converter is connected to the grid side converter through dc-link capacitor. The control of the overall system has been done through the generator side converter and the grid side converter.

MPPT algorithm has been achieved through controlling the generator side converter using FLC. The grid-side converter controller maintains the dc-link voltage at the desired value by exporting active power to the grid and it controls the reactive power exchange with the grid.

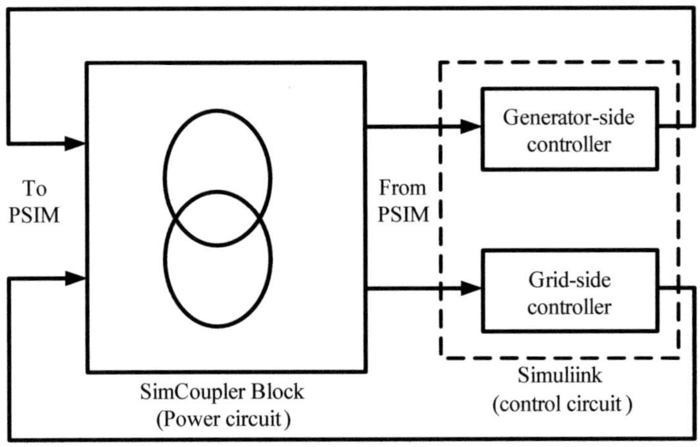

Figure 22. Co-simulation block of wind energy system interfaced to electric utility.

4.1.1. Wind turbine model

Wind turbine converts the wind power to a mechanical power. This mechanical power generated by wind turbine at the shaft of the generator can be expressed as:

$$P_m = \frac{1}{2} C_P(\lambda, \beta) \rho A u^3 \tag{30}$$

where ρ is the air density (typically 1.225 kg/m^3), β is the pitch angle (in degree), A is the area swept by the rotor blades (in m^2); u is the wind speed (in m/s) and $C_p(\lambda, \beta)$ is the wind-turbine power coefficient (dimensionless).

The turbine power coefficient, $C_p(\lambda, \beta)$, describes the power extraction efficiency of the WT and is defined as the ratio between the mechanical power available at the turbine shaft and the power available in wind. A generic equation shown later in Equation (3) is used to model $C_p(\lambda, \beta)$. C_P is a nonlinear function of both tip speed ratio, λ and the blade pitch angle, β. The tip speed ratio, λ is the ratio of the turbine tip speed, ω_m*R to the wind speed, u. This tip speed ratio, λ, is defined as [28]:

$$\lambda = \frac{\omega_m {}^* R}{u} \tag{31}$$

Where ω_m is the rotational speed and R is the turbine blade radius, respectively.

For a fixed pitch angle, β, C_P becomes a nonlinear function of tip speed ratio, λ, only. According to Equation (29), there is a relation between the tip speed ratio, λ and the rotational speed, ω_m. Hence, at a certain wind speed, the power is maximum at a certain ω_m called optimum rotational speed, ω_{opt}. This speed corresponds to optimum TSR, λ_{opt} [15]. The value of the TSR is constant for all maximum power points. So, to extract maximum power at variable wind speed, the WT should always operate at λ_{opt} in speeds bellow the rated speed. This occurs by controlling the rotational speed of the WT to be equal to the optimum rotational speed. Figure 23 shows that the mechanical power generated by WT at the shaft of the generator as a function of the rotational speed, ω_m. These curves have been extracted from PSIM support for the wind turbine used in this study. It is clear from this figure that for each wind speed the mechanical output power is maximum at particular rotational speed, ω_{opt} as shown in Figure 23.

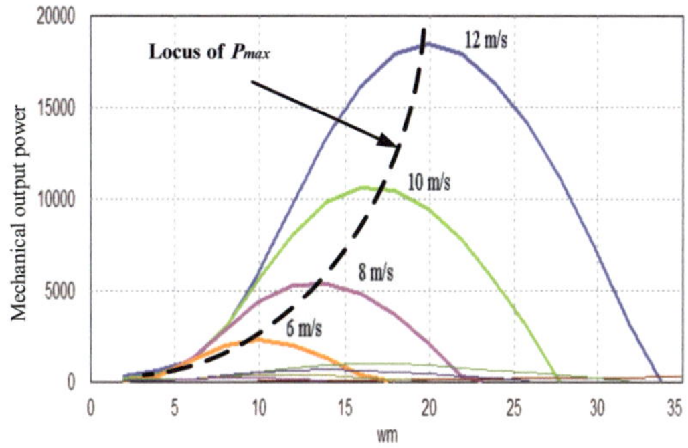

Figure 23. Typical output power characteristics.

4.1.2. PMSG model

The generator is modeled by the following voltage equations in the rotor reference frame (dq axes) [29]:

$$\begin{aligned} v_{sd} &= R_s i_{sd} + \frac{d\lambda_{sd}}{dt} - \omega_r \lambda_{sq} \\ v_{sq} &= R_s i_{sq} + \frac{d\lambda_{sq}}{dt} + \omega_r \lambda_{sd} \end{aligned} \tag{32}$$

Where λ_{sq}, and λ_{sd} are the stator flux linkages in the direct and quadrature axis of rotor which in the absence of damper circuits can be expressed in terms of the stator currents and the magnetic flux as [29].

$$\begin{aligned} \lambda_{sd} &= L_s i_{sd} + \psi_F \\ \lambda_{sq} &= L_s i_{sq} \end{aligned} \tag{33}$$

Where ψ_F is the flux of the permanent magnets.

The electrical torque T_e of the three-phase generator can be calculated as follows [29, 30]:

$$T_e = \frac{3}{2} P \left[\lambda_{sd} i_{sq} - \lambda_{sq} i_{sd} \right] \tag{34}$$

Where P is the number of pole pairs. For a non-salient-pole machine the stator inductances L_{sd} and L_{sq} are approximately equal. This means that the direct-axis current i_{sd} does not contribute to the electrical torque. Our concept is to keep i_{sd} to zero in order to obtain maximal torque with minimum current. Then, the electromagnetic torque results:

$$T_e = \frac{3}{2} P \psi_F i_{sq} = K_c i_{sq} \tag{35}$$

K_c is called the torque constant and represents the proportional coefficient between T_e and i_{sq}.

4.2. Control of the generator side converter (PMSG)

The generator side controller controls the rotational speed to produce the maximum output power via controlling the electromagnetic torque according to Equation (33), where the indirect vector control is used. The proposed control logic of the generator side converter is shown in Figure 24. The speed loop will generate the q-axis current component to control the generator torque and speed at different wind speed via estimating the references value of i_α, i_β as shown in Figure 24. The torque control can be achieved through the control of the i_{sq} current as shown in Equation (33). Figure 25 shows the stator and rotor current space phasors and the excitation flux of the PMSG. The quadrature stator current i_{sq} can be controlled through the rotor reference frame (α, β axis) as shown in Figure 25. So, the references value of i_α, i_β can be estimated easily from the amplitude of i_{sq*} and the rotor angle, Θ_r. Initially, to find the rotor angle, Θ_r, the relationship between the electrical angular speed, ω_r and the rotor mechanical speed (rad/sec), ω_m may be expressed as:

$$\omega_r = \frac{P}{2} \omega_m \tag{36}$$

So, the rotor angle, Θ_r, can be estimated by integrating of the electrical angular speed, ω_r. The input to the speed control is the actual and reference rotor mechanical speed (rad/sec) and the output is the (α, β) reference current components. The actual values of the (α, β) current components are estimated using Clark's transformation to the three phase current of PMSG. The FLC can be used to find the reference speed along which tracks the maximum power point.

http://dx.doi.org/10.5772/54675

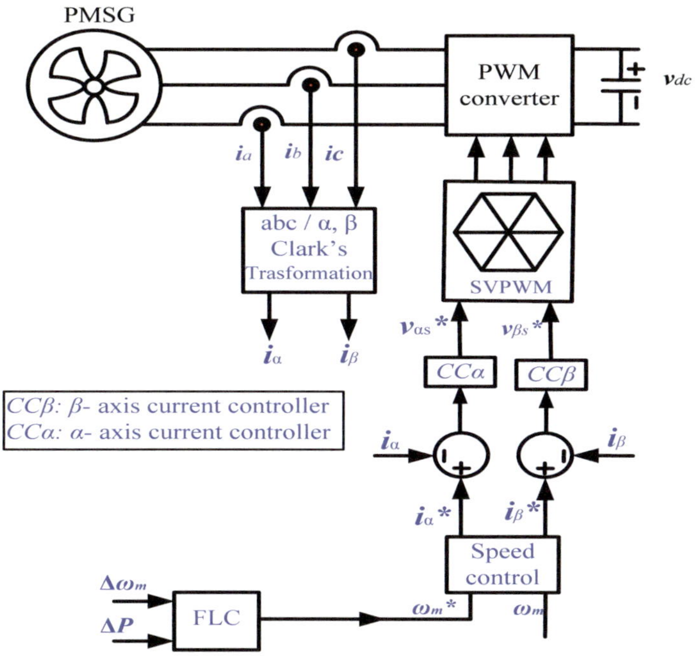

Figure 24. Control block diagram of generator side converter.

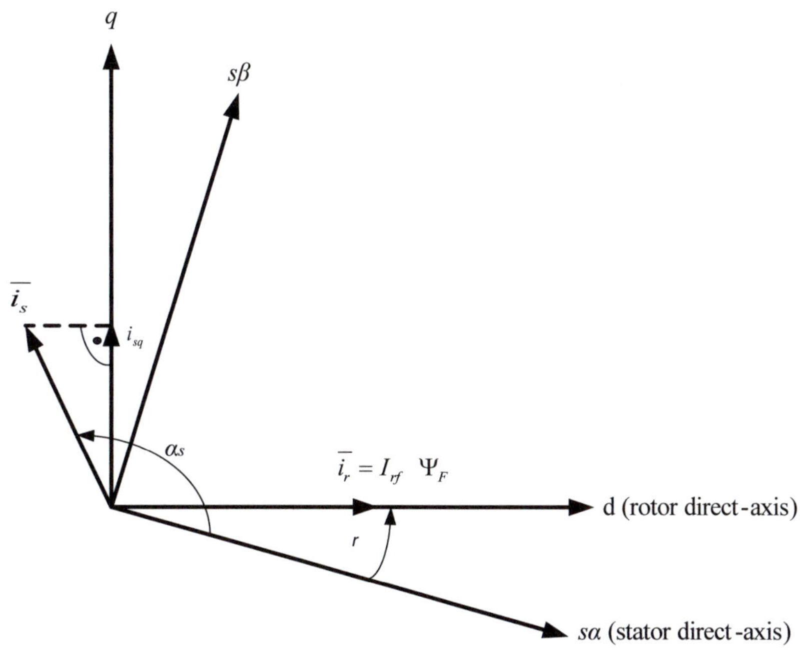

Figure 25. The stator and rotor current space phasors and the excitation flux of the PMSG [29].

4.3. Fuzzy logic controller for MPPT

At a certain wind speed, the power is maximum at a certain ω called optimum rotational speed, ω_{opt}. This speed corresponds to optimum tip speed ratio,λ_{opt} [15]. So, to extract maximum power at variable wind speed, the turbine should always operate at λ_{opt}. This occurs by controlling the rotational speed of the turbine. Controlling of the turbine to operate at optimum rotational speed can be done using the fuzzy logic controller. Each wind turbine has one value of λ_{opt} at variable speed but ω_{opt} changes from a certain wind speed to another. From Equation (29), the relation between ω_{opt} and wind speed, u, for constant R and λ_{opt} can be deduced as follow:

$$\omega_{opt} = \frac{\lambda_{opt}}{R} u \tag{37}$$

From Equation (35), the relation between the optimum rotational speed and wind speed is linear. At a certain wind speed, there is optimum rotational speed which is different at another wind speed. The fuzzy logic control is used to search (observation and perturbation) the rotational speed reference which tracks the maximum power point at variable wind speeds. The fuzzy logic controller block diagram is shown in Figure 26. Two real time measurements are used as input to fuzzy (ΔP, and $\Delta\omega_m$*) and the output is ($\Delta\omega_{m\text{-}new}$*). Membership functions are shown in Figure 27. Triangular symmetrical membership functions are suitable for the input and output, which give more sensitivity especially as variables approach to zero value. The width of variation can be adjusted according to the system parameter. The input signals are first fuzzified and expressed in fuzzy set notation using linguistic labels which are characterized by membership functions before it is processed by the FLC. Using a set of rules and a fuzzy set theory, the output of the FLC is obtained [22]. This output, expressed as a fuzzy set using linguistic labels characterized by membership functions, is defuzzified and then produces the controller output. The fuzzy logic controller doesn't require any detailed mathematical model of the system and its operation is governed simply by a set of rules. The principle of the fuzzy logic controller is to perturb the reference speed ω_m* and to observe the corresponding change of power, ΔP. If the output power increases with the last increment, the searching process continues in the same direction. On the other hand, if the speed increment reduces the output power, the direction of the searching is reversed. The fuzzy logic controller is efficient to track the maximum power point, especially in case of frequently changing wind conditions [22].

Figure 27 shows the input and output membership functions and Table 2 lists the control rule for the input and output variable. The next fuzzy levels are chosen for controlling the inputs and output of the fuzzy logic controller. The variation step of the power and the reference speed may vary depending on the system. In Figure 27, the variation step in the speed reference is from -0.15rad/s to 0.15rad/s for power variation ranging over from -30W to 30W. The membership definitions are given as follows: N (negative), N++ (very big negative), NB (negative big), NM (negative medium), NS (negative small), ZE (zero), P (positive), PS (positive small), PM (positive medium), PB (positive big), and P++ (very big positive).

http://dx.doi.org/10.5772/54675

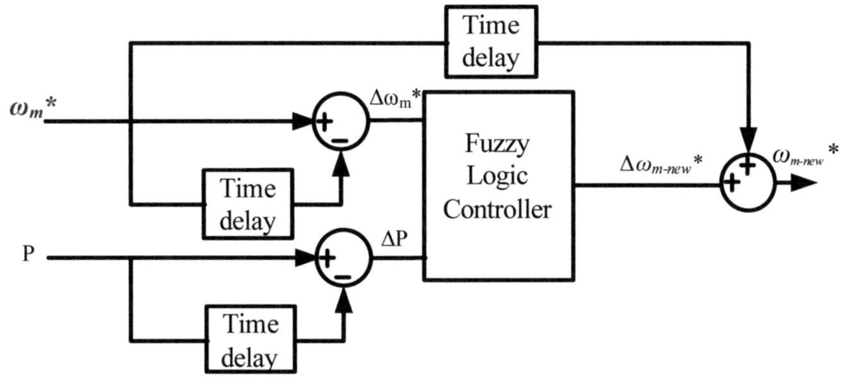

Figure 26. Input and output of fuzzy controller.

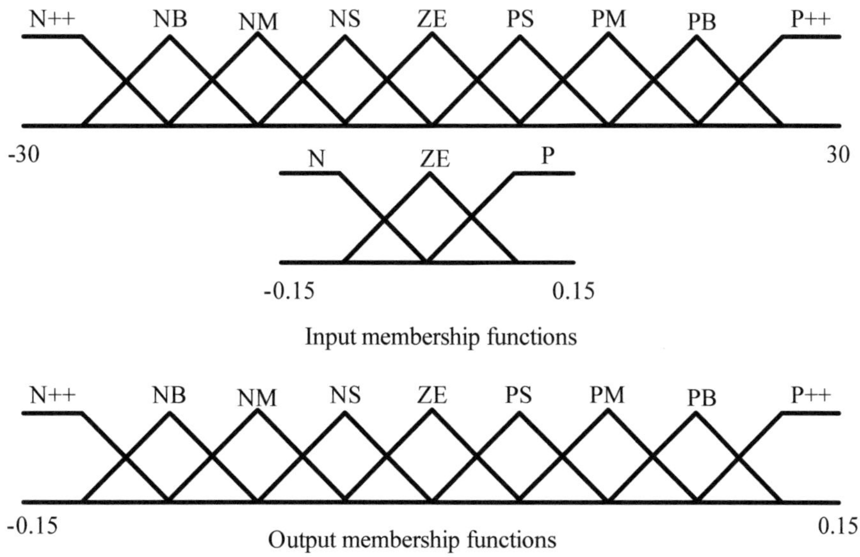

Figure 27. Membership functions of fuzzy logic controller

ΔP / $\Delta\omega_m$	**N++**	**NB**	**NM**	**NS**	**ZE**	**PS**	**PM**	**PB**	**P++**
N	P++	PB	PM	PS	ZE	NS	NM	NB	N++
ZE	NB	NM	NS	NS	ZE	PS	PM	PM	PB
P	N++	NB	NM	NS	ZE	PM	PM	PB	PB

Table 2. Rules of fuzzy logic controller

4.4. Control of the grid side converter

The power flow of the grid-side converter is controlled in order to maintain the dc-link voltage at reference value, 600v. Since increasing the output power than the input power to dc-link capacitor causes a decrease of the dc-link voltage and vise versa, the output power will be regulated to keep dc-link voltage approximately constant. To maintain the dc-link voltage constant and to ensure the reactive power flowing into the grid, the grid side converter currents are controlled using the d-q vector control approach. The dc-link voltage is controlled to the desired value by using a PI-controller and the change in the dc-link voltage represents a change in the q-axis (i_{qs}) current component. Figure 28 shows a control block diagram of the grid side converter.

The active power can be defined as;

$$P_s = \frac{3}{2}\left(v_{ds}i_{ds} + v_{qs}i_{qs}\right) \tag{38}$$

The reactive power can be defined as:

$$Q_s = \frac{3}{2}\left(v_{qs}i_{ds} - v_{ds}i_{qs}\right) \tag{39}$$

By aligning the q-axis of the reference frame along with the grid voltage position v_{ds}=0 and v_{qs}= constant because the grid voltage is assumed to be constant. Then the active and reactive power can be obtained from the following equations:

$$P_s = \frac{3}{2}v_{qs}i_{qs} \tag{40}$$

$$Q_s = \frac{3}{2}v_{qs}i_{ds} \tag{41}$$

http://dx.doi.org/10.5772/54675

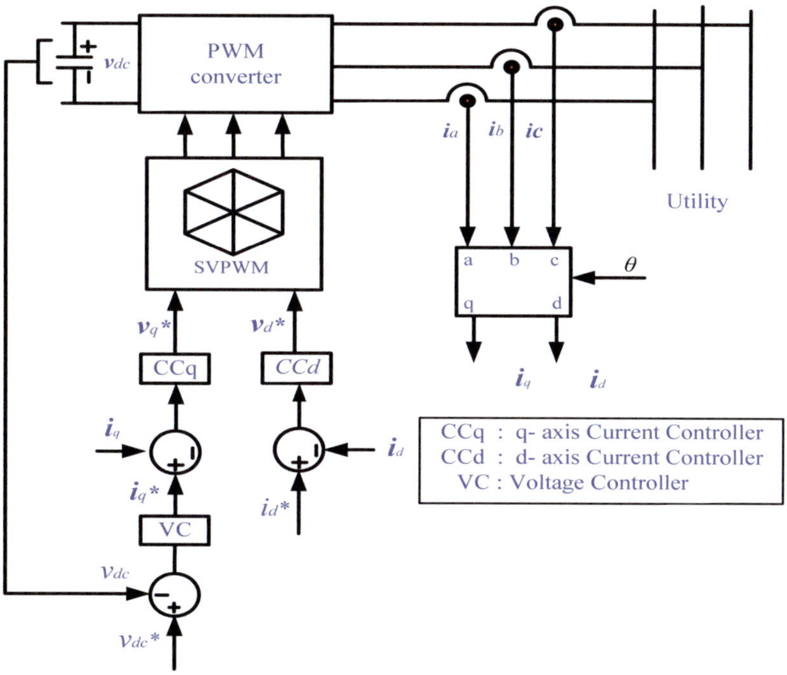

Figure 28. Control block diagram of grid-side converter.

4.5. Simulation results

A co-simulation (PSIM/Simulink) program has been used where PSIM contains the power circuit of the WECS and Matlab/Simulink has the whole control system as described before. The model of WECS in PSIM contains the WT connected to the utility grid through back–to-back bidirectional PWM converter. The control of whole system in Simulink contains the generator side controller and the grid side controller. The wind turbine characteristics and the parameters of the PMSG are listed in Appendix. The generator can be directly controlled by the generator side controller to track the maximum power available from the WT. The wind speed is variable and changes from 7 m/s to 13 m/s as input to WT. To extract maximum power at variable wind speed, the turbine should always operate at λ_{opt}. This occurs by controlling the rotational speed of the WT. So, it always operates at the optimum rotational speed. ω_{opt} changes from a certain wind speed to another. The fuzzy logic controller is used to search the optimum rotational speed which tracks the maximum power point at variable wind speeds. Figure 29 (a) shows the variation of the wind speed which varies randomly from 7 m/s to13 m/s. On the other hand, Figure 29 (b) shows the variation of the actual and reference rotational speed as a result of the wind speed variation. At a certain wind speed, the actual and reference rotational speed have been estimated and this agree with the power characteristic of the wind turbine shown later in Figure 23. I.e. the WT always operates at the optimum rotational speed which is found using FLC; hence, the power extraction from wind is maximum at variable wind speed. It is seen that according to the wind speed variation the generator speed varies and that its output power is produced corresponding to the wind speed variation. The fuzzy logic controller works well and it gives the good tracking performance for the maximum output power point. The fuzzy logic controller makes WT always operates at the optimum rotational speed. On the other hand, the grid-side controller maintains the dc-link voltage at the desired value, 600v, as shown in Figure 29 (c). The dc-link voltage is regulated by exporting active power to the grid as shown in Figure 29 (d). The reactive power transmitted to the grid is shown in Figure 29 (e).

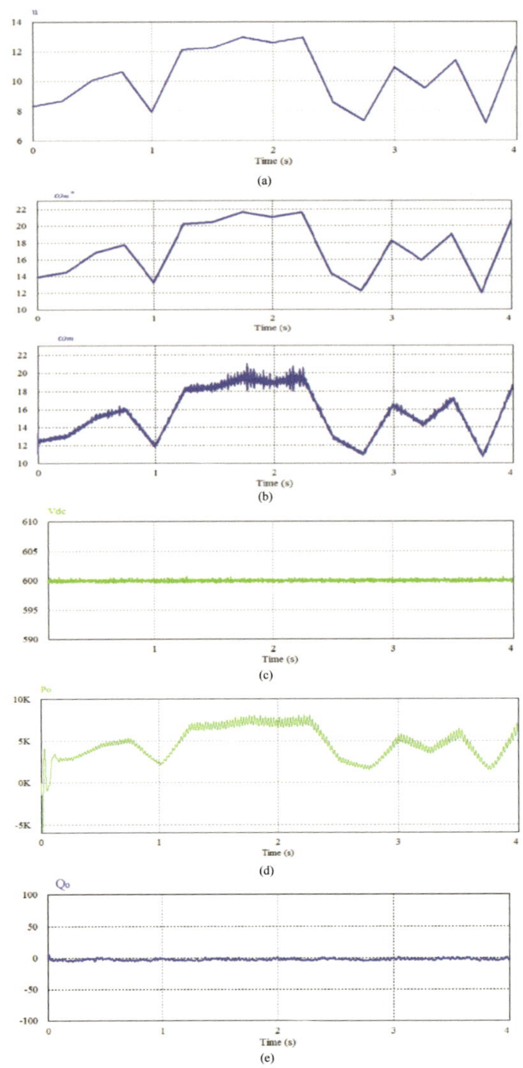

Figure 29. Different simulation waveforms: (a) Wind speed variation (7-13) m/s. (b) Actual and reference rotational speed (rad/s). (c) dc-link voltage (v). (d) Active power (watt). (e) Reactive power (Var).

http://dx.doi.org/10.5772/54675

5. Conclusions

Wind energy conversion system has high priority among the various renewable energy systems. Maximum power extraction from wind energy system became an important research topic due to the increase in output energy by using this technique. Wind speed sensorless MPPT control has been a very active area of research. In this study, a concise review of MPPT control methods has been presented for controlling WECS. On the other hand, there is a continuing effort to make converter and control schemes more efficient and cost effective in hopes of developing an economically viable solution of increasing environmental issues. Wind power generation has grown at a high rate in the past decade and will continue with power electronic technology advanced. The survey of MPPT algorithms have been classified into MPPT algorithms with wind speed sensor and MPPT algorithms without wind speed sensor. A co-simulation (PSIM/Simulink) program has been proposed for WECS where PSIM contains the power circuit of the WECS and Matlab/Simulink has the control circuit of the system. The WT is connected to the grid via back–to-back PWM-VSC. The generator side controller and the grid side controller have been done in Simulink. The main function of the generator side controller is to track the maximum power from wind through controlling the rotational speed of the turbine using fuzzy logic controller. The fuzzy logic algorithm for the maximum output power of the grid-connected wind power generation system using a PMSG has been proposed and implemented above. The PMSG was controlled in indirect-vector field oriented control method and its speed reference was determined using fuzzy logic controller. The grid-side converter controls the dc-link voltage at a desired value, 600V, for the proposed system. Active and reactive power control has been achieved by controlling q-axis and d-axis grid current components respectively. The d-axis grid current is controlled to be zero for unity power factor and the q-axis grid current is controlled to deliver the power flowing from the dc-link to the grid. The simulation results prove the superiority of FLC and the whole control system.

Appendix

Wind turbine		PMSG	
Nominal Output Power	19kw	R_s (stator resistance)	1m
Wind speed input	7:13 m/s (saw tooth)	L_d (d-axis inductance)	1m
Base Wind Speed	12 m/s	L_q (q-axis inductance)	1m
Base Rotational Speed	190 rpm	No. of Poles P	30
Moment of inertia	1m	Moment of inertia	100m
Blade pitch angle input	0°	Mech. Time Constant	1

Table 3. Parameters of wind turbine model and PMSG

Acknowledgements

The authors acknowledge the National Plan for sciences and Technology program (Project No.: ENE226-02-08) by King Saud University for the financial support to carry out the research work reported in this chapter.

Author details

Ali M. Eltamaly, A. I. Alolah and Hassan M. Farh

Department of Electrical Engineering, College of Engineering, King Saud University, Riyadh, Saudi Arabia

References

[1] Vahid, O, & Hassan, N. Maximum power extraction for a wind-turbine generator with no wind speed sensor. in Proc. on IEEE, Conversion and Delivery of Electrical Energy in the 21st Cen. (2008)., 1-6.

[2] Thomas, A, & Lennart, S. An overview of wind energy status (2002). Renewable and sustainable energy reviews 2002, 6:67-128.

[3] DuboisOptimized permanent magnet generator topologies for direct-drive wind turbines. Ph.D. dissertation, Delft Univ. Technol., Delft, The Netherlands; (2004).

[4] Anders, G. Design of direct-driven permanent-magnet generators for wind turbines. Ph.D. dissertation, Chalmers Univ. Technol., Goteborg, Sweden; (1996).

[5] Torbjörn, T, & Jan, L. Control by variable rotor speed of a fixed pitch wind turbine operating in a wide speed range. IEEE Trans. on Energy Conversion, EC-8, (1993)., 520-526.

[6] BuehringFreris. Control policies for wind energy Conversion System. Proc. Inst. Elect. Eng. C, (1981)., 128, 253-261.

[7] Erimis, H. B, Ertan, E, & Akpinar, F. Ulgut, Autonomous wind energy conversion systems with a simple controller for maximum power transfer, Proc. Inst. Elect. Eng. B, (1992)., 139, 421-428.

[8] ChedidMrad, Basma. Intelligent control of a class of wind energy conversion systems. IEEE Trans. on Energy Conversion, EC-14, (1999)., 1597-1604.

[9] Marcelo, S, Bimal, B, & Ronald, S. Fuzzy logic-based intelligent control of a variable speed cage machine wind generation system. IEEE Trans. on Power Electron., PE-12, (1997)., 87-94.

[10] EnslinWyk. A study of a wind power converter with micro-computer based maximum power control utilizing an over- synchronous electronic Scherbius cascade. Renewable Energy World (1993). , 2(6), 551-562.

[11] Quincy, W, & Liuchen, C. An intelligent maximum power extraction algorithm for inverter-based variable speed wind turbine systems. IEEE Transactions on Power Electronics (2004). , 19(5), 1242-1249.

[12] Quincy, W. Maximum wind energy extraction strategies using power electronic converters. PhD. dissertation, Univ. of New Brunswick, Canada, (2003).

[13] Hui, L, Shi, K. L, & Mclaren, P. G. Neural-network-based sensorless maximum wind energy capture with compensated power coefficient. IEEE Trans. Ind. Appl., (2005). , 41(6), 1548-1556.

[14] RajuFernandes, Chatterjee. A UPF power conditioner with maximum power point tracker for grid connected variable speed wind energy conversion system. Proc. of 1st Int. Conf. on PESA, Bombay, India, (2004). , 107-112.

[15] Majid, A. A, Yatim, A. H. M, & Chee, W. Tan. A Study of Maximum Power Point Tracking Algorithms for Wind Energy System. Proc. of 1st IEEE Conf. on Clean Energy and Technology CET, (2011). , 321-326.

[16] Mónica, C, Santiago, A, & Juan, C. B. Control of Permanent-Magnet Generators Applied to Variable-Speed Wind-Energy Systems Connected to the Grid. IEEE Trans. Energy Conversion, 21 March (2006). , 130-135.

[17] Seung-ho, S, Shin-il, K, & Nyeon-kun, H. Implementation and control of grid connected AC-DC-AC power converter for variable speed wind energy conversion system. IEEE, (2003). , 154-158.

[18] Ali, M. E. Modelling of wind turbine driving permanent Magnet Generator with maximum power point tracking system. J. King Saud Univ., Riyadh,(2007). , 19(2), 223-237.

[19] Mahmoud, M. H, Mohamed, O, Mahrous, E. A, & Mahmoud, A. S. Simple sensorless control technique of permanent magnet synchronous generator wind turbine. Proc. of IEEE Int. Conf. on Power and Energy,PEC201),Kuala Lumpur, Malaysia, (2010). , 512-517.

[20] Swagat, P, Mohanty, K. B, & Benudhar, S. Performance Comparison of a Robust Self Tuned Fuzzy Logic Controller used for Power Control in Wind Conversion Systems. Proc. of Modern Electric Power Systems, MEPS'10, Wroclaw, Poland, September (2010). , 20-22.

[21] Xingjia, Y, Changchun, G, Zuoxia, X, Yan, L, & Shu, L. Variable Speed Wind Turbine Maximum Power Extraction Based on Fuzzy Logic Control. Proc. of Int. Conf. on Intelligent Human-Machine Systems and Cybernetics, IEEE; (2009). , 202-205.

[22] Ahmed, G. A, Dong-choon, L, & Jul-ki, S. Variable speed wind power generation system based on fuzzy logic control for maximum power output tracking. in Proc. 35th Annual IEEE Power Electron. Specialists Conf., PESC, Aachen, Germany, (2004). , 3, 2039-2043.

[23] Tanaka, T, & Toumiya, T. Output control by hill-climbing method for a small scale wind power generating system. Renewable Energy, (1997). , 12(4), 387-400.

[24] Tan, K, & Islam, S. Optimum control strategies in energy conversion of PMSG wind turbine system without mechanical sensors. IEEE Trans., (2004). , EC-19, 392-399.

[25] Datta, R, & Ranganathan, V. T. A method of tracking the peak power points for a variable speed wind energy conversion system. IEEE Trans. on Energy Conversion, March (2003). , EC-18, 163-168.

[26] Brendan, F, Damian, F, Leslie, B, Nick, J, David, M, Mark, O, Richard, W, & Olimpo, A. Wind power integration: connection and system operational aspects. Proc. IET Power and Energy Series, (2007). , 50, 77-85.

[27] Siegfried, H. Grid Integration of Wind Energy Conversion Systems. John Wiley & Sons, Germany; (2006).

[28] Gary, L. J. Wind Energy Systems. Prentice Hall, Englewood cliffs; (2003).

[29] Peter, V. Sensorless Vector and Direct Torque Control. Oxford science publications,New York, (1998).

[30] Orlando, M, Liserre, V. G, & Monopoli, A. Dell'Aquila. Speed Sensorless Control of a PMSG for Small Wind Turbine Systems. Proc. of IEEE International Symposiumon Industrial Electronics, ISIE, Seoul, Korea, July (2009). , 5-8.

Chapter 9

Comparative Analysis of Endowments Effect Renewable Energy Efficiency Among OECD Countries

Tser-Yieth Chen, Tsai-Lien Yeh and Yi Hsuan Ko

Additional information is available at the end of the chapter

http://dx.doi.org/10.5772/52020

1. Introduction

With the rapid development of the global economy, more people are living in urban than rural areas, thereby contributing to a significant increase in demand for energy, especially in emerging countries [1] [2] [3] [4]. The realization that fossil fuel resources required for energy generation are becoming scarce and that climate change is related to carbon emissions into the atmosphere has increased interest in energy conservation and environmental protection [5].

Among other recent issues, climate change, energy demand and fluctuations in international oil prices have become the focus of global attention. Renewable energy sources are now the fastest-growing sector of the energy mix and offer significant potential to address issues of energy security and sustainability [6]. All countries seeking to achieve the Kyoto Protocol target of reducing greenhouse gases, have renewable energy as the focus of their energy policy, and some have even become the mainstream of energy development. The energy we currently use is subject to unrestricted exploitation, not only about to run out of the global stock of face time, energy conversion process produce tangible and intangible waste, that have a significant effect on the global environment. To ensure a stable energy supply, enhance our energy supply security, reduce dependence on fossil fuels, and reduce greenhouse gas emissions, developing renewable energy sources has also become our current economic development and environmental resources for the biggest issue.

Thus, in face of the current trends, the demand for energy is rising. In addition to focusing on the power generation efficiency of power plants, we should also consider each unit of electricity efficiency to achieve an effective energy-saving effect in the pursuit of sustainable resource use. Therefore, we adopt a two-stage data envelopment analysis (DEA) [7] and incorporate two sub-processes into a DEA efficiency model to evaluate the level of management performance within renewable energy in the OECD countries. We measure managerial efficiency in two phases: operating efficiency (OE) and the energy density efficiency (DE).

This method is different from those of previous studies that focused primarily on assessing OE [8] [9]. We divide the efficiency of energy plants into two components. Management performance is no longer constrained with production efficiency but constitutes a broader dimension that covers operating activities and the efficiency of energy use. Compared to the traditional single-efficiency model, the sub-processes model is more suitable for evaluating the usage performance because of energy industry characteristics.

This evaluation model is useful for energy managers and current policy-makers. For managers, it provides a more detailed performance evaluation process including two essential operational elements in the energy generation industry; for policy- makers, it offers a complete measurement of efficiency and is based on variable combinations of these two dimensions; policy-makers can identify the most suitable policy (e.g., a subsidy) and develop the most effective strategy.

Taiwan is an island, country that is extremely lacking in energy and is more than 98% dependent on imported energy. Taiwan is also influenced by political and geographical constraints; therefore, the capacity to acquire energy is difficult compared to other countries. Thus, implementation of renewable energy and the abolition of nuclear power generation is a potential policy priority for Taiwan. Seeking the most cost-effective strategy, Taiwan's national conditions, if we can use the experience of other countries, will become Taiwan's development of a great help. OECD countries including highly and lowly developed countries, especially developing countries, is from the energy consumption, low efficiency and serious pollution to the economic development mode shift to energy efficient, less polluting economic development mode. In this study, we discuss and compare 34 OECD countries' renewable energy OE and DE by DEA. Finally, we present our conclusions and provide suggestions for renewable energy development in Taiwan. β

2. Literature review and hypothesis setting

Numerous studies related to the efficiency evaluation of renewable power plants have focused primarily on single efficiency and have assisted in the selection of input and output variables used in this study [10] [9]. First, Criswell and Thompson [11] applied DEA with a sample of large-scale commercial power systems for earth in global. They used three input and three output variables are exogenously fixed for the research. Azadeh, Ghaderi and Maghsoudi [12] used data from 25 cities in Iran with six regions within each city. Four types of input variables and two types of output variables were used in their analysis. More recently, Madlener, Antunes and Dias [13] justified the use of DEA logically and systematically in 41 agricultural biogas plants situated in Austria. They used three input and two output variables and identified that DEA offers considerable potential and advantages for seeking accurate evaluate productivity. Iglesias, Castellanos and Seijas [14] evaluated the performance of a group of 57 Spanish wind farms located in the region of Galicia by using three input and two output variables. Azadh, Ghaderi and Nasrollahi [15] measured the efficiency of wind power plants with the lowest possible costs using DEA, with data collected from 25 cities in Iran with 5 regions within each city using DEA with four input and two outputs.

In this section, we propose several hypotheses. Considering Sahelian countries, energy access remained relatively low until recently, despite the abundance of renewable resources such as wind and solar energy. The abundance of renewable resources assumes that access to renewable technologies could increase and improve energy access in remote rural areas [16]. They are compatible with local conditions and resource endowment. Research on regional development specifically related to China's Western Development Program by the China Energy Strategic Research Group and Fan, Sun & Ren [17] discussed sustainable development issues for economically disadvantaged areas such as the ecological deterioration and sustainable livelihoods of rural households, and suggested reasonable approaches to address energy problems in these areas, such as the use of rich natural resources (endowment), development of renewable energy, and developing a moderate centralized energy supply that considers local energy endowment conditions. Shi [18] supported a similar type of energy development because a region's unique energy endowments reflect it is energy developmental differences. Chen and Zhu [19] specifically used resource endowment, zoning separation of wind power and solar power resources, the classification results for the preliminary study on China's energy and economic regionalization. Chen and Zhu argued that there is little evidence on whether the impact of economic development on the electricity mix is affected by energy resource endowments [19]. Marcotullio and Schulz [20] provided evidence of endowment's heterogeneity in energy mix transitions across countries. Therefore, we present the following hypotheses:

H1a: Endowment and OE are positively causal related

H1b: Endowment and DE are positively causal related

British Petroleum discussed China and India's rapid increase in energy use because they represent approximately one-third of the global population, the expected depletion of oil resources in the near future, and the effect of human activities on global climate change. Bettencourt [2] indicated that as economies and populations continue to grow rapidly, energy and power consumption also increase at the same rate. The Empresa de Pesquisa Energética (EPE) indicated that because of population growth, urbanization and higher income, annual electricity consumption in the residential sector is growing steadily from 4.7% in 2003 up to 6.2% in 2009. The International Energy Agency (IEA) [6] and United Nations (UN) [4] stated that approximately 4.9 billion people (80 % of the global population) lived in developing countries as of 2001. The current annual population growth rate is approximately 1.5 % in developing countries. However, despite the lower living standards and lower per capita energy use in developing countries, total energy use in developing countries is increasing fairly rapidly. Crane and Kinzig [4] indicated that many countries in the pursuit of economic development, the population increase rapidly as the same time, but also face a requirement to increase energy. There is a growing need to implement energy efficiency. Therefore, we present the following hypotheses:

H2a: Population and OE are positively causal related

H2b: Population and DE are positively causal related

Because energy efficiency improvement relies on total-factor productivity improvement [21], the technical efficiency (TE) index is computed to analyze the energy efficiencies of economies. The TE index incorporates energy, capital, and labor as multiple inputs for production. They use DEA to find the TE of each economy. Chien and Hu [22] stated that it is possible that capital inputs may increase energy generations. From an economic production perspective, these practices imply that energy savings as well and emission reduction can be achieved by means of factor substitution between energy and capital [16] [23] [24]. This effectively mitigates the dependence of economic growth on energy input and environmental capacity; in other words, it improves the aggregated energy and environmental efficiency (AEEE). Hudson and Jorgenson [25] stated that intensity effects in the industrial sector might depend on three strong interactions. Energy and capital are both, substitutes for labor, whereas capital and energy are complements. In other words, capital and energy can be increased simultaneously. Turner [26] proposed another factor of production that is critical in determining substitution and other effects driving economy-wide responses. Specifically, rebound effects, from increased energy efficiency are capital. Therefore, we present the following hypotheses:

H3a: Capital and OE are positively causal related

H3b: Capital and DE are positively causal related

The renewable energy-developing indicators of an economy are obtained from Renewables Energy Information [5] and have been published by the IEA since 2002. Indicators such as household consumption, capital formation, trade balance, energy imports, and gross domestic product (GDP) are obtained from the world energy development. Anderson and Leach [27] also indicated that if renewable energy technologies supply a significant share of total energy supply, then the energy storage problem must be solved in advance. First, the manner in which GDP affects the promotion of energy policies must be studied. Bettencourt [2] indicated that there seems to be a long way to go to fully use renewable resources. Until the early 1980s, changes in the energy–GDP ratio were the subject of many studies. Questions were raised as to how the ratio would evolve over time if a country experiences different stages of economic development. Understanding such trends provides indicators for how future energy demand would evolve. A number of studies have suggested that as the process of industrialization advances, with agriculture replaced by manufacturing, energy consumption tends to increase more rapidly than GDP, creating an increasing value of the energy–GDP ratio. Among the theories on the relationship between energy consumption (or energy-related environmental indicators) and GDP, the most famous is the environmental Kuznets curve. A recent overview was provided by Ang and Liu [28]. With the GDP measured in common units, comparisons can be made between countries. Cross-country variations in the energy–GDP ratio have been studied for industrialized countries and for developing countries [27] [29].

Therefore, we present the following hypotheses:

H4a: GDP and OE are positively causal related

H4b: GDP and DE are positively causal related

Some researchers have reached an opposing conclusion that energy subsidy reform would produce positive results. Steenblik and Coroyannakis [30] used the computable general equilibrium (CGE) model to simulate the positive effects of removing coal subsidies in Western European countries, such as promoting the industrialization of the power sector and increasing coal production and exports. United Nations [3] concluded that cutting energy subsidies could have significant impacts on residents, although this requires a more in-depth analysis in the future. Conversely, some researchers believed that fossil energy reform would increase energy use efficiency and household income levels. Choi, Roh and Yoon [9] indicated that increase in energy price could improve energy efficiency significantly. Thus, the energy price mechanism is at the core of energy reform, and energy subsidies are crucial determinant of energy prices.

Anderson and Leach [27] showed that energy subsidies in the United States would impede the use of new energy and reduce energy use efficiency. Shah and Larsen [31] showed that if the total energy subsidies worth almost $230 billion in 1990 could be removed, CO_2 emissions worldwide would decrease by 9.5%. Using the global coal model, Lam and Shiu [32] analyzed coal subsidy reform in Japan; the results showed that removing the coal subsidies in the power supply and industrial boiler sector would reduce global CO_2 emissions by 0.2%. The IEA [5] also indicated that global CO_2 emissions would decrease by more than 6% by 2010 if the fossil energy subsidies in the power sector were removed. We use these research data to test and verify these countries, and the relationship between subsidies policy and efficiency. Therefore, Hypotheses 5a and 5b are as follow:

H5a: Verify that causal relationship between subsidy and OE

H5b: Verify that causal relationship between subsidy and DE

3. Research methodology

3.1. Two-phase data envelopment analysis framework

We adopted a two-stage DEA [7] to evaluate the level of management performance in renewable energy industries in OECD countries. These two types of efficiency are based on sub-processes that detail the two essential phases of a country's renewable power plants: outputs provided and use generation. We then followed the approach by Seiford and Zhu [33], who divided the entire production activity into two sub-production processes. Fuel, labor, generating capacity, and operating expenses were the original input variables, whereas total primary renewable energy supply (TPES)/GDP ratio, TPES/population ratio, and grid were final output variables. Medial input variables included electricity-only plants (EOP), combined heat and power plants (CHP) of electricity, and CHP of heat, and heat-only plants (HOP). Figure 1 shows this process.

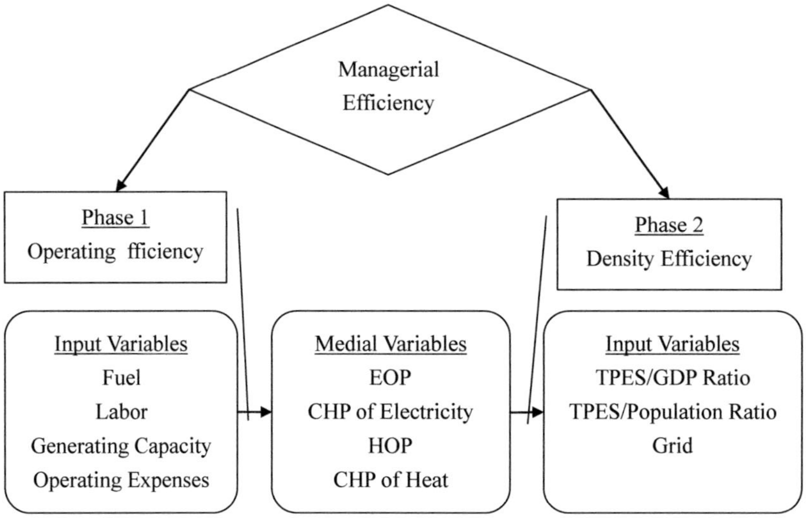

Figure 1. Two Phases DEA Model

3.2. Input and output variables

Donthu, Hershberger, and Osmonbekok [10] emphasized the significance of variable selection because the research outcome is heavily dependent on the input and output variables used in the model. Their arguments led researchers to believe that there should be a more rigorous method than those of previous studies for selecting input and output variables for efficiency assessment.

Phase I Input Variables: Selection of input variables is critical task for performance analysis, and the choice of variables depends on the selected methodology and technical requirements, the availability and quality of data, and on countries' individual socio-economic structures [34]. In this study, we use fuel, labor, generating capacity and operation expenses as our input:

1. *Fuel:* According to the IEA, renewable energy is divided into three categories of: (1) hydro fuel; (2) geothermal, solar, tidal, and wind fuel; and (3) combustible renewable energy and waste. The three categories of energy are all different in nature and cost [35]. On this basis, we discuss four renewable power sources: (1) Solar radiation: Glaser [36] provided a critical insight for a new source of solar energy. He proposed that large satellites be placed in geosynchronous orbit around Earth. These solar power satellites (SPS) would continually face the sun. Each SPS would convert a steady stream of sunlight to electric power, transform the electric power to microwave energy, and then transmit the microwaves in a tight beam to a receiver (rectenna) on Earth; (2) Wind speed: Boud and Thorpe [37] and, Bedard et al. [38] suggested that progress ratios from the wind and offshore engineering industries may be expected within the renewable energy industry; (3) Wave energy: Reviews of wave energy technologies are presented by

Thorpe [39], among others. Wave energy conversion devices have been classified according to numerous features including their relative location to the shore, the wave mode that energy is captured from, or the device operational type; and (4) Bio-energy: Many studies have stated that the substitution of conventional fossil fuels with biomass for energy production results in a net reduction of greenhouse gas emissions and in the replacement of non-renewable energy sources [40] [41] [42].

2. *Labor:* Adjaye [43] and Ghosh [44] indicated that the relationship among output, energy use, and labor employment are built on an econometric framework. From a policy viewpoint, the direction of causality between these variables has important implications. Bettencourt [2] proposed the primary reason for the continued use of labor as an input was because labor cost is a significant cost in many industries. Dugan and Autor [45] and Morey [46] indicated that electric power production is a comprehensive process that includes generation, transmission, distribution, and retailing, involving large amounts of capital, labor, and financial resources.

3. *Generating Capacity:* Electric power production is a comprehensive process that includes generation, transmission, distribution, and retailing, involving large amounts of capital, labor, and financial resources [45] [46]. Furthermore, major infrastructure facilities, such as electric power and transport systems, have been improved [21].

4. *Operation Expenses:* Many studies on operational processes have been produced within the energy industry [47] a greater energy density of renewable at the design sites schemes increases the importance of efficient operations and maintenance (O&M) planning. Marcotullio and Schulz [20] indicated that controlling the operating costs results in achieving specific renovation and maintenance (R&M) program, adopting better maintenance practices and promoting greater plant utilization.

3.3. Medial input/output variables: The medial outputs of phase I and also the medial inputs of phase II

1. *Electricity-Only Plants (EOP):* EOP refers to plants that are designed to produce electricity only [48]. The electric power business is separated into the following four functions: generation, transmission, distribution and retailing. Numerous previous have studies applied DEA to evaluate the performance of electricity generation facilities in many industrialized nations [32] [49] [50] [51].

2. *Heat-only Plants (HOP):* HOP refers to plants designed to produce heat only [48]. The names used below for each model originate from the study by Agrell and Bogetoft [52]. The heat output used by Agrell and Bogetoft is the production at the plant and not the quantity sold to heat customers.

3. *Combined Heat and Power Plants (CHP):* CHP refers to plants designed to produce heat and electricity, occasionally referred to as co-generation power stations [48]. If possible, fuel inputs and electricity/heat outputs are on a unit basis rather than on a plant basis. However, if data are unavailable on a unit basis, the convention for defining a CHP plant is adopted [5].

3.4. Output variables

Phase II Output Variables in: The results of recent studies have contributed to energy efficiency or environment efficiency evaluation problems that consider total production activity factors. Ramanathan [53] proposed an overall efficiency index that combined energy inputs, desirable outputs and undesirable outputs using DEA to study the relationships among global GDP, energy consumption, and carbon dioxide emissions. The final outputs used in this research, are as follows:

1. *TEPS/GDP ratio:* TPES per US. $1000 of GDP. The ratios are calculated by dividing each country's annual TPES by their annual GDP expressed in constant prices and converted to US. dollars using purchasing power parities (PPPs) (www.OECD-iLibrary.org). TPES consists of primary energy production adjusted for net trade, bunkers and stock changes.

2. *TPES/Population ratio:* TPES includes the sum of the unexplained statistical differences for individual fuels, as they appear in basic energy statistics [48]. TPES per population, ratios are calculated by dividing each country's TPES by unit people (www.OECD-iLibrary.org). This ratio can be used to obtain the electricity consumption of residents for each country.

3. *Grid:* The energy's transport length of line required to moor multiple devices is dependent on the spacing between devices and the array configuration. The length of cable required depends on the array configuration, although groups of devices are typically interconnected in series and each group is connected to a hub [38].

We used the intermediation approach to view the energy industry as intermediaries, and summarized the major input and output variables in Table 1.

3.5. Research resource and sample

This research is based on DEA of operating procedures. Though data collection and literature review on performance measurement of renewable power, we can understand the differences in renewable energy efficiency among 34 OECD countries (Table 2) and provide suggestions for Taiwan. The data obtained for this analysis were gathered from many relevant data resources, including the IEA, Renewable Information, World Bank, and other energy indices of a representative sample from 2007 to 2009. However, the data obtained from Renewable Energy Information [48] are used account for the full range of statistics collected from the Annual Renewables and Waste Questionnaire. This database of annual statistics for OECD countries covers hydroelectricity, solid biofuels, geothermal, renewable municipal waste, wind, gas from biomass, liquid biofuels, solar photovoltaic, solar thermal, tide/wave/ocean, non-renewable municipal waste and industrial waste. It includes EOP and HOP from renewable sources and supply/demand balances of renewable and waste products. The primary data from this system are from IEA annual publications.

Input Variables (Phase I)	
1. Fuel	IEA; Glaser, 1977; Thorpe, 1999; Boud and Thorpe, 2003; Schneider and McCarl, 2003;Owen, 2004; Bedard et al., 2005; Previsic et al., 2005; Dowaki and Mori, 2005; Caputo et al., 2005
2. Labor	Buonafina, 1992; Adjaye, 2000; Morey, 2001; Ghosh, 2002; Dugan et al., 2002;
3. Generating Capacity	Dorian, 1998; Morey, 2001; Dugan et al., 2002
4. Operation Expenses	Kannan and Pillai, 2000; Herman, 2002; AMEC, 2004
Medial Input/Output Variables (Output Phase I and Input Phase II)	
1. EOP	Olatubi and Dismukes, 2000; Lam and Shiu, 2001; Nag, 2006; Pombo and Taborda, 2006; Sueyoshi and Goto, 2010; Renewables Information, 2011
2. HOP	Agrell and Bogetoft, 2004; Renewables Information, 2011
3. CHP	IEA; Renewable Information, 2011
Output Variables (Phase II)	
1. TEPS/GDP ratio	PPPs (www.OECD-iLibrary.org)
2. TEPS/Population ratio	Renewable Information, 2011; PPPs
3. Grid	Halcrow, 2005; Bedard et al., 2005
Note. Source from this study	

Table 1. Input and Output Variables

3.6. DEA model

Data Envelopment Analysis (DEA) is a method for measuring the performance efficiency of decision units, characterizing by multiple input and output variables [8]. The DEA technique uses linear programming to estimate the maximum potential efficiency for various levels of inputs based on each firm's actual inputs and output. DEA includes two major models, the CCR model, and the BCC model. Charnes, Cooper and Rhodes [54] proposed a model under the assumption of constant return to scale (CRS), called the CCR model. This model is only appropriate when all DMUs are operating at an optimal scale. Banker, Charnes and Cooper [55] extended the CCR model to include the variable returns to scale named the BCC model, which can further decompose the TE into two components: pure technical efficiency (PTE) and scale efficiency (SE). The problem of calculating efficiency can be formulated as a fractional linear programming problem as below:

$$
\begin{aligned}
& Max\ \mathrm{E}_j = \sum_{n=1}^{N} U_n Y_{jn} - u_0 \,/\, \sum_{m=1}^{M} V_m X_{jm} \\
& s.t.\ \sum_{n=1}^{N} U_n Y_{jn} - u_0 \,/\, \sum_{m=1}^{M} V_m X_{jm} \le 1;\ \forall \mathrm{r} \\
& \mathrm{U}_n,\ \mathrm{V}_m \ge 0\ \ \mathrm{m=1,2,...,M;\ n=1,2,...N;} \\
& r = 1,2,...,j,...R
\end{aligned} \tag{1}
$$

We utilized the BCC input-oriented model to measure phase I and II to find a maximum output with certain medial output.

DMUs	Country Name	DMUs	Country Name
D1	Australia	D18	Japan
D2	Austria	D19	Korea
D3	Belgium	D20	Luxembourg
D4	Canada	D21	Mexico
D5	Chile	D22	Netherlands
D6	Czech Republic	D23	New Zealand
D7	Denmark	D24	Norway
D8	Estonia	D25	Poland
D9	Finland	D26	Portugal
D10	France	D27	Slovak Republic
D11	Germany	D28	Slovenia
D12	Greece	D29	Spain
D13	Hungary	D30	Sweden
D14	Iceland	D31	Switzerland
D15	Ireland	D32	Turkey
D16	Israel	D33	United Kingdom
D17	Italy	D34	United States

Note. Source from this study

Table 2. Country Names of each DMU

4. Empirical analysis

First, multi-collinearity analysis was employed to examine the correlation coefficient between input and input variables, and then between output and output variables [56]. We used isotonicity diagnosis to examine positive correlation coefficients between input and output variables [57]. We then used sensitivity analysis to sequentially increase or reduce the input or output variables to examine variation of efficiency [8]. The obtained sensitivity analysis result does not consider the operating expenses because of their highly correlation. Additionally, we also test the rule of thumb issued by Golany and Roll [58]. The four tests are all hold.

Tables 3 and 4 report the BCC efficiency scores of OE and DE for the 34 OECD countries from 2007 to 2009. Table 3 shows the comparison of the main goal in phase I to evaluate how efficiently countries use their resources; in other words, to identify any inefficiency result from PTE or SE. The resource inefficiency 2007, 2008, and 2009 is primarily pure technical efficiency (0.729, 0.704, and 0.727, respectively). In other words, the inefficiency is a result of inappropriate input and output configuration, rather than inappropriate scale. Table 4 shows a comparison of the main goal in Phase II to evaluate how efficiency energy is used to identify inefficiency resulting from PTE or SE. The resource inefficiency during 2007, 2008, and 2009 is primarily scale efficiency (0.439, 0.431, and 0.45, respectively). In other words, the inefficiency is result of inappropriate scale.

				Phase 1					
		2007			**2008**			**2009**	
DMUs	TE	PTE	SE	TE	PTE	SE	TE	PTE	SE
Ave.	0.666	0.729	0.917	0.5	0.704	0.848	0.597	0.727	0.812
SD	0.26	0.247	0.148	0.266	0.255	0.176	0.274	0.247	0.183
No. of Efficient DMUs Efficient DMUs	9	11	11	6	10	7	7	13	7
				Phase 2					
		2007			**2008**			**2009**	
DMUs	TE	PTE	SE	TE	PTE	SE	TE	PTE	SE
Ave.	0.387	0.735	0.439	0.373	0.726	0.431	0.367	0.718	0.45
SD	0.445	0.293	0.474	0.441	0.291	0.48	0.407	0.285	0.431
No. of Efficient DMUs Efficient DMUs	9	15	12	8	14	12	7	12	11

Note. Source from this study

Table 3. BCC-efficiency Scores for operating efficiency for each year

We employed the Mann-Whitney U-Test, a non-parameter statistical method, to test the same mean between two groups. The results in Table 4 show that the OE and DE for all cases do not achieve a level of significance (p >.05) for all compared years. Therefore, these 3-years are suitable for the DEA model using 102 DMUs to determine if there is a significant difference between OE and DE (Table 5).

Phase 1				
Case	**Test Value**	**TE**	**PTE**	**SE**
Between 2008 and 2007	Z Test	-0.21	-0.372	-0.501
	p-vale	0.834	0.71	0.617
Between 20.50.58 and 20.50.57	Z Test	-1.349	-0.604	-1.571
	p-vale	0.177	0.546	0.077
Phase 2				
Case	**Test Value**	**TE**	**PTE**	**SE**
Between 2008 and 2007	Z Test	-0.98	-0.234	-0.5.057
	p-vale	0.327	0.815	0.29
Between 20.50.58 and 20.50.57	Z Test	-0.433	-0.179	-0.345
	p-vale	0.665	0.858	0.73

Note. Source from this study

Table 4. Results of Mann – Whitney U Test

	2007		2008		2009	
DMUs	**OE**	**DE**	**OE**	**DE**	**OE**	**DE**
Ave.	0.666	0.387	0.5	0.373	0.597	0.367
SD	0.26	0.445	0.266	0.441	0.274	0.407
No. of Efficient DMUs	9	9	6	8	7	7

Note. Source from this study

Table 5. Bcc-efficiency score for OE and DE for each year

The Mann-Whitney U-Test is also used to determine if there is a significant difference between OE and DE before and after 2008 (Table 6). The results show that the global financial crisis did not influence OE and DE. Because OE and DE are non-significant, we can assert that the data are consistent and that renewable energy capital investments in each country have a certain proportion; thus, 2008 financial crisis did not have s significant influence on renewable energy development. This implies that the development of renewable energy is crucial. Furthermore, we want to determine if there is a significant difference between OE and DE.

Case	Test Value	OE	DE	OEvsDE
Between 2008 and 2007	Z Test	-0.006	-0.98	-2.812
	p-vale	0.995	0.327	0.005**
Between 20.50.58 and 20.50.57	Z Test	-1.258	-0.5.433	-2.819
	p-vale	0.208	0.665	0.005**

Note. Source from this study

Table 6. Results of Mann – Whitney U Test of OE and DE

Tobit regression analysis was conducted to determine whether the efficiency scores are related to characteristics such as GDP, population, capital, endowment and subsidy (Table 7). Furthermore, a dummy variable was included to evaluate the renewable energy subsidies in OECD countries. The function of a regression model can be expressed as: $Y = a + bX$, where Y represents the dependent variable and X represents regression form to a logistic probability function because the efficiency ranges from zero to one. The transformed regression function is expressed as:

$$\ln(\frac{Y}{1-Y}) = a + bX, \tag{2}$$

derived from:

$$Y = F(a + bX) = \frac{1}{1 + \exp(-a - bx)}. \tag{3}$$

The Tobit regression analysis result shows that endowment, population, and capital all have high positive significance with OE and DE. Thus, H1a, H1b, H2a, H2b, H3a, and H3b are supported. However, GDP has a non-significant negative correlation with OE and DE. Thus, H4a and H4B are rejected. GDP, capital, trade balance, household consumption, and energy imports are critical factors for measuring renewable energy indicators [5]. Our finding in H4a and H4b is that the GDP and OE are negatively correlated, and GDP and DE are also negative correlated. This is potentially because countries did not allocate the use of renewable energy in accordance with GDP degree. For example, compared to poorer countries, wealthy countries must improve the relatively large number of renewable energy use to achieve the target.

Specially, subsidies are significant with OE but not with DE. In other words, the subsidy is positively correlated with OE but negatively correlated with DE. Thus, H5a is supported and H5b is rejected. Some researchers have reached an opposing conclusion that subsidies and OE are positively correlated and that energy subsidy reform would produce positive results. Promoting "subsidy" policies can reduce industrial production costs. However, if they are implemented inefficiently without carefully assessing the cost-efficiency and associated financial risks, a "free rider" phenomenon is created with consequent disadvantages; thus, the subsidies do not have a positive benefit. Therefore, renewable energy subsidies and DE may be negatively correlated [59].

Model		**Phase 1**			**Phase 2**	
Dependent Variables		**OE**			**DE**	
	β-value coe.	t-value	p-value	β-value	t-value	p-value
Independent Variables						
Endowment	3.131	-4.7	0.013**	-14.294	2.54	0.000**
Population	-0.004	4.43	0.006**	0.13	-2.83	0.000**
Capital	0.001	-3.3	0.018**	-0.007	2.4	0.001**
GDP	-2.3	0.39	0.961	0.001	-0.05	0.695
Subsidy	0.17	-0.26	0.017**	-0.029	2.42	0.797
r^2		0.337			0.264	
F-value		4.04			8.69	
P-value		0.000**			0.000**	

Note. Here is the efficiency scores derived from operating efficiency (OE) and density efficiency (DE). The observation is 102. ** represents significant at 0.05 level and * represents significant at 0.1 level.

Table 7. Estimated Results of the Tobit Regression Analysis

5. Conclusion

Numerous DEA studies have incorporated the concept of production activities with multiple phases. They subsequently divided the DEA model into several sub-processes [60] [55] [33] [61] [62] [19]. Our first finding in this research is that there is a significant difference between OE and DE. In other words, sub-process DEA model is suitable for measuring management performance because of the characteristic of production activities in renewable energy industries [61].

http://dx.doi.org/10.5772/52020

OECD countries in response to United Nations climate Change Framework Convention and the relevant provisions of the Kyoto Protocol, In addition to adjusting the energy supply and demand side policies, and with the greenhouse gas performance of fiscal policy (subsidy) to promote energy conservation and reduce dioxide emissions [63]. Our second finding is that subsidies are positively correlated with OE and negatively correlated with DE. Promoting subsidy policies can reduce industrial production cost. However, if they are implemented inefficiently without carefully assessing the cost-efficiency and associated financial risks, a "free rider" phenomenon is created with consequent disadvantages; thus, the subsidies do not have a positive benefit. Therefore, renewable energy subsidies and DE may be negatively correlated [59].

In our study, we attempted to measure OECD countries' renewable OE and DE simultaneously, to examine the OECD renewable energy's promote, employ, and the relevance to the development of research, and provide feasible suggestions for a renewable energy development strategy in Taiwan. For example, because Taiwan is an island country, and resources are difficult to obtain, efforts should be made to actively develop renewable energy technology to replace traditional energy sources. In addition, policy-makers should assess renewable energy subsidy programs, promote renewable energy industry research and development, assist the industry in developing cost-efficient production technologies, and develop a new energy market. Furthermore, strengthen the use of renewable energy demonstration and propaganda work, and to enhance the efficiency of the client to use.

Finally, this study has several limitations that require discussion. First, only 34 OECD samples were selected that could provide the data required to conduct this study. Future research could include more countries, especially developing countries, such as Taiwan, China, and India to achieve more precise results. Second, non-financial data such as output quality and investment of renewable land were not included in our model. These variables are also critical factors for the evaluation of energy industry performance. Future research could include this as an additional evaluation variable. Finally, in this study, we independently tested and verified the two phases of efficiency. However, future research could use a supply chain model that assumes that the two phases of efficiency are dependent and further evaluate the real scores of management efficiency.

Author details

Tser-Yieth Chen[1*], Tsai-Lien Yeh[2] and Yi Hsuan Ko[1]

*Address all correspondence to: chenty@mail.ntpu.edu.tw

1 Graduate Institute of International Business, National Taipei University, New Taipei City, Taiwan, R.O.C.

2 Department of International Business, Ming Chuan University, Taipei, Taiwan, R.O.C.

References

[1] Marshall. J.D. Urban Land Area and Population Growth: A New Scaling Relationship for Metropolitan Expansion. Urban Studies 2007; 44(10) 1889-1904.

[2] Bettencourt L.M.A. Growth, Innovation, Scaling and the Pace of Life in Cities. PNAS 2007; 104(17) 7301-7306.

[3] United Nations. UN World Urbanization Prospects: The 2003 Revision. United Nations: New York; 2004.

[4] Crane P. and Kinzig A. Nature in the Metropolis. Science 2005; 27(1) 225-308.

[5] Vine E. and Hamrin J. Energy Savings Certificates: A Market-Based Tool For Reducing Greenhouse Gas Emissions. Energy Policy 2008; 36(1) 456-476.

[6] International Energy Agency, http://www.iea.org/ ; 2011.

[7] Chen Y., Cook W.D., Li N., Zhu J. Additive Efficiency Decomposition in Two Stage DEA. European Journal of Operational Research 2009; 196(5) 1170–1176.

[8] Donthu N. and Yoo B. Retail Productivity Assessment: Using Data Envelopment Analysis. Journal of Retailing 1998; 74(1) 89-105.

[9] Choi K.W., Roh Y.S., and Yoon J.H. An Empirical Examination of Productivity of a Chain Restaurant Using Data Envelopment Analysis (DEA). International Journal of Quality and Productivity Management 2007; 7(1) 47-67.

[10] Donthu N., Hershberger E.K., and Osmonbekok T. Benchmarking Marketing Productivity Using Data Envelopment Analysis. Journal of Business Research 2005; 58(11) 1474-1482.

[11] Criswell D.R. and Thompson R.G. Data Envelopment Analysis of Space and Terrestrially-Based Large Scale Commercial Power Systems for Earth: A Prototype Analysis of Their Relative Economic Advantages. Solar Energy 1996; 56(1) 119-131.

[12] Azadeh A., Ghaderi S.F. and Maghsoudi A. Location Optimization of Solar Plants by an Integrated Hierarchical DEA PCA Approach. Energy Policy 2008; 36(10) 3993-4004.

[13] Madlener R., Antunes C.H. and Dias L.C. Assessing the Performance of Biogas Plants with Multi-Criteria and Data Envelopment Analysis. European Journal of Operational Research 2009; 197(3) 1084-1094.

[14] Iglesias G., Castellanos P. and Seijas A. Measurement of Productive Efficiency with Frontier Methods: A Case Study for Wind Farms. Energy Economics 2010; 32(5) 1199-1208.

[15] Azadeh A., Ghaderi S.F. and Nasrollahi M.R. Location Optimization of Wind Plants in Iran by an Integrated Hierarchical Data Envelopment Analysis. Renewable Energy 2011; 36(5) 1621-1631.

[16] Maiga A.S., Chen G.M., Wang Q. and Xu J.Y. Renewable Energy Options for a Sahel Country: Mali. Renewable and Sustainable Energy Reviews 2008; 12(2) 564-574.

[17] Fan J., Sun W. and Ren D.M. Renewables Portfolio Standard and Regional Energy Structure Optimization In China. Energy Policy 2005; 33(8) 1095.

[18] Shi P. Planning of Wind Farm Development in China. World Renewable Energy Congress VI. 2007; 1176-1178.

[19] Chen Y. and Zhu J. Measuring Information Technology's Indirect Impact on Firm Performance. Information Technology and Management 2004; 5(1-2) 9-22.

[20] Marcotullio P.J. and Schulz N.B. Urbanization and Global Environmental Change. International Working Paper Series. 2007; 1011-1025.

[21] Boyd G.A. and Pang J.X. Estimating the Linkage between Energy Efficiency and Productivity. Energy Policy 2000; 28(5) 289-296.

[22] Chien, T.C. and Hu J.L. Renewable Energy and Macroeconomic Efficiency of OECD and Non-OECD Economies. Energy Policy 2007; 35(12) 3606-3615.

[23] Gerlagh R. and Zwaan, B.C.C. A Sensitivity Analysis of Timing and Costs of Greenhouse Gas Emission Reductions. Climatic Change 2004; 65(1-2) 39-71.

[24] Smyth R. Substitution between Energy and Classical Factor Inputs in the Chinese Steel Sector. Applied Energy 2011; 88 (1) 361-367.

[25] Hudson E.A. and Jorgenson D.W. Economic Analysis of Alternative Energy Growth Patterns, 1975-2000. Growth. Econometric General Equilibrium Modeling, Cambridge, Mas. The MIT Press 1998; 1(1) 191-220.

[26] Turner K. Negative Rebound and Disinvestment Effects in Response to an Improvement in Energy Efficiency in the UK Economy. Energy Economics 2009; 31 (5) 648-666.

[27] Anderson D. and Leach M. Harvesting and Redistributing Renewable Energy: On the Role of Gas And Electricity Grids to Overcome Intermittency Through The Generation and Storage of Hydrogen. Energy Policy 2004; 32(14) 1603-1614.

[28] Ang B.W. and Liu N. A Cross-Country Analysis of Aggregate Energy and Carbon Intensities. Energy Policy 2006; 34(5) 2398-2404.

[29] Ang B.W. A Cross-Sectional Analysis of Energy—Output Correlation. Energy Economics 1987; 9(4) 274-286.

[30] Steenblik, R. and Coroyannakis P. Reform of Coal Policies in Western and Central Europe. Energy Policy 1995; 23(6) 537-553.

[31] Shah A. and Larsen B. Carbon Taxes, the Greenhouse Effect, and Developing Countries. Policy Research Working Paper Series 957. The World Bank; 1992.

[32] Lam P. and Shiu A. A Data Envelopment Analysis of the Efficiency of China's Thermal Power Generation. Utilities Policy 2001; 10(1) 75–83.

[33] Seiford L.M. and Zhu J. Marketability and Profitability of the Top 55 US Commercial Banks. Management Service 1999; 45(9) 1270-1288.

[34] Jamasb T. and Pollitt M. Benchmarking and Regulation: International Electricity Experience. Utilities policy 2001; 9(3) 107-130.

[35] Owen A.D. Oil Supply Insecurity: Control versus Damage Costs. Energy Policy 2004; 32(16) 1879-1882.

[36] Glaser P.E. Editorial. Solar Energy 1977; 19(1) 1.

[37] Boud R. and Thorpe T. Wavenet: Section D Financing and Economics, available at: www.wave-energy.net/Library/ ;2003.

[38] Bedard R., Hagerman G., Previsic M., Siddiqui O., Thresher R. and Ram B. Final Summary Report Project Definition Study Offshore Wave Power Feasibility Demonstration Project. Electric Power Research Institute Inc.; 2005.

[39] Thorpe, T.W. Wave Energy Potential in the UK. Report Number AEAT-3028 for the DTI. ; 1999.

[40] Akbarzadeh A., MacDonald R.W.G. and Wang Y.F. Reduction of Surface Mixing In Solar Ponds By Floating Rings. Solar Energy 1983; 31(4) 377-380.

[41] Richter W.F. and Schneider U.A. Energy Taxation: Reasoning for Discriminating in Favor of the Production Sector. European Economic Review 2003; 47(3) 461-476.

[42] Caputo A.C., Palumbo M., Pelagagge P.M. and Scacchia F. Economics of Biomass Energy Utilization in Combustion and Gasification Plants: Effects of Logistic Variables. Biomass and Bio-energy 2005; 28(1) 35-51.

[43] Adjaye, J.A. The Relationship between Energy Consumption, Energy Prices and Economic Growth: Time Series Evidence from Asian Developing Countries. Energy Economics 2000; 22(4) 615-625.

[44] Ghosh J. Macroeconomic Reforms and a Labor Policy Framework for India. International Labor Office, Employment Strategy Papers; 2002.

[45] Dugan M.G. and Autor D.H. The Growth in the Social Security Disability Rolls: A Fiscal Crisis Unfolding. Journal of Economic Perspective 2002; 20(3) 71-96.

[46] Morey E. A Simple Model of Industrial Pollution. Environment Economics 2001; 10(1) 85-94.

[47] Herman A. Three-Dimensional Structure of Wave Induced Momentum Flux in Ir-rotational Waves in Combined Shoaling-Refraction Conditions. Coastal Energy, 2002; 53(2) 545-555.

[48] Renewable Energy Information. Renewable Energy http://www.guardian.co.uk/environment/renewableenergy ; 2011.

[49] Olatubi W.O. and Dismukes D.E. A Data Envelopment Analysis of the Levels and Determinants of Coal-Fired Electric Power Generation Performance. Utilities Policy 2000; 9(2) 47-59.

[50] Pombo C. and Taborda R. Performance and Efficiency in Colombia's Power Distribution System: Effects of the 1994 Reform. Energy Economics 2006; 28(3) 339-369.

[51] Sueyoshi T. and Goto M. DEA Approach for Unified Efficiency Measurement: Assessment of Japanese Fossil Fuel Power Generation. Energy Economics 2010; 33(2) 292-303.

[52] Agrell P.J. and Bogetoft P. Economic and Environmental Efficiency of District Heating Plants. Energy Policy 2004; 33(10) 1351-1362.

[53] Ramanathan, P. Energy Efficient Transmission Scheme for Data-Gathering in Mobile Sensor Networks. 2006; In Proceedings of IEEE SECON.

[54] Charnes A.C., Cooper W.W. and Rhodes E. Measuring Efficiency of Decision Making Units. European Journal of Operations Research 1978; 2(6) 429-444.

[55] Banker R.D., Charnes A.C. and Cooper W.W. Models for Estimation of Technical and Scale Inefficiencies in Data Envelopment Analysis. Management Science 1984; 30(9) 1078-1092.

[56] Charnes A., Cooper W.W., Gollany B., Seiford L. and Stutz J. Foundations of Data Envelopment Analysis for Pareto-Koopmans Efficient Empirical Productions Functions. Journal of Econometrics 1985; 30(1) 91-107.

[57] Bowlin W.F. Evaluating the Efficiency of U.S. Air Force Real-Property Maintenance Activities. Journal of the Operational Research Society 1987; 38(1) 127-135.

[58] Golany B. and Roll Y. An Application Procedure of DEA. OMEGA 1989; 17(3) 237-250.

[59] Taiwan Research Institute (TRI). http:// www.tri.org.tw; 2011.

[60] Farrell M.J. The Measurement of Productive Efficiency. Journal of the Royal Statistical Society 1957; 120(3) 253-290.

[61] Chen T.Y. Measuring Operation, Market and Financial Efficiency in the Measurement of Taiwan's Banks. Services Marketing Quarterly 2002; 24(2) 15-28.

[62] Keh H.T. and Chu S. Retail Productivity and Scale Economies at the Firm Level: A DEA Approach. Omega 2003; 31(2) 75-82.

[63] Environmental Energy Technologies Division (EETD). http://eetd.lbl.gov/ ; 2011.

Section 3

Wind Power

Chapter 10

Wind Speed Regionalization Under Climate Change Conditions

Masoomeh Fakhry, Mohammad Reza Farzaneh, Saeid Eslamian and Rouzbeh Nazari

Additional information is available at the end of the chapter

http://dx.doi.org/10.5772/55985

1. Introduction

Green energy and renewable energy are one of the most essential fundamentals for future developments of countries all over the world. Running out of fossil fuels in near future, makes the use of renewable energies almost inevitable. Evaluation of the capability of using these types of energies is an essential issue at present. Extreme dependency of mankind to the limited energy resources is both dangers and unsustainable and finding a way out is one of the most important challenges that we face. Achieving some unlimited energy sources has been human's dream. Daily increase of energy demands and limitation of fossil energy resources from one side, and increase of environmental pollutants caused by using these resources in the other side, has made application of renewable energies more essential and widespread. The wind, as one of the climatological factors has a wide effects on agriculture, transport, pollution, energy, manufacturing and industrial plans. Wind power is one of the first energy sources discovered by humans. It is applied for different purposes such as power source for ship movement, irrigation and milling in some countries such as Iran and China. However, construction of wind power plants would make this type of energy more applicable. Nowadays, coal, oil and gas are the main resources to provide energy. In recent years, an increasing trend in price of the mentioned materials has become vivid due to the globalization and political events and mostly because of water crises. These increasing prices make us to find an appropriate solution to decrease the expenses and increasing the stability. Furthermore, renewable energies are considered as available, exchangeable and inexhaustible resources. New energies are available until these resources exist. Wind energy is used in two ways, first in direct way in which the wind is applied for drying and ventilation, and secondly, the wind is utilized indirectly for milling the grains, to pump water to the fields and to generate electricity. The evidence shows that in some countries such as Iran, Iraq, Egypt, China, Italy, Spain, the wind energy has been used for milling and irrigation. According to IPCC special report on renewable energy sources and decreasing the climate change damages, there is an increasing trend in the magnitude of the wind power plants from 17 meters high and 75 kilowatts in 1980 decade to 80 meters high and 1800 kilowatts in 2005-2010. It is predicted that the size of these generators should be 250 meters high and 20000 kilowatts in future years which is clarifying the increasing trend of application of this green energy. An important challenge in front of this kind of power plants is the uncertainty in accessible capacity of electrical power. This problem has been caused by random nature of effective factors such as random variation in mechanical forces generating wind power. In other words, due to the continuous variations in meteorological and climatological conditions, the wind speed, duration, density and power are randomly changing. Thus, for using its power, it is necessary to study the windy conditions in the area and statistical data reported by meteorology centers. However, the analysis of such large amount of data recorded in meteorology centers to estimate the mechanical power input of wind power plants is not possible except using some applied methods. Obviously, the study of the behavior and speed of winds will lead to more accurate estimation of accessible capacity in wind power plants. Moreover, spatial analysis of this climatological phenomenon will provide some essential knowledge about the areas with potential capacity of constructing wind power plants. The frequency analysis is an operative tool in assessing this aim. This study is going to describe these materials and their effects on regionalization of wind speed under effect of climate change.

2. Previous investigations on wind prediction

There are some studies reported in the literature about wind speed in future periods which some of them are summarized in Table 1.

Results	Paper Type	Location	Scenarios	Model	References
Historical wind time series for future application					Harmsen et al. (2009)
Historical wind time series for future application					Doria et al. (2006)
Wind speed change is stronger than the changes obtained for the same region on daily precipitation and temperature	Case Study	Nebraska	2 scenarios	MPI	Bogardi and Matyasovszky (1996)
Dynamical Downscaling	Case study	German	2 scenarios	ECHAM	Hoyme and Zieleke (2001)
The GCM models of climate change become more reliable and as tools are refined for improving results at regional scales, it will be desirable to	Case study	USA	2 scenarios	HADCM2, CGCM1	Breslow and Sailor (2002)

Results	Paper Type	Location	Scenarios	Model	References
include improved estimates of vulnerabilities in the wind power site selection decision process.					
-ANN and statistical downscaling -A2 is preferred to A1b changes in daily mean wind speeds at each location and they are presented and discussed with respect to potential implications for wind power generation.	Case study	USA	A1b, A2	4 GCM	Sailor et al. (2000)
-Impact of the climate change scenarios on wind power may be as high as a 40% reduction in summer time generation potential.	Case study	Texas and California	2 scenarios	CCM	Sailor et al. (2008)
In order to create the scenarios of changes in power density under climate change, a method is developed for mapping the daily-resolution downscaled GCM output to the hourly level. The model results for either of the SRES scenarios were similar. In most cases, the impacts of the A2 scenario were slightly larger than those for the A1B scenario, but these inter-scenario differences were smaller than the inter-model differences, even after downscaling.					
-The average of wind speed will increase significantly towards the end of this century under the changing climate impacts.	Case study	Finland	A2	HWIND and SIMA	Peltola et al. (2010)
- Use of NCEP and ANN for downscaling. - For the two locations considered, the increase in the 100-year wind was found to be varying from 44% to 74%.	Case study	Indian	A2	CGCM3	Deepthi and Deo (2010)
-By the end of the twenty-first century, there is an evidence for small magnitude changes in the wind resource and increases in extreme wind speeds, and the declines in sea ice and icing frequencies.	Review Paper	-Some changes associated with climate evolution will likely benefit the wind energy industry while other changes may negatively have an impact on wind energy developments, with such 'gains and losses' depending on the region under consideration. - Gumbel is presented for the probability distribution of extreme wind speeds			Pryor and Barthelmie (2010)
15 – 30 % wind speed growth	Case study	Brazil	A1B	HADCM3	Pereira et al. (2012)

Table 1. Wind speed investigations in future period

3. Climate change impacts

Fast development of industries and its outcome as increasing of the emission of greenhouse gases, has led to destruction of climatic equilibrium of the earth. This phenomenon is called "Climate Change" (IPCC 2007, Leander et al. 2006). The research is indicating the negative impacts of this phenomenon on different systems such as water resources, agriculture, environment, health, industry, and economy. The importance and hazardous of climate change has been emphasized in different international communities such as the group of eight (G8) which is a forum for the governments of eight of the world's largest economies and some of its facing solutions to save water resources, agriculture, and environmental resources have been suggested. As the water is an important resource, which is extremely under effect of climate change, the analysis of its changes in future years can provide a very useful key for future droughts, floods, evapotranspiration and etc.

The first step in the study of climate change impacts on future resources is to simulate the behavior of climatological factors under the effect of greenhouse gases. A general circulation model (GCM) is a three dimensional mathematical models of the general circulation of a planetary atmosphere or ocean. Atmospheric and oceanic GCMs (AGCM and OGCM) are key components of global climate models, which are systems of differential equations. Using such models, scientists divide the atmosphere, hydrosphere, geosphere, cryosphere, and biosphere of the planet into a 3-dimensional grid, apply the basic equations, and evaluate the results. Atmospheric models calculate winds, heat transfer, radiation, relative humidity, and surface hydrology within each grid and evaluate the interactions with neighboring points. Different greenhouse gases emission scenarios such as A1, B1, A2, and B2 are going to be used during the simulation process.

4. Generation of climatic scenarios

The aforementioned 3-dimensional joint atmospheric-oceanic general circulation models (AOGCM) are used in this study among different available methods for the generation of climatic scenarios. The GCM models have a physical basis presented by mathematical relations. They are going to be solved in a 3-dimensional grid all over the planet. In order to simulate the climate of the planet, the fundamental climatic processes in the atmosphere, hydrosphere, geosphere, cryosphere, and biosphere will be simulated in separate secondary models. Then, these atmospheric and oceanic secondary models are joining together to form AOGCMs. To study the condition of climate in the past periods, the observed values of greenhouse gases, solar radiation changes, and volcanic eruption aerosols until the 2000 are entered as input to the GCM models and the climatic variables are simulated as time series. After simulation of these variables in the past periods by using the GCM models, the introduction of future greenhouse gas conditions is necessary for simulation of these variables in future periods. For this purpose, at first the amounts of emitted greenhouse gases presented by emission scenarios (which are always until 2100) are transformed to concentrate and then

http://dx.doi.org/10.5772/55985

to the amount of solar radiation and these values would be the input of the GCM models. The results obtained from the GCM models under emission scenarios will form the time series of climatic variables up to 2100.

5. Downscaling

One of the main challenges using the output of the AOGCM models is the spatial scale of their calculation cell and the downscaling method is used to solve this challenge. These methods are generally consists of two main groups of dynamic and statistical ones. In these methods, the downscaling procedure is done by using the observed meteorological data. A considerable point in application of the final outputs is the different sources of uncertainty, which can be evaluated using the Bootstrap method (Efron, 1993) in each confidence level.

6. Frequency analysis

The magnitude of an extreme event has an inverse relation with its frequency. In other words, the higher magnitude is the event, the less is its frequency of occurrence. The primary objective of frequency analysis is to relate the magnitude of extreme events to their frequency of occurrence through application of probability distributions (Chow et al., 1988). The first assumption in this manner is that under study data are independent and identically distributed and their underlying system is random and is spatially and temporally independent. This would be available when there is no correlation between observations. In application, these conditions can be achieved by using annual maximum values noting the independency of events between years. However, the wind speed parameter has been rarely examined by this method among other meteorological parameters; therefore, the studies in this field are at the beginning.

7. Probability distribution function and frequency formula

To describe the probability distribution of a random variable X, a cumulative distribution function (CDF) is used. The value of this function $F(x)$ is simply the probability P of the event that the random variable takes on value equal to or less than the argument:

$$F(x)=P[X \leq x] \tag{1}$$

This is the probability of the random variable X, it will not exceed x and is shown by the non-exceedance probability $F(x)$.

The occurrence of extreme events is not according to a constant regime or with a fixed magnitude and the time interval between two such events is variable. Thus, the return period defined as the average inter-arrival time between two extreme events is an applicable tool in such cases. An extreme event x_T with return period T can occur more than one time in a year and its exceedance probability can be expressed by:

$$P[X > x_T] = \frac{1}{T} \tag{2}$$

Thus, the above mentioned non-exceedance probability can be presented as follows:

$$F(x_T) = P[X \leq x_T] = 1 - P[X > x_T] = 1 - \frac{1}{T} \tag{3}$$

Equation (3) is presenting the magnitude of an extreme event correspond to a return period T. For a preselected value of return period and corresponding non-exceedance probability ($p = 1 - \frac{1}{T}$), it is possible to determine the quantile x_p using the inverse function of F:

$$x_p = F^{-1}(p) = F^{-1}\left(1 - \frac{1}{T}\right) \tag{4}$$

which gives the value of x_p corresponding to any particular value of p or T.

8. Selection of probability distribution

The probability distribution is a function for describing the probability of occurrence of a random event. Large amount of statistical information will be summarized in the distribution and its parameters by fitting a probability distribution on a set of hydrologic data. The most important and widely used methods for estimation of distribution parameters from data samples are method of moments, linear moments, and maximum likelihood which will be described more in next paragraph.

9. Method of moments

The method of moments was first introduced by Pearson (1902). He found that the appropriate estimations of the parameters of a probability distribution are those which their moments match with corresponding sample moments in the best way. In this method, general formula for calculation of moments of order r of the distribution $f(x)$ around the mean is:

$$\mu^r = \int_{-\infty}^{\infty} x^r f(x) dx \tag{5}$$

The method of moments is describing the relation between moments and distribution parameters. The most important moments around the mean are the mean, variance, skewness, and kurtosis, which are the one to four order moments, respectively.

10. Maximum likelihood method

Maximum-likelihood estimation was recommended, analyzed and vastly popularized by R. A. Fisher between 1912 and 1922 (Aldrich, John 1997). He argued that the best value of a parameter of a probability distribution should be one, which maximizes the likelihood or joint probability of occurrence of the sample. Assume that the sample space is divided into parts with length dx and x_i is selected from independent and identically distributed observations $x_1, x_2, \ldots, x_n$. The probability density for x_i can be denoted by $f(x_i)$. The probability that a random event occur in a distance consisting x_i would be equal to $f(x_i)\,dx$. As the samples are independent, the joint density function for all observations can be calculated by:

$$f(x_1)\,dx.f(x_1)\,dx\ldots f(x_n)\,dx=\left(\prod_{i=1}^{n} f(x_i)\right)(dx)^n \tag{6}$$

As the distance dx is constant, the maximization of above joint density function is equivalent to maximization of likelihood function defined as:

$$L\;(x_1,\,x_2,\,\ldots,\,x_n)=f(x_1,\,x_2,\,\ldots,\,x_n)=\prod_{i=1}^{n} f(x_i) \tag{7}$$

The maximum likelihood method is theoretically the most accurate method in estimation of the parameters of probability distributions. In fact, it is estimating those parameters with minimum average error with respect to correct parameters.

11. L-moments method

Probability weighted moments (PWMs) are defined by Greenwood et al. (1979) as

$$\beta_r=E\left(x\{F(x)\}^r\right) \tag{8}$$

Which can be rewritten as:

$$\beta_r=\int_0^1 x(F)F^{\,r}dF\;,\qquad r=0,\;1,\;\ldots,\;s \tag{9}$$

Where $F=F(x)$ is CDF of x and $x(F)$ is its inverse. In the case of $r=0$, β_r is equal to the mean of the distribution $\mu=E(x)$. PWMs are precursors of L-moments and developed more by the works of Hosking (1986, 1990) and Hosking and Wallis (1991, 1993, 1997). Hosking defined L-moments as:

$$\lambda_{r+1}=\sum_{k=0}^{r}\beta_r(-1)^{r-k}\binom{r}{k}\binom{r+k}{k} \tag{10}$$

The first four order L-moments can be calculated as bellow:

$$\lambda_1 = \beta_0 \tag{11}$$

$$\lambda_2 = 2\beta_1 - \beta_0 \tag{12}$$

$$\lambda_3 = 6\beta_2 - 6\beta_1 + \beta_0 \tag{13}$$

$$\lambda_4 = 20\beta_3 - 30\beta_2 + 12\beta_1 - \beta_0 \tag{14}$$

L-moment ratios, which are analogous to conventional moment ratios, are defined by Hosking (1990) as:

$$\tau = \lambda_2 / \lambda_1 \tag{15}$$

$$\tau_r = \lambda_r / \lambda_2 \quad , \quad r \geq 3 \tag{16}$$

where λ_1 is a measure of location, τ is a measure of scale and dispersion (L-CV), τ_3 is a measure of skewness (L-skew), and τ_4 is a measure of kurtosis (L-kurt.). Analogous to conventional (product) moments, the L-moments of order one to four characterize location, scale, skewness and kurtosis, respectively (Karvanen, 2006). In order to estimate the distribution parameters in the method of L-moments, similar to other methods, sample L-moment ratios are calculated by replacing distribution L-moments λ_r by their sample estimates.

However, L-moments have significant advantages over PWM's, specially their ability to summarize a statistical distribution in a more meaningful way. Since L-moment estimators are linear functions of the ordered data values, they are virtually unbiased and less influenced by outliers. Also they have relatively small sampling variance and the bias of their small sample estimates remains quite small. L-moments have become popular tools for solving various problems related to parameter estimation, distribution identification, and regionalization in different fields such as hydrology, water resources, and especially in regional analysis of rainfall and floods.

12. At-station goodness of fit tests

After estimation of the parameter of the prescribed distributions, the usual question is the selection of the best fitted distribution to the observation sample. For this aim, the goodness-of-fit tests are used to compare fitted theoretical distributions and observations. Two very common goodness-of-fit tests with wide application in the literature are Chi-square and Kolmogorov-Smirnov tests. Another extensively used test is root mean square error (RMSE) which estimates the root of the square differences between observed values and calculated ones divided by the sample size. Among theoretical probability distributions, some of them are selected by researchers for describing wind speed, which are presented in Table 2.

Year	Distribution function	Scientists
1940 to 1945	Pearson Type III	Putnum*
1951	Pearson Type III	Sherlock*
1950 to 1970	Bivariate Distributions of Two Components	Essenwanger*
	Two-Parameter Normal	Crutcher and Bear*
1970s	Isotropic Gaussian Model of McWilliams et al.	Justus and Koeppl*
1974, 1976, 1977	Three-Parameter Log-normal	Luna and Church* Kaminsky, Justus et al.
1976 to 1977	Square-root Normal Model	Winger*
1978	Three-parameter Weibull	Stewart and Essenvanger*
1980	Three-Parameter Generalized Gamma	Auwera*
1980	Inverse Gaussian	Bardsley*
1983	Pearson Type I (Beta)	Lavagnini et al.*
1994	Weibull	Stelios Pasardes*
1996	Log-Normal	Bogardi and Matyasovski
2009	Beta	Carta et al.
2009	Two-Component Weibull	Akdog et al.*
2010	Gumbel and Weibull	Deepthi and Deo

*by Carta et al. 2009

Table 2. Probability distribution function of wind speed in the literature

13. Regional frequency analysis

One of the main problems in frequency analysis is the lack of adequate long time data in locations under study, which beside insufficient accuracy in data recording in at-site estimates, has caused regional frequency analysis to be more applicable in such studies. Steps of regional frequency analysis are presented below.

14. Discordancy test

For the screening of similar sites, the discordancy measure, in terms of the sample L-moment ratios (L-CV, L-skew, and L-kurt.) of the gauging sites' observed data is suggested by Hosking and Wallis (1997). The aim of the data screening performed using L-moments based on a discordancy measure D_i is to identify data that are grossly discordant with the group as a whole for the regional flood frequency analysis. Hosking and Wallis (1997) defined the discordancy measure (D_i) considering that there are N sites in the sample. Let $u_i = \left(t^{(i)}, t_3^{(i)}, t_4^{(i)}\right)$ be a vector containing the sample L-moment ratios for site $i(i=1, 2, \ldots, N)$ (Hosking and Wallis, 1993, 1997). The classical discordancy measure for any gauging site i is calculated as follows (Hosking and Wallis 1997):

$$D_i = \frac{1}{3}\left(u_i - \bar{u}\right)^T S^{-1}\left(u_i - \bar{u}\right) \tag{17}$$

$$\bar{u} = \frac{1}{N}\sum_{i=1}^{N} u_i \tag{18}$$

$$S = \frac{1}{N-1}\sum_{i=1}^{N}\left(u_i - \bar{u}\right)\left(u_i - \bar{u}\right)^T \tag{19}$$

Where N is the number of stations, u_i is the vector of L-moments, $\bar{u}$ and S are the sample mean and covariance matrix respectively, and T denotes the transposition of a vector or matrix. Large values of D_i indicate the sites that are the most discordant from the group as a whole and are the most suitable for investigating the existence of data errors (Hosking and Wallis 1993). Generally, if a site's D statistic exceeds three when the number of sites in one region is greater than 15, its data are considered to be discordant from the rest of the regional data (Hosking and Wallis 1997). Large values of D_i indicate that cautious investigation of the ith site should be carried out to detect the presence of data errors.

15. Heterogeneity test

L-moment heterogeneity tests allow assessing whether a group of sites may reasonably be treated as a homogeneous region. The heterogeneity measure compares the between-site variations in sample L-moments for a group of sites with what would be expected for a homogeneous region (Hosking and Wallis, 1993).

The homogeneity test used in this study is the test that is proposed by Hosking and Wallis (1993) and is based on various orders of sample L-moment ratios. It is particularly based on the variability of three different levels of tests: a test based on the L-CV only; a test based on the L-CV and L-skew; and a test based on the L-skew and L-kurt. These tests are called the V-statistics and are respectively defined as:

$$V_1=\left\{\sum_{i=1}^{N} n_i\left(\tau^{(i)}-\tau^{R}\right)^2 \Big/ \sum_{i=1}^{N} n_i\right\}^{0.5} \tag{20}$$

$$V_2=\sum_{i=1}^{N} n_i\left\{\left(\tau^{(i)}-\tau^{R}\right)^2+\left(\tau_3^{(i)}-\tau_3^{R}\right)^2\right\}^{0.5} \Big/ \sum_{i=1}^{N} n_i \tag{21}$$

$$V_3=\sum_{i=1}^{N} n_i\left\{\left(\tau_3^{(i)}-\tau_3^{R}\right)^2+\left(\tau_4^{(i)}-\tau_4^{R}\right)^2\right\}^{0.5} \Big/ \sum_{i=1}^{N} n_i \tag{22}$$

The heterogeneity measure is then defined as:

$$H=\frac{\left(V_i-\mu_{V_i}\right)}{\sigma_{V_i}} \quad ; \quad i=1,\ 2,\ 3 \tag{23}$$

where μ_{V_i} and σ_{V_i} are the mean and the standard deviation of the simulated value of the V, respectively. The regional averages L-moment ratios are determined by following equations:

$$\tau^{R}=\sum_{i=1}^{N} n_i\tau^{(i)} \Big/ \sum_{i=1}^{N} n_i \tag{24}$$

$$\tau_3^{R}=\sum_{i=1}^{N} n_i\tau_3^{(i)} \Big/ \sum_{i=1}^{N} n_i \tag{25}$$

$$\tau_4^{R}=\sum_{i=1}^{N} n_i\tau_4^{(i)} \Big/ \sum_{i=1}^{N} n_i \tag{26}$$

where N is the number of sites, n_i is the record length at site i, and $\tau^{(i)}$, $\tau_3^{(i)}$, and $\tau_4^{(i)}$ are the sample L-moment ratios at site i. After fitting a Kappa distribution to the regional average L-moment ratios, a large number of realizations N_{sim} of a region with N sites, each having the Kappa distribution is simulated. The simulated regions are homogenous and have no cross-correlation or serial correlation.

Declare the region to be heterogeneous if H is sufficiently large. Hosking and Wallis (1993) suggested that the region be regarded as "acceptably homogeneous" if $H<1$, "possibly heterogeneous" if $1\leq H<2$, and "definitely heterogeneous" if $H\geq 2$.

In order to achieve reliable estimates μ_V and σ_V, Hosking and Wallis (1993) recognized that the value of $N_{sim}=500$ should usually be adequate. Furthermore, they judged that larger values may need to resolve H values very close to 1 or 2.

Hosking and Wallis (1993) revealed that H statistics based on V_2 and V_3 lacks a power to distinguish between homogeneous and heterogeneous regions and the H statistics based on V_1 has much better discriminatory power and thus, it is suggested as heterogeneity measure.

16. Estimation of the parameters of regional frequency distribution

Four first orders L-moments for each site inside a homogeneous region is making dimensionless by dividing them by the average of the data. Weighted values of dimensionless L-moments are used to calculate standardized regional L-moments.

$$\lambda_r^R = \frac{\sum_{i=1}^{N} n_i \lambda_r^{(i)}}{\sum_{i=1}^{N} n_i} \tag{27}$$

where λ_r^R is the regional standardized L-moment of order r, $\lambda_r^{(i)}$ is the standardized L-moment of order r in site i, n_i is the number of years in site i, and N is the number of sites in the homogenous region. The parameters of best fitted distribution are estimated using the relation between distribution parameters and L-moments presented by Hosking (1989). Then the quantile values corresponding to different return periods are estimated for under study variable as regional quantile. The quantiles for the sites in each sub-region are determined by multiplying the regional quantile with the site's mean. At any site, the ith quantiles for the subregions are calculated using Eq. (34)

$$Q_i(F) = \lambda_1^{(i)} q(F) \tag{28}$$

Where $Q(F)$ and $q(F)$ are the at-site i and regional quantiles with non-exceedance probability, respectively.

17. Spatial interpolation

In classic statistics, the samples derived from a population are usually considered as random sets and the recorded value of a particular variable in an individual sample cannot present any details about the value of that variable in another sample with specific distance. In geostatistics, it is possible to link the values of one variable in a population and distance and direction of samples relating to each other. Furthermore, in classic statistics, it is assuming that the variables are randomly changing, while in geostatistics, some parts of the variable is random and some other parts have structure and is a function of distance and direction. Thus, using geostatistics, first the existence or the absence of a spatial structure between data is considered and then, in presence of a spatial structure, the data will be analyzed. It is possible to adjacent data to be spatially dependent together in a certain distance. In such cases, as in the presence of spatial structure, the variations in a certain space have more chances to be effective on near spaces with regard to more far ones, it is clear that variables are maybe more similar in closer samples.

Geostatistics is a field of statistics with the base of "local variables theory". Any variable distributed in 3d space with spatial dependence is called local variable and can be studied and analyzed in geostatistical studies. Some method of geostatistical studies are Inverse Distance Weighting (IDW), Global Polynomial (GP), Local Polynomial (LP), Radial Basis Functions (RBF), Kriging (Simple, Ordinary, Universal, Disjunctive and CoKriging).

Applying these methods needs to parallel application of spatial and statistical analysis, which is possible only in some environments like ArcMap. To visualize the above process, the steps are summarized as follows in Figure 1.

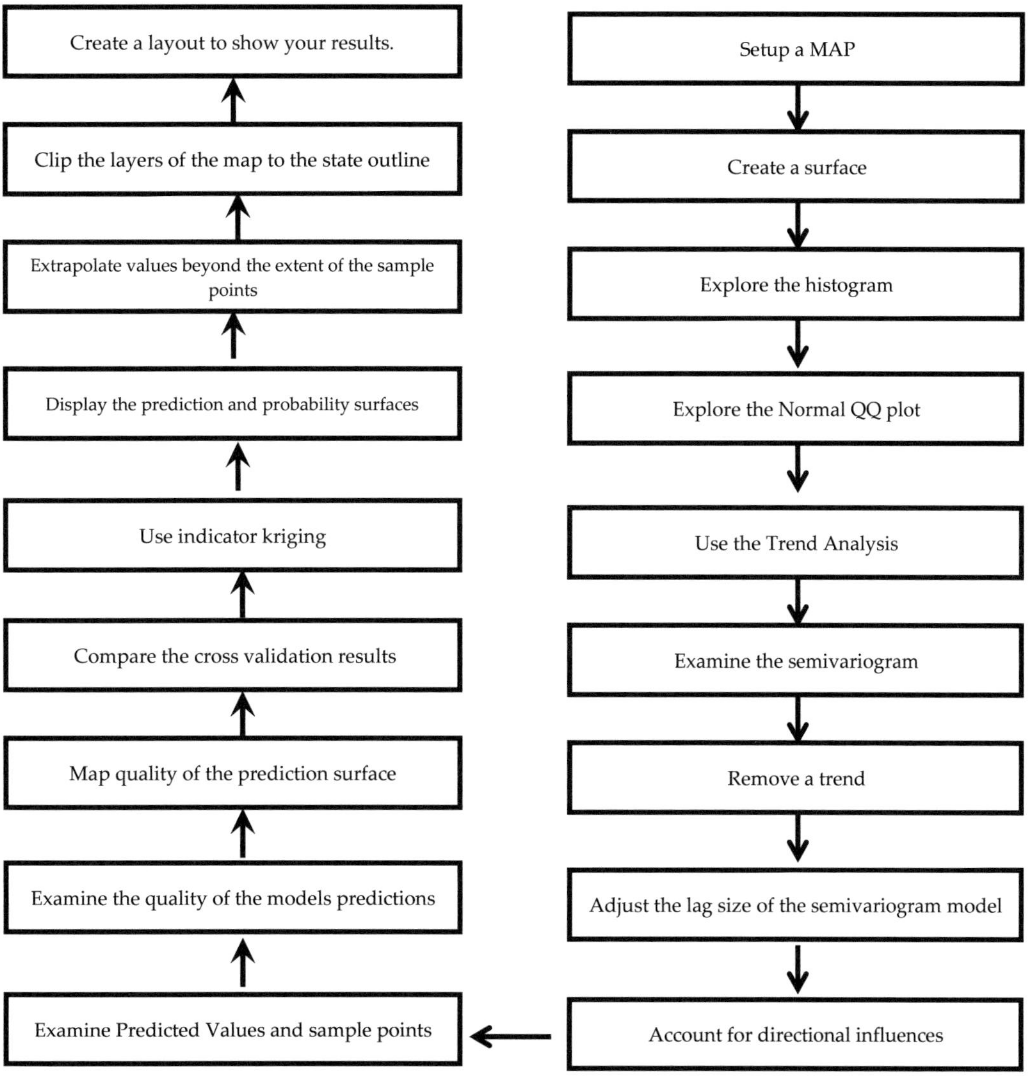

Figure 1. Spatial interpolation steps

18. Comparison of different methods

In above sections, all methods are separately presented but their comparison is very important to choose the most appropriate method for analysis. However, a wrong selection in this step will lead to large amount of uncertainty in output results.

Model	Type	Output Surfaces	Speed	Exact Interpolation	Flexibility	Advantage	Disadvantages	Assumptions
IDW	Deterministic	Prediction	Fast	Yes	Little flexibility, few parameter decision	Few decisions	NO assessment of prediction errors, bull's-eyes around data location	None
GP	Deterministic	Prediction	Fast	No	Little flexibility, few parameter decision	Few decisions	NO assessment of prediction errors, may be too smooth, edge points have large influence	None
LP	Deterministic	Prediction	Fairly fast	No	Some flexibility, more parameter decision	Flexible	NO assessment of prediction errors, may be hard to choose a good local neighborhood	None
RBF	Deterministic	Prediction	Fairly fast	Yes	Flexibility, more parameter decision	Flexible	NO assessment of prediction errors, may be too automatic	None
Kriging	Stochastic	Prediction. Standard Error, Probability, Quintile	Fairly fast	Yes without measurement error, No with measurement error	Very flexibility, assess spatial autocorrelation, obtain standard errors, many decisions more parameter decision	Flexible with modeling tools; prediction standard errors	Many decision on transformations trends, models, parameters, and neighborhoods	Stationary, some methods require a normal data distribution
Cokriging	Stochastic	Prediction. Standard Error, Probability, Quintile	Fairly fast	Yes without measurement error, No with measurement error	Very flexibility, assess spatial autocorrelation, obtain standard errors, very many decisions	Flexible with modeling tools; prediction standard errors	Many decision on transformations trends, models, parameters, and neighborhoods	Stationary, some methods require a normal data distribution

*by compus.esri.com

Table 3. Summarized properties of the interpolators

19. A case study

To make the above discussions more clear, a case study including all steps needed for assessment of climate change effects on positioning of wind power plant station is briefly presented. Figure 2 shows these steps applied in the assessment.

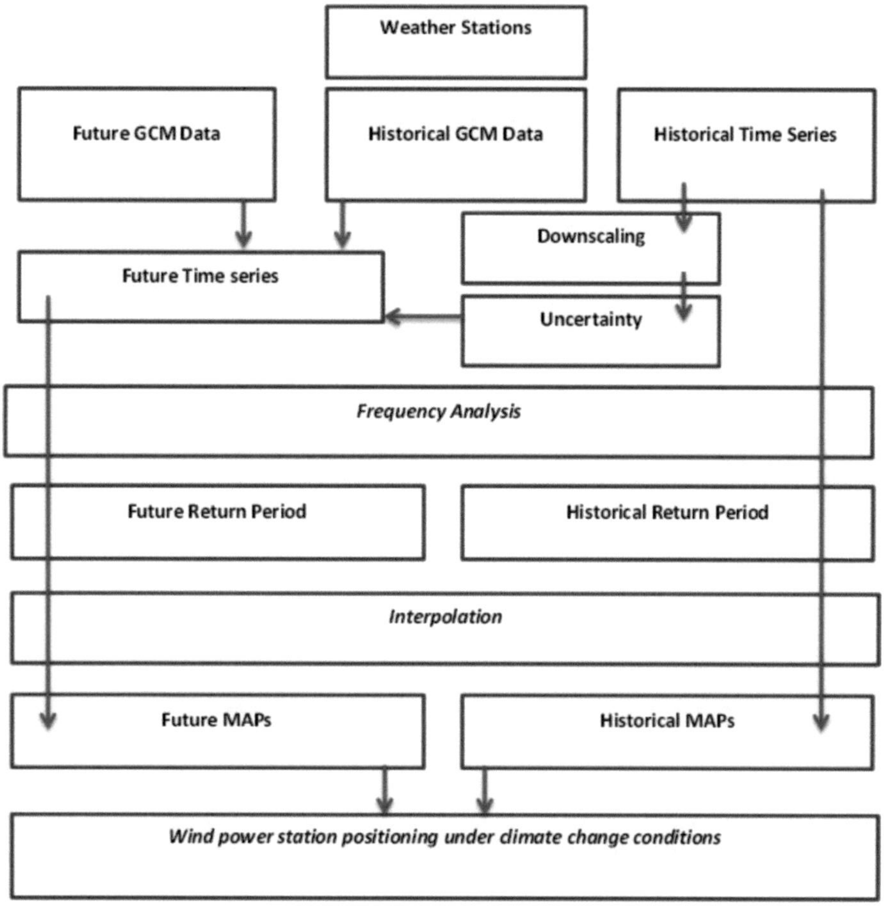

Figure 2. Methodology of wind power positioning under climate change conditions

For this purpose, a region in Southern Khorasan, Iran is chosen. Five synoptic stations are considered as reference stations as displayed in Figure 3.

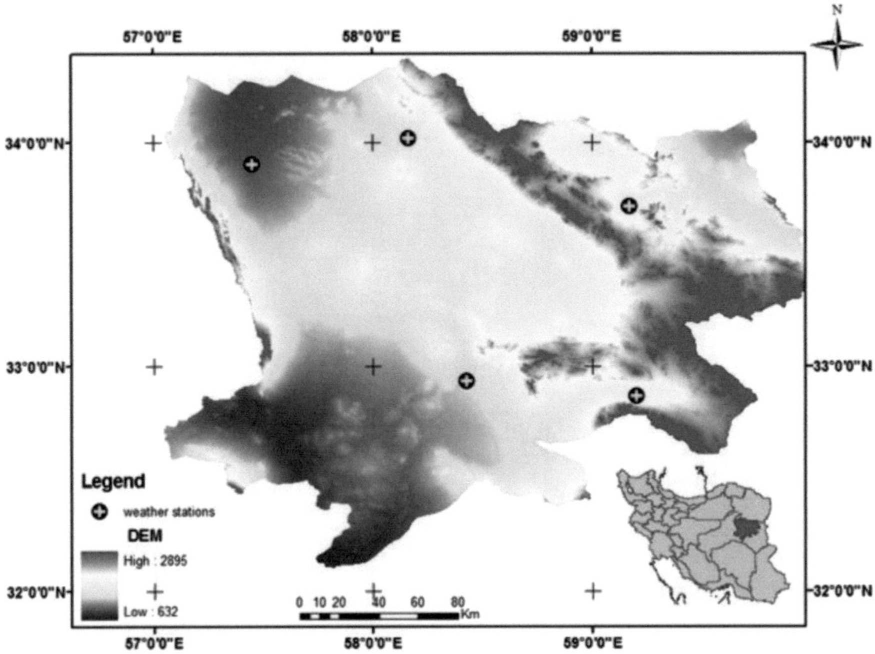

Figure 3. Under study region

Considering high level of sensitivity of the results to different sources of uncertainty in this study, uncertainty analysis was applied with bootstrap method at 95% confidence interval on the results after downscaling steps (Efron, 1993; Khan, 2006; Fakhry, 2012a; Fakhry, 2012b). Figure 3 shows the results of uncertainty analysis of downscaled data.

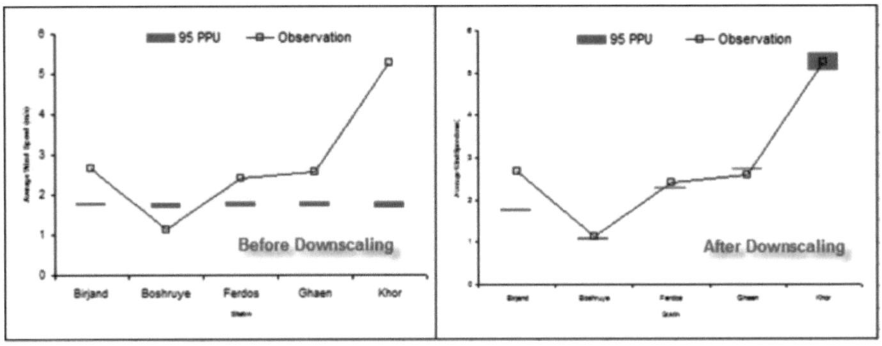

Figure 4. Uncertainty band before and after downscaling

After preparing average wind speed parameter for each station, frequency analysis is performed and the Weibull distribution is selected using L-moments method. Finally, the quantile values corresponding to each return period were derived. Then, geostatistics method was applied for local interpolation and final map for the historical period was prepared.

As the average wind speed was the single used parameter, only one map was drawn. In the case of many existed maps, final positioning could be possible by weighting, according to importance of the maps. Figure 4 shows precedence of potential locations to install wind power station in the region according to long duration of historical records. It is clear from the Figure that the regions with the highest and lowest potential of wind power plant construction are respectively located in the south eastern and north western parts.

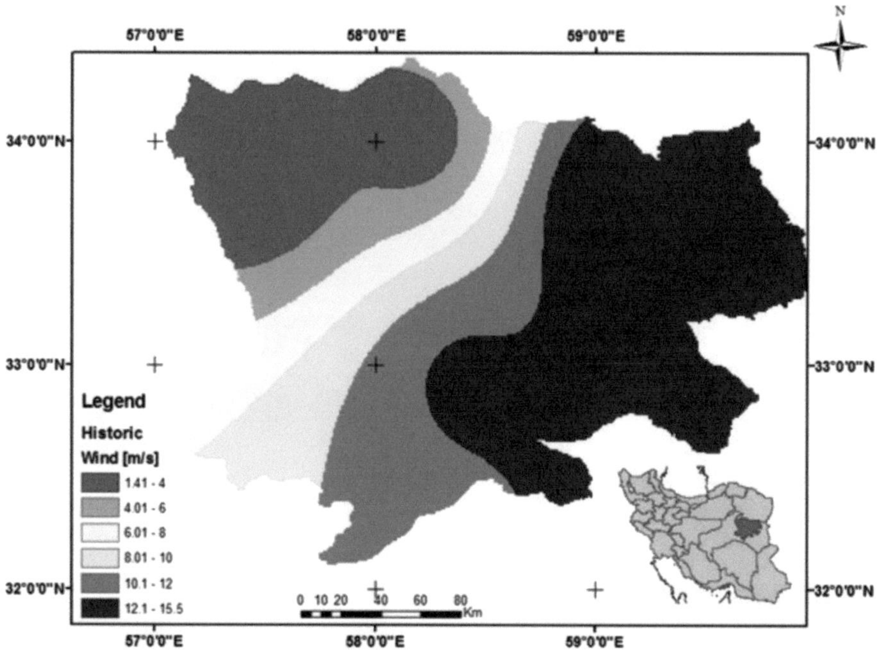

Figure 5. Final historical wind map

To investigate the effect of climate change on positioning process, first the output of HADCM3 model under A2 emission scenario is derived from IPCC website and downscaled by using statistical downscaling techniques. In this study, linear regression method is used and by making wind speed time series of under study stations for near future period (2010-2039), the frequency analysis is performed again and the future map is presented.

As presented in Figure 6, climate change impacts on wind power plant construction in the regions with low priority will be negligible but in high priority ones, the best points are concentrated in the north eastern part.

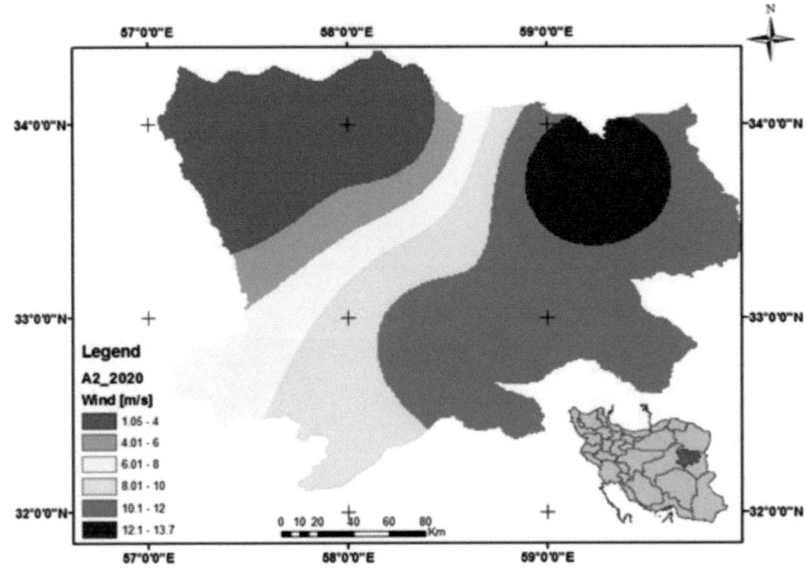

Figure 6. Final future wind map under climate change conditions

20. Conclusions

This chapter made an attempt to initiate discussions on the impact of the climate change on construction of wind farms through various cases that have been presented. Climate change impact studies on wind farms should consider the following principle steps.

This study suggests that the main focus should be on using observational data, time series and frequency analysis and utilizing spatial interpolation methods to create the initial maps, which depict the wind velocity, direction and the average power generation could be calculated. The decision making of the wind farm location could start based on these initial maps and power analysis, which takes various factors into account. The following step would involve down-scaling process of the GCM model for initially selected sites. Comparing the GCM model output for future conditions to the historical maps drawn based on existing data. This process will be instrumental in helping to create a path to coupe with a changing climate and its impact on the wind farms.

Author details

Masoomeh Fakhry[1*], Mohammad Reza Farzaneh[2], Saeid Eslamian[3] and Rouzbeh Nazari[4]

*Address all correspondence to: Masoomeh.fakhri@gmail.com; m.farzaneh@modares.ac.ir; saeid@cc.iut.ac.ir; rnazari@citytech.cuny.edu

1 Department of Hydraulic Structure, Faculty of Water Science Engineering, Shahid Chamran University, Ahwaz, Iran

2 Department of Water Engineering, Tarbiat Modares University, Tehran, Iran

3 Department of Water Engineering, Isfahan University of Technology, Isfahan, Iran

4 Department of Construction Management and Civil Engineering Technology, City University of New York, USA

References

[1] Ackermann Th(2005). Wind power in power system, Royal Institute of Technology, Stockholm, Sweden.

[2] Akday, S. A, Bagiorgas, H. S, & Mihalakakou, G. (2010). Use of Two-component weibull mixtures in the analysis of wind speed in the eastern, Applied Energy: , 2560-2573.

[3] Aldrich, John ((1997). R. A. Fisher and the making of maximum likelihood 1912-1922. Statistical Science, , 12(3), 162-176.

[4] Asif, M, & Muneer, T. (2007). Energy supply, its demand and security issues for developed and emerging economies, Renewable and Sustainable Energy Reviews,, 11, 1388-413.

[5] Bogardi, I, & Matyasovszky, I. (1996). Estimating daily wind speed under climate change, Solar Energy , 57(3), 239-248.

[6] Bonham-carter, G. F. (1994). Geographic Information Systems for Geoscientists: Modelling with GIS, 1st Ed. Pergamon Press, Oxford, UK.

[7] Breslow, P. B, & Sailor, D. J. (2002). Vulnerability of wind power resources to climate change in the continental United States, Renewable Energy: , 27, 585-598.

[8] Carta, J. A, & Ramirez, P. Vela´zquez, S. ((2009). A review of wind speed probability distributions used in wind energy analysis Case studies in the Canary Islands. Renewable and Sustainable Energy Reviews.13. , 933-955.

[9] Chow, V. T, Maidment, D. R, & Mays, L. W. (1998). Applied Hydrology, McGraw-Hill.

[10] Cunnane, C. (1989). Statistical distributions for flood frequency analysis, world meteorological organization operational hydrology, Report WMO-NO.718, Geneva, Swiczerland.(33)

[11] Deepthi, R, & Deo, M. C. (2010). Effect of climate change on design wind at the Indian offshore locations, Ocean Engineering , 37, 1061-1069.

[12] Diakoulaki, D, & Karangelis, F. (2007). Multi-criteria decision analysis and cost-benefit analysis of alternative scenarios for the power generation sector in Greece, Renewable and Sustainable Energy Reviews;, 11, 716-27.

[13] Doria, R, Madramooto, C, & Mehdi, B. B. (2006). Estimation of Future Crop Water Requirements for 2020 and 2050, Using CROPWAT. IEEE.

[14] Efron, B, & Tibshrani, R. J. (1993). An Introduction to the Bootstrap, Monographs on Statistics 57 & Applied Probability, Chapman & Hall/CRC.

[15] Enio, B. Pereira, Fernando R. Martins, Marcelo P. Pes, Eliude I. da Cruz Segundo, André de A. Lyra. ((2012). The impacts of global climate changes on the wind power density in Brazil, Renewable Energy :, 1-4.

[16] Eslamian, S. S, & Feizi, H. (2007). Maximum Monthly Rainfall Analysis Using L-Moments for an Arid Region in Isfahan Province, Iran. J. Appl. Meteorol. Clim. , 46, 494-503.

[17] Fakhry, M, Farzaneh, M, Eslamian, R, & Khordadi, S. S. M, J. ((2012a). Uncertainty assessment of downscaled rainfall to investigate impact of climate change on the probability of flood, Journal of Flood Engineering, , 3(1)

[18] Fakhry, M, Farzaneh, M, Eslamian, R, & Khordadi, S. S. M. J. ((2012b). Confidence interval assessment to estimate dry and wet spells under climate change in Shahrekord Station, Iran, ASCE, Journal of Hydrologic Engineering, (In Press).

[19] Fisher, R. A. (1912). On an absolute criterion for fitting frequency curves. Messenger of Mathematics 41 155 160. CP1 Z. in Bennet 1971, Z., 1

[20] Fisher, P. F. (1989). Knowledge-based Approaches to Determining and Correcting Areas of Unreliabilityin Geographic Databases, Goodchild, M., Gopal, S.: The Accuracy of Spatial Databases,, 45-54.

[21] Greenwood, J. A, Landwehr, J. M, Matalas, N. C, & Wallis, J. R. (1979). Probability Weighted Moments: definition and relation to parameters of several distributions expressible in inverse form, Water Resources Research, , 15(5), 1049-1054.

[22] Harmsen, E, Miller, N. L, Schlegel, N. J, & Gonzalez, J. E. (2009). Seasonal climate change impacts on evapotranspiration, precipitation deficit and crop yield in Puerto Rico. Agricultural Water Management , 96, 1085-1095.

[23] Hiremath, R. B, Shikha, S, & Ravindranath, N. H. (2007). Decentralized energy planning; modeling and application review, Renewable and Sustainable Energy Reviews;, 11, 729-52.

[24] Hosking, J. R. (1986). The theory of probability weighted moments. Res. Rep. RC12210, IBM Research Division, Yorktown Heights, New York 10598.

[25] Hosking, J. R. and estimation of distributions using linear combinations of order statistics. J. Royal Stat. Soc. B , 52, 105-124.

[26] Hosking, J. R. (1991). Approximations for Use in Constructing L-Moment Ratio Diagrams. Res. Rep. RC 16635, IBM Research Division, Yorktown Heights, New York 10598.

[27] Hosking, J. R, & Wallis, J. R. (1991). Some Statistics Useful in Regional Frequency Analysis. Res. Rep. RC 17096, IBM Research Division, Yorktown Heights, New York 10598.

[28] Hosking, J. R, & Wallis, J. R. (1993). Some statistics useful in regional frequency analysis. Wat. Resour. Res. , 29, 271-281.

[29] Hosking, J. R, & Wallis, J. R. (1997). Regional Frequency Analysis: An Approach based on Lmoments. Cambridge University Press, Cambridge, UK.

[30] Hoyme, H, & Zielke, W. (2001). Impact of climate changes on wind behaviour and water levels at the German North Sea coast. Estuarine, Coastal & Shelf Science , 53(4), 451-458.

[31] IPCC(2012). The IPCC Special Report on Renewable Energy Sources and Climate Change Mitigation.

[32] IPCC(2007). in: Alfsen, K., Barrow, Bass, E.B, Dai, Desanker, X.P., Gaffin, Giorgi, S.R.F., Hulme, M.M., Lal, L.J., Mata, L.O., Mearns, J.F.B., Mitchell, T., Morita, R., Moss, D., Murdiyarso, J.D., Pabon-Caicedo, J., Palutikof, M.L., Parry, C., Rosenzweig, B., Seguin, R.J., Scholes, D., and Whetton, P.H., General guidlines on the use of scenario data for climate impact and adaptation assessment. Cambridge University Press, UK.

[33] IPCC(2001). in: Watson, R.T., Zinyowera, M.C., Moss, R.H., Dokken, D.J. (Eds.)., Sepecial Report on The Regional Impacts of Climate Change, An Assessment of Vulnerability. Cambridge University Press, UK.

[34] Jose, A. cart, Penelope Ramirez, Celia Bueno. ((2008). A joint probability density function of wind speed and direction for wind energy analysis, Energy Conversion and Managemant, , 1309-1320.

[35] Karvanen, J. (2006). Estimation of quantile mixtures via L-moments and trimmed L-moments, Computational Statistics & Data Analysis , 51(2), 947-959.

[36] Khan, M. S, Coulibaly, P, & Dibike, Y. (2006). uncertainity analysis of statistical downscaling methods,. journal of Hydrology, , 319, 357-382.

[37] Leander, R, & Bouishand, T. A. (2006). Resampling of regional climate model output for the simulation of extreme river flows. Journal of Hydrology, , 332, 487-496.

[38] Pearson, K. (1902). On the systematic fitting of curves to observations and measurements, Biometrika, , 1, 265-303.

[39] Peltola, H, Ikonen, V, Gregow, P, Strandman, H, Kilpelainena, H, Venalainenb, A, & Kellomaki, A. S. ((2010). Impacts of climate change on timber production and regional risks of wind-induced damage to forests in Finland, Forest Ecology and Management , 260, 833-845.

[40] Pereira Enio BMartins Fernando R., Marcelo P. Pes, Eliude I. da Cruz Segundo, André de A. Lyra. ((2012). The impacts of global climate changes on the wind power density in Brazil, Renewable Energy :, 1-4.

[41] Pryor, S. C, & Barthelmie, R. J. (2010). Climate change impacts on wind energy: A review, Renewable and Sustainable Energy Reviews , 14, 430-437.

[42] Sailor, D J, Smith, M, & Hart, M. (2008). Climate change implications for wind power resources in the Northwest United States, Renewable Energy , 33, 2393-2406.

[43] Sailor, D. J, Hu, T, Li, X, & Rosen, J. N. (2000). A neural network approach to local downscaling of GCM output for assessing wind power implications of climate change, Renewable Energy , 19, 359-378.

[44] Samadi, S. Z, & Mahdavi, M. Sharifi Forood. Bihamta, M.R)(2009). Methodology for selecting the best predictor for climate change impact assessment in Karkheh Basin, Iran, Journal of Environment Engineering and Science, 51(4), 249-256.

[45] Valle Costa CLa Roverea E., Dirk A. ((2008). Technological innovation policies to promote renewable energies: Lessons from the European experience for the Brazilian Case, Renewable and Sustainable Energy Reviews;, 12, 65-90.

Section 4

Biomass

Chapter 11

Biomass Conversion to Energy in Tanzania: A Critique

Mashauri Adam Kusekwa

Additional information is available at the end of the chapter

http://dx.doi.org/10.5772/ 52956

1. Introduction

The United Republic of Tanzania (URT) is the largest country in East Africa in terms of size and population. It is made up by Tanzania Mainland and the island of Zanzibar. It is bordered by the Democratic Republic of Congo, Rwanda and Burundi in the west; Zambia, Malawi and Mozambique in the Southern part, Uganda and Kenya in the Northern side and the Indian Ocean on the East. The country lies between meridians 30^0E and 40^0E and parallels 1^0S and 12^0S.

It has an area of 945,000 Square Kilometres.While about 62,000 square Kilometres of the land is covered by water, including three fresh trans-boundary lakes of Victoria, Tanganyika and Nyasa. Woodlands accounts for 33,500 square Kilometres [1] and arable land suitable for agriculture is concentrated in central part and Southern Highlands of the country, covering about 44 million hectares.

According to URT, Economic survey report 2009 [2], the Tanzanian population was estimated to be 41,915,799 of which 21,311,150, that is about 50.8 percent, were female, while 20,604,730 about 49.2 percent were male. Tanzania mainland had an estimated population of 40,683,294, while Zanzibar had an estimated population of 1,232,505. The population distribution indicated that 31,143,439 of people, about 74.3 percent live in rural areas, while 10, 772, 360 people about 25.7 percent live in urban areas. These estimates are based on the population growth rate of 2.9 percent per annum established out of the Population and Housing Census of the years 2002. [3]

Hydropower, Coal and Petroleum are Tanzania's main source of commercial energy. However, solid biomass energy such as agro residue, forestry residue and wood fuels are used throughout the country and they account for 88 percent of total energy consumption in rural and semi-urban areas [4] while modern commercial energy contribute about 2 percent. Of the 24 mainland regions, Dar es Salaam region has the greatest access to electricity; however, less than 50 percent of all households in the regions are connected.

The conventional energy sector, and in particular the electricity sector has not lived up to expectations of the Tanzanians. The sector is mainly characterized, among other problems, by unreliability power supply, low access levels at about 15% [5] low capacity utilization and availability factor; deficient maintenance of generation transmission facilities and equipment; poor procurement of spare parts due to liquidity problems; and high transmission and distribution losses at 20% [6] are typical problems.

Provision of electricity is largely confined to urban middle and upper income groups as well as the formal commercial and industrial sub-sector. At the moment the energy sector is characterized by large and increasing import of liquid petroleum products, which account for significant proportions of export earnings. Liquid petroleum is used in electricity generation and in the transport sector. The transport sector is the major consumer of liquid petroleum product accounting to about 60% of total consumption. The high liquid petroleum products import bill expose the country's energy sector to the external energy price shocks. Renewable energy such as ethanol would assist in mitigating the negative impact of high liquid petroleum fuel imports

Tanzania is endowed with substantial renewable energy resources [4]. The renewable resource potential in the country has not been fully exploited, but only to limited investment level;. Rural Energy Agency (REA) is making initiatives to disseminate information as the importance of renewable energy in the energy balance of the country. In addition, technical and financial barriers have contributed to the low levels of uptake of Renewable Energy Technologies (RETs) in the country. However, these constrains are being addressed by REA.

2. The Introduction and the Problem

The objective of this chapter is to discuss the potential and implementation of biomass conversion to energy in Tanzania. Generally, the feasibility as well as suitability of the various categories of biomass to energy conversions in the country is presented. Detailed descriptions of potential conversion routes are included with their possible and existing scope of implementation. The most recent statistical data of food, commercial agricultural crops as source of biomass energy are reported from the available sources. Tanzania has abundant and diverse indigenous energy resources which are yet to be fully exploited. The sources include; hydropower, mini-hydro, natural gas, coal, petroleum, wind, solar, and geothermal.

2.1. Hydropower sources

The generation capacity of electricity was on a 60:40 hydro/thermal proportion before 2005. Following introduction and expanded use of natural gas usage in power generation, the hydro-thermal mix is now standing at 41:59 (including the emergency plants).Up to June 2012, electricity installed capacity is about 1,375.74 [7]of which represent about 41 percent is from hydropower sources. The other percentage is from thermal and oil.

http://dx.doi.org/10.5772/ 52956

Out of Tanzania's 41.9 million inhabitants, so far only 14 percent of urban and 2 percent of rural areas are electrified [5], which means that less than 12 percent have access to grid-based electricity or other forms of commercial electricity. The national electricity connectivity is about 14%; though, it is expected that electricity demand will triple by 2020[8]. On the supply side, TANESCO increased connections by almost 66,000 in 2010 bringing the total number of its customers to 868,953 by the end of 2010. REA currently (2011) implements grid extension projects initially benefitting 20,000 new customers [9]. In the current setting demand will therefore even more outpace supply.

Moreover, Tanzania's electricity sector faces another important challenge since it is heavily dependent on hydropower, which means that energy provision cannot be ascertained in times of drought. This was clearly visible in the years 2010/2011, where re-curing droughts effectively removed around 420 MW from a system of around 900 MW, forcing the country to endure a programme of load shedding coupled with unplanned outages. With consequently suppressed sales the countries' utility TANESCO financial situation became increasingly parlous (on top of economic losses for non-productivity to the country as a whole).

This led to the design of an 572 MW Emergency Power Plan at the end of 2011, to be fully fuelled by liquid fossil fuels (HFO, JetA1, diesel) at (fuel) costs varying from 30-43 ct/ kWhThese are to be financed by TANESCO tariff revenues and through government guaranteed loans, leading to an increasing weaker financial position of TANESCO. There is quite some critique on the EPP because it does not take into account planned natural gas supply projects, dispersed capacity owned by the private sector, and power projects to be commissioned already in the short and medium term. In general there is a disconnect between expected power demand (both unconstrained -1089 MW- and constrained) and the total generation capacity (1855 MW) proposed by the EPP.[9]

2.2. Micro/Mini-hydro

It is estimated that 32 GWh per year [10] is generated from smaller systems, many of which are private schemes run by religious missionaries. The potential for micro/mini-hydro is large [4]; however, exploitation is still low because of barriers hindering full exploitation of these potentials.

2.3. Natural Gas

Tanzania has so far made five onshore and shallow water discoveries of natural gas fields in the vicinity of SongoSongo Island, Mnazi bay, Mkuranga, Kiliwani North and Nyuni. Out of the five discoveries, only two gas fields, SongoSongo and Mnazi bay are producing. Mkuranga and Nyuni gas fields have not been assessed. It is estimated that about 27 trillion [11] cubic feet gas is available in the country. Natural gas is expected to become a reliable and economical source of energy to replace petroleum in the near future.

2.4. Coal

Coal reserves in Tanzania are estimated at about 1,200 million tonnes of which 304 million tonnes are proven [12]. Coal sites include Kiwira, Mchuchuma/Katewaka on the south east of Lake Nyasa, and Ngaka in Ruvuma region. Coal has been used in limited quantities for electricity generation as well as in some industries such as cement factories. Low coal consumption in the country is due to part to huge investment costs and quality of the coal itself. However, there is a plan to generate 600 MW from Mchuchuma coal mine in the near future.

2.5. Petroleum

Imported petroleum and related products are widely used in the transport and industrial sector. It also used to generate electricity in isolated grid-diesel power stations that have and installed capacity of 33.8 MW [10]. Petroleum and related by-products are imported by a single company and regulated by EWURA, which controls the price and standard.

2.6. Wind

Based on the available information much of the wind resources in Tanzania is located in the central part of the country, North-East part, and Southern part [4]. Currently wind energy is used to pump water for irrigation and to meet domestic and livestock water needs [10]. Very limited number of attempts has been made to install wind turbine for electricity generation. However, efforts are underway to utilize wind energy in electricity generation. Several companies like Geo-Wind Power (T) limited, Wind East Africa/ Six Telecoms, and Sino-Tanzania Renewable Energy Limited have been licenced to generate electricity of about 500 MW from wind [7].

2.7. Solar

Solar has not been utilized fully as energy sources even though the country being one of the solar belts, that is being a county with 2800-3500 hours of sunshine per year and a global radiation of 4-7 kWh/m^2 per day. Despite the huge solar potential, solar energy has predominately been used only for drying process. In the recent years solar PV technology has been promoted as an energy solution especially in rural areas where there is no access to the national grid. It is anticipated that in the near future, solar as a source of energy will play a great role in rural electrification.

2.8. Geothermal

The country is endowed with a huge geothermal potential, which has not yet exploited [13]. Geothermal power is a reliable, low-cost, environmental friendly, alternative energy supply, indigenous, renewable energy source and suitable for electricity generation. Estimations by analogue method [14] show that the geothermal potential in the country is about 650 MW of which most of the prospects are located within the East African Rift Valley system. Geothermal resource exploitation is a capital intensive investment; hence, private investment is not expected to come in before obtaining detail information on the resources, particularly on their economic viability as potential geothermal energy resource. It is anticipated that the country in order to move from surface assessments to further detailed investigation, public and donor fund will be required.

Other energy sources are petroleum, which makes up 8 percent of total primary energy consumption, natural gas 2.4 percent, and hydropower 1.2 percent. About 6.6 percent of primary energy needs to be imported, primarily from Uganda (8 MW) and Zambia (5 MW) [15]

2.9. Wood fuel and other biomass fuels

Tanzania's energy supply depends mainly on biomass. Since 85-88 percent of the population are not connected to the electricity grid, the overwhelming majority of households use fire wood and charcoal for cooking. As a total, biomass makes up to 88 percent [4] of the total primary energy consumption in Tanzania. Unfortunately, this leads to the deforestation of 100,000 ha per year, of which is very serious since only about a quarter of the Tanzanian land is re-forested

About 50% of the population lives in poverty, out of which 35% is unable to access all of the basic needs including energy services. The poor spend about 35% of their household income on energy while the well-off spends only 14%. Lack of access to modern energy services creates a vicious cycle of poverty for rural communities due to continued limited production opportunities and social facilities. This situation creates a very big challenge to the country. There is a need; to look for an alternative means for assisting the rural poor to have opportunities of accessing to modern energy for reason of alleviating poverty. This chapter is proposing biomass to be one of the alternatives of energy resource which can be employed in modern form to change the situation.

3. Literature review

Biomass is a term used to define all organic matter that is derived from plants as well as animals. Biomass resources include wood and wood waste, agricultural crops and their waste products, municipal and city solid waste, and wastes from food processing, aquatic plants and algae.

Biomass is mainly composed of cellulose suitability of a particular biomass as a potential for energy generation depends on such characteristic; moisture content, calorific value, fixed carbon, oxygen, hydrogen, nitrogen volatiles, as contents, and cellulose/lignin ratio. Generally, cellulose is the largest fraction and constitutes about 38-50% of the biomass by weight. These characteristics are important to determine efficient biomass utilization and are provided in the paper.

Biomass is considered to be one of the key renewable energy resources of the future at both small- and large-scale levels. It already supplies 14 per cent of the world's energy, and if many future projects being assessed, could be implemented, increase the role of biomass in the overall energy system. On average, biomass produces 38 per cent of the primary energy in developing countries (90 per cent in some countries), where it is the largest single energy source like Tanzania. Biomass energy is likely to remain an important global energy source in the next century

Biomass is generally and wrongly regarded as a low-status fuel, and rarely finds its way into energy statistics. Nevertheless, biomass can lay claim to being considered as a renewable equivalent to fossil fuels. It offers considerable flexibility of fuel supply due to the range and diversity of fuels which can be produced. It can be converted into liquid and gaseous fuels and to electricity via gas turbines; it can also serve as a feedstock for direct combustion in modern devices, ranging from very-small-scale domestic boilers to multi-megawatt size power plants.

Biomass-energy systems can increase the energy available for economic development without contributing to the greenhouse effect since it is not a net emitter of CO_2 to the atmosphere when it is produced and used sustainably. It also has other benign environmental attributes such as lower sulphur and NO_x emissions and can help rehabilitate degraded lands.

Despite its wide use, biomass is usually used so inefficiently like firewood (Figure 1) that only a small percentage of useful energy is obtained. The overall energy efficiency in traditional use is only about 5-15 per cent, and biomass is often less convenient to use compared with fossil fuels. It can also be a health hazard in some circumstances; for example, cooking stoves can release particulates, CO, NO_x, formaldehyde, and other organic compounds in poorly-ventilated homes. Furthermore, the traditional uses of biomass energy, i.e., burning fuel wood, animal dung and crop residues, are often associated with the increasing scarcity of hand-gathered wood, nutrient depletion, and the problems of deforestation and desertification

Figure 1. Women carrying firewood in rural Tanzania

http://dx.doi.org/10.5772/ 52956

Figure 2. Biomass heaps (bagasse) in rural areas

There is an enormous biomass potential in the country such (Figure 2) as bagasse that can be tapped by improving the utilization of existing resources and by increasing plant productivity. Bioenergy can be modernized through the application of advanced technology to convert raw biomass into modern, easy-to-use energy carriers (such as electricity, liquid or gaseous fuels, or processed solid fuels). Therefore, much more useful energy could be extracted from biomass. The present lack of access to convenient energy sources limits the quality of life of millions of people, particularly in rural areas. Since biomass is a single most important energy resource in these areas its use should be enhanced to provide for increasing energy needs. Growing biomass is a rural, labour-intensive activity, and can, therefore, create jobs in rural areas and help to reduce rural-to-urban migration, whilst, at the same time, providing convenient energy carriers to help promote other rural industries.

Enhanced biomass availability on a sustainable basis requires support and development of new biomass systems in which production, conversion and utilization are performed efficiently in an environmentally sustainable manner. Efforts to modernize biomass energy should concentrate on those applications for which there are favorable prospects of rapid market development, e.g., biogas, the generation of electricity from residues and biomass plantations through the gasifier/dual-fuel engines route or using advanced gas turbines fired by gasified biomass, and the production of alcohol fuels from sugarcane.

4. Methodology

The methodology used towards accomplishing the project on biomass conversion to energy in Tanzania involved analytical approach, data collection, and analysis.

4.1. Analytical approach

A comprehensive study and review of documents relevant to biomass resources, conversion and application in the country, and other African countries, Asia, Latino America, United Stated of America (USA), Europe, the Middle East and China were made. The aims of the study were to find available and valuable information on the subject. In addition, situation analysis and brain storming on application of biomass, conversion methods with biomass entrepreneurs were discussed.

4.2. Data collection and analysis

Questionnaires were prepared and used in data collection. Interviews were held with leaders at the Regional, City, municipal, wards and subward levels on biomass conversion to modern energy instead of using raw biomass. Interviewees at this level were held with, Mayors, Municipal Directors, and City and Municipal Solid Waste Management experts from all the city councils. Others included relevant Ministries of Health, Lands and Human Settlements Development and Industries. The interviews were undertaken during day times in weekdays and weekends. Interviews were conducted in a tranquil and friendly atmosphere. The information disclosed by the interviewees was treated as confidential.

5. Results

The following findings were obtained from the analytical approach and the interviews:

5.1. Biomass potential

5.1.1. Agricultural biomass

Agriculture is the mainstay of the economy; it employs about 80% of the work force and account for over 50% of gross domestic product (GDP) at factor cost and over 50% of foreign exchange earnings. It is also the major source of food supply and raw materials for industrial sector. Furthermore, it provides the market for industrial sector. Agricultural sector development has been undertaken with the objective of increasing production of food and cash crops in order to improve food security generate and raise income levels to alleviate poverty. Major food crops cultivated in the country include maize, rice, sorghum, cassava, groundnuts, cowpeas, banana, soya beans and sweet potatoes. A part from food crops, commercial crops cultivated include: cotton, sisal, coffee, coconut pineapples, palm oil, cocoa, sugarcane about 80% of farms in the country are less than 2 % owned by small scale holders. All of these crops contribute to biomass potential in the country.

5.1.1.1. Cash crops

Production of main cash crops has been fluctuating depending on whether conditions, availability and usage of agricultural inputs and fluctuation in the World Market. Table 1 gives the trend of cash crops produced in the past 4 years.

CROP	YEARS			
	2006	2007	2008	2009
Cotton	130,565,000	199,954,000	200,662,000	267,004,200
Tobacco	50,617,400	50,784,000	55,356,000	60,990,000
Sugar	290,863,000	279,494,000	276,605,000	279,850,000
Tea	31,348,000	34,763,000	34,770,000	33,160,000
Pyrethrum	2,046,800	1,000,000	1,500,000	3,320,000
Coffee	1,049,900	33,708,000	58,053,000	40,000,000
Sisal	30,847,000	33,039,000	33,000,000	26,363,000
Cashew nut	88,213,000	92,573,000	99,017,000	74,169,000

Table 1. Production of Cash crops in kg (*Source: Ministry of Agriculture and Cooperatives (URT)*)

5.1.1.2.Food Crops

Production of main food crops has been fluctuating depending on whether conditions, availability and usage of agricultural inputs and fluctuation in the World Market. Table 2 gives the trend of food crops produced in the past 4 years.

CROP	YEARS			
	2006	2007	2008	2009
Maize	2,423,000	3,302,000	3,555,000	3,324,200
Rice	805,400	872,000	875,000	885,610
Wheat	109,500	83,000	92,000	93,690
Millet	941,500	1,165,000	1,064,000	203,580
Cassava	2,052,800	1,733,000	1,797,000	1,758,790
Beans	1,049,900	1,156,000	1,125,000	1,183,880
Banana	1,169,200	1,027,000	982,000	990,540
Sweet potato	1,396,400	1,322,000	1,379,000	1,381,120

Table 2. Production of food crops in kg (*Source: Ministry of Agriculture and Cooperatives (URT)*)

5.1.1.3.Oilseed crops

Production of various important oil seeds such as Simsim, groundnuts, sunflower, palm oil and soya for food and Jatropha for petroleum producing continued to be emphasized. Already a policy on Jatropha production for biofuel is in place [4]. Table 3 gives production of oil seed crops for the past 5 years.

CROP	YEARS			
	2006	2007	2008	2009
Sunflower	373,391,000	369,803,000	418,317,000	466,831,000
Groundnuts	783,775,000	408,058,000	396,769,000	385,480,000
Simsim	221,421,000	155,794,000	46,767,000	115,895,000
Soya	5,000,000	3,000,000	3,450,000	3,900,000

Table 3. Production of oilseed crops

There are factors, which determine whether a crop is suitable for energy use. The main material of interest during processing as an energy source relate to moisture contents, calorific value, proportions of fixed carbon dioxide and volatiles, ash content, alkali metal content and cellulose-to-lignin ratio.

5.1.2.Agriculture crop residues

A large amount of agricultural residues are produced in the country. These constitute a potential biomass feedstock for energy conversion. Generally agricultural residue is used to describe all organic materials which are produced as by-product from the harvesting or processing of agricultural crops. These residues can be further categorized into two groups. The first group consisting of the residues which are generated in the field at the time of harvest or field based residues such as rice straw, sugar cane tops etc.

The second group is of those residues that are co-produced during processing or well known as processing based residues e.g. rice husk, cashewnut husk, coffee husk, bagasse, etc. the availability of the first category residue for energy conversion or application is usually low since collection is difficult and they have other uses as fertilizer, animal feeds, etc. Experience has shown that most of the first category residues are left or burnt in the farms. However, the secondary category residues are usually available in relatively large quantities at the processing site or mill and may be used as captive energy source for the same processing mill involving no or little transportation and handling costs. Figure 3 gives selected agricultural residue estimated potential in the country.

These residues have a high potential for energy production and therefore contribute to the energy balance of the country. Major residues generated from harvesting and processing of maize/corn is potential biofuel feedstock. Similarly, the stalk of sorghum which is rich in sugar is a potential feedstock for ethanol production. Figures 4-7 show some of the crops and their corresponding residues available in the country at the moment

Oil palm plantations can be found in Kigoma Region, along the shore of Lake Tanganyika, Western Tanzania. There are three main residues from oil palm processing, namely: empty fruit bunches, shells (Figure 5) and fronds. Empty fruit, bunches are rich in potassium and they can be used as fertilizer. The shells can be used for production of carbon and heating. The fronds are usually used for mulching.

Coffee production in the country is increasing. The husk (Figure 6), which is the main residue generated during processing, can be utilized as an organic fertilizer as well as a source of energy. When compressed it can be used in modern energy generation; at the moment coffee husks are disposed by burning.

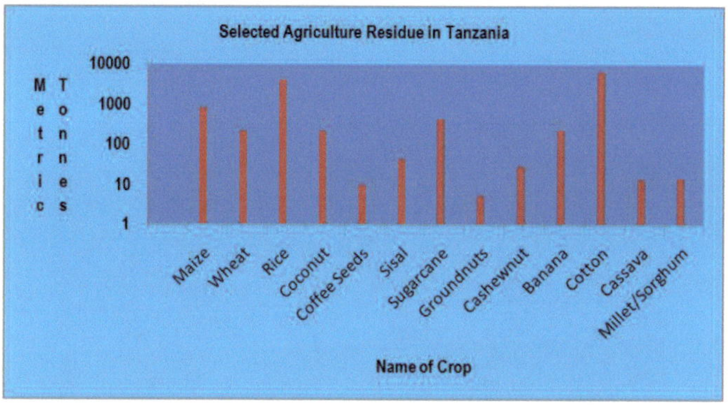

Figure 3. Selected agriculture residue potential in Tanzania (*Source: National Bureau of Statistics, 2006)*

Figure 4. Rice husk (Crop residue)

Figure 5. Palm Oil Shell

Figure 6. Coffee husk (Source of energy)

Figure 7. Coconut shells

The residue from coconut harvestings are mainly the husk and shells. These residues (Figure 7) are a potential source of energy generation. Large plantation of coconut tree can be seen in the coastal area of Tanzania and in Zanzibar

The main residues generated from harvesting and processing sugarcane namely, the tops, baggase and molasses are also sources of energy. However, only bagasse is utilized in co-generation of electricity. The tops and molasses are underutilized even though they are potential source of energy. Biomass contains sugars that are deemed uneconomical to remove. The recent and development in technology can assist in reducing sugar concentrate and hence the residue can be potential source of energy generation.

Rice husk is underutilized, efforts are underway to use rice husk in briquette production. At the moment rice husk are used by brick markers to burn their bricks. Moving across the country in particularly in rice growing areas, large amount of rice husk are seen unutilized. Rice straws are virtually unutilized and could serve as major source of energy in the country.

During harvesting period, most of agricultural residues are burnt on the farms to facilitate the harvesting process or as pest a control measure e.g. cotton some of the residues are also used as a substitute for firewood. However, at the moment there is no adequate information about the share of the agricultural residue in modern energy generation.

5.1.3.Residue chemical composition

Residue chemical composition of some of the agriculture residue was analysed, and the aim of the analysis was to establish if these residues have the characteristics as feedstock for conversion to usable energy using gasification method. The chemical compositions used in the analysis were: proximate analysis (%) dry basis, ultimate analysis (%) dry basis and heating value. Tables 4, 5 and 6 give the proximate, ultimate analyses and higher heating values.

Other agro-residues: Cotton stalks, cassava stalks and straws are mainly used as local fuels in rural areas. Besides, the residues can be used in production of ethanol. Table 7 gives cellulose, Hemicelluloses, lignin and theatrical yield (litres/tones) from some of agro-residue.

Experience gained from agriculture sector, particularly agricultural activities show that the agricultural crops generate considerable amount of residue which can be harnessed for modern energy generation. Annual evaluation of total amount of residue that originates from agricultural activities (Figure 3) is about 13 million tonnes; residues with higher potential for modern energy generation are:

- Maize stalk and straw about 8 million tones
- Rice husk and straw about 4.1 million tones
- Wheat straw about 232,400 tonnes
- Sisal tole and flume 46,080 tonnes
- Bagasse is 447,030 tonnes

Condition	Residue type								
	Rice husk	Rice Bran	Coffee husk	Sisal leaf	Sisal pole	Sisal boles	Palm branch	Palm Stem	Sugarcane Bagasse
Moisture	8.80	7.80	10.10	8.50	10.10	7.50	8.10	9.10	9.00
Volatile Matter	59.20	64.60	83.20	80.20	79.30	84.10	79.60	81.20	75.8
Fixed Carbon	14.60	14.20	14.30	12.60	14.60	12.80	12.60	15.30	20.1
Ash Content	26.20	21.20	2.50	7.20	6.10	3.10	7.80	3.50	4.2

Table 4. Proximate analysis (%), dry basis –Source CEET (2008)

Elements	Residue type								
	Rice husk	Rice Bran	Coffee husk	Sisal leaf	Sisal pole	Sisal boles	Palm branch	Palm Stem	Sugarcane Bagasse
C	35.60	37.80	49.40	47.00	47.00	48.00	45.60	47.50	48.10
H	4.50	5.00	6.10	5.70	6.00	6.00	5.60	5.90	5.90
N	0.19	0.55	0.81	0.14	1.66	0.10	0.19	0.28	0.15
O	33.40	35.40	41.20	39.90	39.10	42.70	39.30	42.50	42.40
Cl	0.08	0.09	0.03	0.04	0.05	0.06	1.33	0.18	0.07
S	0.02	0.05	0.07	0.03	0.13	0.03	0.16	0.13	0.02

Table 5. Ultimate analysis (%), dry basis- Source CEET (2008)

Elements	Residue type								
	Rice husk	Rice Bran	Coffee Husk	Sisal Leaf	Sisal Pole	Sisal Boles	Palm Branch	Palm Stem	Sugarcane Bagasse
HHV [MJ/kg]	13.24	13.93	18.34	17.23	17.20	17.35	16.24	17.38	17.33

Table 6. Heating values

Agro Residues	Cellulose	Hemicellulose	Lignin	Theoretical ethanol yield per litres
Rice straw	41.36	20.36	12.06	39.75
Rice Husk	44.06	17.85	17.20	392.33
Maize cob	36.80	27.90	11.15	418.21
Maize husk	43.39	19.58	22.82	407.03
Ground nut straw	36.55	13.94	31.28	319.96
Sugarcane baggage	33.60	29.00	18.50	404.64

Table 7. Other chemical component

http://dx.doi.org/10.5772/ 52956

5.1.4. *Economic and environmental implications of biomass*

Advantages of using crop residues as energy source are twofold: economic and environmental.

Economic: For the farmer, agricultural residues can be a cash crop. Traditionally, farmers burnt agricultural residue or left them in the field. The market ability of crop residues will boost local/village economies by providing jobs and services.

Environment: the burning of agricultural residues causes air pollution, soil erosion and a decrease in soil biological activity, which eventually leads to low yields. Burning of agricultural residues yield smoke and other pollutants, which adversely affect air quantity, visibility and human and environmental health

5.1.5. *Challenges with agricultural residues are*

- Agricultural resources in the country are largely un- organized, scattered and not evenly distributed.
- Collection, storage, pre-processing and distribution are a big challenge because of low land holdings and low level of mechanization in complete value chain.
- Exact data on consumption of agriculture- residues in the country is not well known. Hence, more studies are needed
- Overall agro-residue conversion is low through existing technology;
- Agriculture-residue is marginalized in the country;

5.1.6. *Agriculture waste*

Apart from agricultural residue from the farms, in Urban and semi-urban areas certain other residues and waste water also constitute a potential source of energy. The agro-processing industries such as fruit processing and vegetable, urban vegetable market places (Figure 8 and 9), road sweepings and road side are areas, which generates significant biomass waste. The management of these wastes are in hands of poor farmers, un-organized sector, rural and semi-urban households. Hence, large amount of these wastes are left to rot in open space resulting into air pollution. Good management of these wastes can contribute to energy generation in the country.

Figure 8. Surplus Tomato Left as Waste

Figure 9. Heaps of Food waste in Urban Tanzania

5.1.7. Forestry residues

Forest residues and wood waste represent a large potential resource for energy generation. They include forest thinning, primary mill residues, (Sao hill forest). Forest residues are left in the forest by harvesting operations (Figure 10) the residue could be collected after timber harvest and used for energy purposes.

The primary advantage of using forest residue for power generation is that an existing collection infrastructure is already set up to harvest wood. Companies that harvest wood already own equipment and transport options that could be extended to collecting forest residue.

Manufacturing operations that produce mill residues usually include sawmills (Urban sawmill), pulp and paper mills (Mufindi paper mills) and other millwork companies involved in producing pulp and other related material.

Primary mill residues are usually in the form of bark, chips, saw dust, slabs etc. These primary mill residues are relatively homogenous and concentrated at one source. Nearly 98 percent of all primary residues in the country are currently used as fuel.

Figure 10. Forest harvesting waste at Sao Hill

There about 80,000 hectares of state owned plantation forest that are mostly linked to state owned wood based panel industry and the pulp and paper industry. It is estimated that there are 25,000 hectares of private owned plantations. In addition, more than 75,000 hectares belongs to villagers, local government, NGOs and civil societies. Hence, the estimate forest residue potential m^3 per year is about 205,400 tonnes. The residue can be used for modern energy conversion.

5.1.8. City and Municipal solid waste (CMW)

Millions of tons of household waste are collected each year with the vast majority, disposed affirm open fields. Table 9 gives quantity of solid waste collected in Tanzania by region. Biomass resource in solid waste comprises paper and plastic. City and Municipal solid waste (CMW) can be converted into energy by direct combustion, or by natural anaerobic digestion in the land fill.

On other land fill sites the gas produced by natural anaerobic digestion, which is approximately 50% methane and 50% carbon dioxide can be collected form the stored material, scrubbed and cleaned before feeding into internal combustion engines or gas turbines to generate heat and power.

The above compiled data clearly shows that large scope exist in the country for the exploitation of bio-crops for their conversion to bio-fuel, e.g. bio-diesel, ethanol, by thermo conversion and bio-chemical conversion routes.

Apart from energy crops, a huge potential exist for energy generation from the various industrial wasters available in the country by bio-chemical routes. Similar, other biomass wastes such as forest residue, crop residue, animal manure and city and municipal waste (Figure 16) also bear a large potential for modern energy generation using bio-chemical as well as thermo-chemical routes. Hence biomass conversion to modern energy such as electricity and fuels may be rewarding for a future developed Tanzania.

5.1.9. Animal Waste

Rural population burn dried animal dung as a fuel, and this is a major source of energy. If a programme is institutes to use the dung as raw material for biogas production, it could benefit the livestock keepers. India for example has pursued a programme to generate biogas from gas with some success. However, at the moment there are more than 6,000 small scale biogas plants operating. The number is small compared to number of livestock the country having (Table 8).

The advantage, from and environmental aspect is that methane that would be released is captured and used to generate heat for cooking purposes. This could reduce the pressure on forests and deforestation. Methane is about twenty times more potent than carbon dioxide as a greenhouse gas and oxidising it while producing usable heat make sense from a climate point of view.

The solid residue remaining from fermentation process could be used as fertilizer in growing other biomass sources such as maize, wheat, cassava, etc.

The challenge is how to quantify the energy potential from animal waste in the country. Advanced investigation is needed. Figure 11 shows cow dung, which is used as source of energy in rural areas.

Figure 11. Cow dung

Decomposition of animal manure can occur either in an aerobic or anaerobic environment. Usually under aerobic condition, carbon dioxide (Co_2) and stabilized organic material are produced. While under anaerobic conditions, methane (CH_4), carbon dioxide (Co_2) and stabilized organic material are produced. Basing on statistical data given in Table 9 the quantity of animal manure produced annually can be substantial for generation of methane (CH_4) and hence energy potential of animal manure is significant. At the moment there are more than 6,000 biogas plants in the country, which use animal manure as their raw material.

Name of Livestock	YEARS			
	2006	2007	2010	2011
Cattle	18,500,000	19,100,000	19,200,000	21,300,000
Goats	13,500,000	13,600,000	13,700,000	14,000,000
Sheep	3,500,000	3,600,000	3,600,000	3,800,000
Chicken	30,000,000	31,000,000	33,000,000	40,000,000

Table 8. Livestock statistic for the country

5.1.10. Food Industry Wastes

The hotels, restaurants, Schools, and community kitchens produce a lot of waste such as vegetable peels, uneaten food, e.g. rice, bread, vegetables, etc., plate and dish washings, fruits and vegetable rejects. Similarly, a huge amount of wastes are generated from confectionary industry. Solid waste from these industries include peelings and scraps from fruits and vegetables, food that does not meet quality control standards, pulp and fibre from sugar and starch extraction, filter sludge and coffee grounds are disposed of and left to lot in the open space. However, all of these wastes make potential feedstock for biogas generation by anaerobic digestion. Usually these wastes are disposed of in landfill dumps [16].

Liquid wastes are generated by washing meat, fruit and vegetables, blanching fruit and vegetables, pre-cooking meats, poultry and fish, cleaning and processing operations and wine making. These wastewaters contain sugars, starches and other dissolved and solid organic matter. There is a potential for these industrial wastes to be anaerobically digested to produce biogas or fermented to produce ethanol.

5.1.11. Industrial Waste

Such waste consists of lawn and tree trimmings, whole tree trunks, wood pallets and any other construction and demolition wastes made from timber (Figure 12). The rejected woody material can be collected after a construction or demolition project and turned into mulch, compost or used to fuel bioenergy plants

Figure 12. Wood waste in Mwanza Municipality

Industrial waste such as bagasse (Figure 13) from sugar plants find application in co-generation process, which generates electricity that is used by the same plant. The excess is supplied to the nation grid

Figure 13. A heap of bagasse at Sugar Factory

6. Benefit of biomass utilization

Biomass is renewable, potentially sustainable and relatively environmentally benign source of energy. It is free from carbon dioxide. Thus, the substitution of fossil fuels for energy generation using biomass will result into a net reduction of greenhouse gas emissions and the replacement of a non-renewable energy source.

Biomass fuels have negligible sulphur content and, hence, do not contribute to sulphur dioxide emissions, which cause acid rain. The combustion of biomass produces less ash than coal combustion, and the ash produced can be used as a soil additive on farm target.

Biomass is a domestic resource, which is not subject to world price fluctuations or supply uncertainties. If well developed in the country, the use of biofuels, such as ethanol and biodiesel, reduces the economic pressures of importing petroleum products.

6.1. Environmental impact of biomass energy

Biomass energy generation systems raise some environmental issues that must be addressed. Issues such as air pollution, impact on forests, and impact due to crop cultivation must be addressed case by case. Produced and consumed in a sustainable fashion, and there is no net contribution of carbon dioxide to global warming when fossil fuels are burnt, carbon dioxide is released that has been stored underground for millions of years, making a net contribution to atmospheric greenhouse gases. Hence, if managed carefully, biomass energy has significant environmental advantage over the use of fossil fuels. An appropriate level of biomass energy use can have less environmental impact than our current use of fossil fuels.

6.2. Barriers

Main barriers to wide spread uses of biomass in the country for modern power generation are:

- Cost
- Low conversion efficiency;
- Feed stock availability;
- Lack of internalization of external costs in power generations;
- Dependence on technology advances from outside instead of development of indigenous technology;
- Competition for feed stock use; and
- Lack of supply logistics.

6.3. Risks

Risks associated with widespread use of biomass are:

- Intensive framing;
- Fertilizers and chemical uses; and
- Biodiversity conservation

7. Biomass conversion

Biomass feedstock can be converted into useful forms of energy using a number of different processes. This is possible in the country because there is potential biomass that could be used for the process. However, before conversion processes can be initiated, factors that influence the choice of conversion [17] have to be established. The critical factors are:-

- The type and quantity of biomass feedstock;
- The desired form of energy i.e. end-use requirements ;
- Environmental standards;
- Economic conditions;
- Project specific factors.

Biomass can be converted into three main products [18].Two related to energy i.e. power or heat generation and transportation, and one as a chemical feedstock. Conversion of biomass to energy is usually undertaken using two main technologies: Thermo – chemical and bio-chemical. Within thermo-chemical conversion four process options are available. The processes are: Direct combustion, Gasification, pyrolysis and liquefaction. Thermo-chemical conversion route is given in Figure 14.

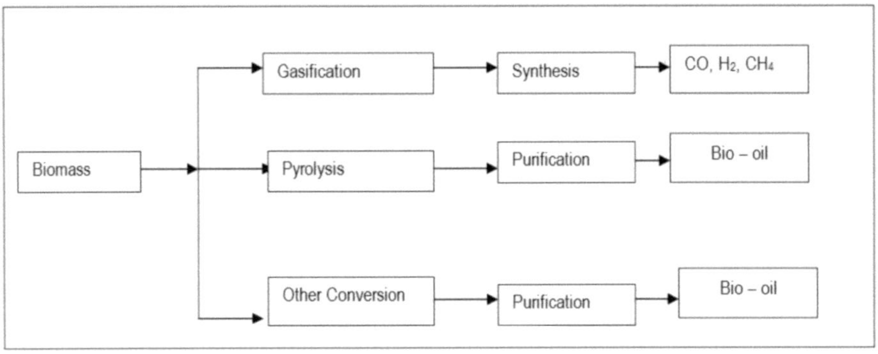

Figure 14. Biomass thermo-chemical conversion route

7.1. Thermo-Chemical conversion

7.1.1. Gasification

Gasification is the conversion of biomass into a combustible gas mixture by the partial oxidation of biomass at high temperature [19] resulting in production of (CO), H_2, and trace of Methane (CH_4). The mixture of these gases is called producer gas. Producer gas can be used to run internal combustion engine, also it can be used as substitute for furnace oil in direct heat applications. The gas can be used to produce methanol-an extremely attractive chemical which is used as a fuel for heat engines as well as chemical feed stock for industries. Since any biomass material can undergo gasification, this process is much more attractive than ethanol production or biogas where only selected materials can produce the fuel. Gasification conversion is suitable for Tanzanian environment.

7.1.2. Pyrolysis

Pyrolysis is the process of converting biomass to liquid termed bio-oil, solid and gaseous fraction, by heating the biomass in the absence of air to around 500°c. Pyrolysis is used to produce predominantly bio-oil. The product i.e. bio-oil can be used in engines and turbine. Obstacle of Pyrolysis is the water dilution [20] of the bio-oil and it's corrosively due to the broad range of organic and inorganic compounds. Hence, the application of bio-oil as a raw material for electricity generation technology is difficult [21].

7.1.3. Other conversions

Other conversions include direct combustion of biomass and liquefaction.

Direct combustion is the process of burning biomass in air. It is used to convert the chemical energy stored in biomass into heat, mechanical power, or electricity using items such a stoves, furnaces, boilers steam turbines, turbo-generators, etc. However, direct combustion does not a fuel suitable for use in gas turbine, etc.

Liquefaction is a process, which tries to clear the large biomass feedstock macro molecules by applying high pressure and how level of heat. Common process parameters of temperature in the range 200-400^{0}C and pressure ranges of 50-200 bar [19, 22], the main products of liquefaction are liquid fuels with similar consistency like that of pyrolysis processes. Given that Liquefaction require high pressure reactor there are only a few commercially available Liquefaction processes. [22].

Ranking of thermo-chemical conversion technology

Table 9 summarizes the findings of performance of thermo-chemical conversion technology and ranking the applicability. The assessments vary from very poor (-), good (+) and very good (+++).

ACTION	PROCESS		
	Gasification	Pyrolysis	Other conversion
Conversion level	+++	++	++
Simplicity	++	++	-
Plant cost	++	++	-
Applicability to scale	+++	++	-
Conversion time	+	+	+

Table 9. Ranking of thermo-chemical conversion Technology

From Table 9, it can be concluded that Gasification process has very good conversion level and applicability. Hence, the process is suitable biomass conversion technology in the country.

7.2. Bio-chemical conversion

Water diluted biomass such as sludge, manure, vegetable waste are difficult to be converted by thermo- chemical conversion process due to difficulties in vaporizing the water present in the biomass. Hence, for feed stock with significantly more than 50% moisture content, it is usually not to apply thermo-chemical technology at comparatively at low temperature is an economic alternative solution.

Two main processes are employed: anaerobic digestion, where biomass is converted by bacteria, and fermentation using yeast to convert biomass. Anaerobic digestion is the standard solution for treating very high dilution levels of biomass, fermentation is used to biomass containing lower amount of water. [20].

7.2.1. Anaerobic digestion

Anaerobic digestion is process of converting of organic material directly into a gas teemed biogas. Biogas is a mixture of methane (CH_4) and carbon dioxide (CO2) with other small quantities of gases such as hydrogen suphide (H_2S) [23]. Anaerobic digestion is a proven technology and is widely used for treating high moisture content organic waste [19]. Biogas a product from anaerobic digestion can be used directly in gas turbine to generate electricity, and can be upgraded to higher quality i.e. natural gas quality by removing carbon dioxide (CO_2).

By- product of anaerobic digestion are settled fibre, which can be used as soil conditioning and liquid fertilizer, which can be used in the farms directly without additional treatment [24-25].

7.2.2. Fermentation

Fermentation process converts biomass into ethanol by the metabolism of microorganisms [26, 20]. The fermentation process is normally anaerobic, but also aerobic process can be feasible. The process consists of two notable steps. First, biomass starch, the sugars are fermented to ethanol using yeast.

The solid residues from fermentation, which still consists of amount of biomass, can then be used for direct combustion or gasification. Typically sugarcane and sugar beet (in Europe) are can theoretically fermented [27].

The final fermentation product allows easier handling and storage when compared to gases produced from anaerobic digestion. However, the intensive feedstock pre-treatment, the necessary temperatures and diluted intermediate product obtained, the fermentation process is complex than anaerobic digestion.

Despite the advantages of storage and transportation, fermentation process is less suitable for micro-scale energy production than gas production technologies. Besides, a major environmental impact of fermentation is the waste water of fermentation process. Treating the waste water can be very energy intensive. The high contents of Nitrate and phosphates in the waste water might influence the development of certain species such as algae.

Ranking of bio-chemical conversion technology

Table 10 summarizes the findings of performance of bio-chemical conversion technology and ranking the applicability. The assessments vary from very poor (-), good (+) and very good (+++).

From the above ranking, it is evident that anaerobic diction is more promising as a biomass conversion technology in the country, especially due to its simplicity.

Gasification and anaerobic digestion are promising conversion technologies in the country. Anaerobic digestion is an excellent technology to produce biogas from wastes in a very small scale i.e. at house hold level. The produced gas (biogas) can be utilized as cooking gas, transportation fuel, and for electricity generation. Gasification is a more demanding technology in small- scale projects with special feed stock requirements.

ACTION	PROCESS	
	Anaerobic digestion	Fermentation
Conversion level	++	++
Simplicity	+++	-
Plant cost	+++	-
Applicability	+++	-
Conversion time	+	++

Table 10. Ranking of bio chemical conversion Technology:

Direct combustion is an ancient technology for heat production purposes. It is a common technology in the country. Pyrolysis is a technology that can be used in large-scale for commercial purposes. The product from pyrolysis i.e. pyrolysis oil, is demanding to upgrade to transport fuel. The pyrolysis oil can be used for combined heat and power generation; however, the pyrolysis process is inefficient.

Fermentation process is a commercial technology but competes with food production. The produced ethanol can be used for heat and power generation and preferably as transportation fuel.

Sensitization on the use of these conversion technologies in the country is required. At the same time training institution should be involved in more research and development aiming at improving the technologies. With this approach, it is clear that the potential of biomass available in the country could contribute to energy mix of the country.

Thermo-chemical and bio-chemical biomass technologies can be summarized in Tables 11 and 12.

	CONVERSION TECHNOLOGY		
	Gasification	Pyrolysis	Direct combustion
Input material	Energy crops, bio waste, Forest, forest product	Energy Crops, mill wood waste, Urban organic waste	Biomass, wood waste
Product	Syngas	Pyrolysis	Heat
By product	Char	Gases and char	Ash
Post treatment	Particulates and tars removal	Oxygen removal	Name
Applications	Synthetic fuel production	Fuel for engine	Heat production, electricity generation.
Direct Environmental Impact	Carbon dioxide, hydrogen Sulphide, NO_X	Carbon dioxide, Carbon monoxide, Methane	Fly and bottom ash, particulates
Indirect environmental Impact	Energy requirement of downstream processes	Energy requirement of downstream processes	Land use impacts of unsustainable forestry practices

Table 11. Summary of conversion technologies

	CONVERSION TECHNOLOGY		
	Anaerobic digestion	**Fermentation**	**Others**
Input material	Biowaste, Wastewaters, by-products, energy crops	Food crops and by-products, forest residues, biowaste, energy crops	-
Product	Biogas	Alcohol, Ethanol	-
By product	Rejects, water	Rejects, water and gases	
Post treatment	Moisture removal	Water removal	-
Applications	Transportation fuel, fertilizer, CHP	Transportation fuel, animal feed, fertilizer	-
Direct Environmental Impact	Energy needs, carbon dioxide emissions	Wastewater, nitrates and phosphates	-
Indirect environmental Impact	Potential water impact of digestate spreading on land	Land use impact	-

Table 12. Summary of conversion Technologies

8. Discussion

A critical analysis of the potential of biomass as an energy source has been presented. The analysis shows that Tanzania has abundant biomass resources for modern power generation. It is evident from the analysis that a large potential exist for exploitation of available biomass to be converted into modern energy. Thus, it is the role of the government, private companies, NGOs, and individuals to increase the share of renewable energies i.e. biomass within the national energy mix. Since biomass is the most important renewable energy source used in the country, the demand for biomass as energy resource will inevitably increase in the near future.

To what extent biomass will penetrate future energy markets in the country; this depends on various aspects; e.g. availability of the resources, the costs of biomass fuels, the development of conversion technologies, cost of converted biomass energy, and social and/or institutional factors. The use of agricultural crop residues, animal waste, and industrial waste as energy sources is a promising opportunity to reduce pressure on energy supply. Since the use is, in the most cases, in compliance with sustainability criteria such as protection of resources, compatibility with environment and climate, social compatibility issues, low risk and error tolerance, and furthermore, it promotes economic efficiency.

Biomass could meet the primary energy demand of the country. And a considerable share could be used for modem power generation within the different conversion routes; the thermal- chemical conversion routes offer opportunities for those residues which are predominantly dry such as rice straw and husk. In view of the availability of waste biomass e.g. food waste, industrial waste, city and municipal solid waste (CMSW) anaerobic digestion is a promising route.

Within the biological conversion technologies, the development of power generation from biogas is at advanced stage. Currently there are more than 6,000 biogas plants in operation. More plants are expected to be in operation in the future. However, awareness on use of biogas in particular to areas with large forks of livestock is still low. This is a challenge to the developers of biogas plants.

There is an increasing interest in gasification technologies for power generation, but a commercial implementation has not yet been received since there are still draw backs such as system reliability, high operation and maintenance cost, which has to be solved first.

Co-generation technology is the only technology at advanced stage of implementation in the country; in particular to sugar processing plants. Electricity generated from these plants is used by the same plants and the excess is supplied to the nation grid. It is anticipated that with "Kilimo Kwanza"[28] initiatives are in the pipe line, production of sugar is expected to increase in the near future; hence more electricity is expected to be generated and supplied into the grid.

Modern energy generation from biomass resources has a great potential in saving for rural energy needs with sustainable benefits. The existing biomass conversion technology such as co-generation, biogas and recently improved thermo-chemical, could be effectively utilized in the process of energy conversion from biomass.

These technologies should be used in the right way to utilize the available biomass energy potential. The power generation from biomass would make the rural areas productivity. The use of local resources would also enhance the employment opportunities and income generation in the rural areas. The available biomass potential in the country should be used to take the nation towards a clear and secure energy source.

9. Conclusion

Biomass is one of the renewable energy sources that can make a significant c contribution to the developing world's future energy supply. Tanzania has a large potential for biomass production. The forms in which biomass can be used for energy are diverse, Optimal resources, technologies and entire systems will be shaped by local conditions, both physical and socio-economic in nature.

Though I have mentioned it numerous times, it bears repeating that the majority of people in the country will continue using biomass as their primary energy source well into the next century. A critical issue for policy-makers concerned with public health, local environmental degradation, and global environmental change is that biomass-based energy truly can be modernized, and that such a transformation can yield multiple socioeconomic and environmental benefits. Conversion of biomass to energy carriers like electricity and transportation fuels will give biomass a commercial value, and potentially provide income for local rural economies. It will also reduce national dependence on imported fuels, and reduce the environmental and public health impacts of fossil fuel combustion. To make progress, biomass markets and necessary infrastructure must be developed with the realization that the large-scale commoditization of biomass resources can have negative impacts to poor households that rely on it for their basic needs. Hence, measures must be taken to ensure that the poor have an opportunity to participate in, and benefit from, the development of biomass markets.

In addition, high efficiency conversion technologies and advanced fuel production systems for methanol, ethanol and hydrogen must be demonstrated and commercialized in the country. Meanwhile, and experiences in industrialized countries should be shared openly. Further, projects must not be concentrated in one region alone. Biomass is obviously a resource that intimately depends on local environmental factors, and experiences gained in other countries will not wholly apply. The benefits of modernized bioenergy systems will only be enjoyed globally if efforts are made to gain experience in a wide variety of ecological and socioeconomic venues.

Biomass can play a major role in reducing the reliance on fossil fuels by making use of thermo-chemical conversion technologies. In addition, the increased utilization of biomass-based fuels will be instrumental in safeguarding the environment, generation of new job opportunities, sustainable development and health improvements in rural areas. The development of efficient biomass handling technology, improvement of agro-forestry systems and establishment of small and large-scale biomass-based power plants can play a major role in rural development. Biomass energy could also aid in modernizing the agricultural economy. A large amount of energy is expended in the cultivation and processing of crops like sugarcane, coconut, and rice which can met by utilizing energy-rich residues for electricity production. The integration of biomass-fuelled gasifier in coal-fired power stations would be advantageous in terms of improved flexibility in response to fluctuations in biomass availability and lower investment costs. The growth of the bioenergy industry can also be achieved by laying more stress on green power marketing.

10. Recommendations

Biomass plays an important role for the energy sustainable development in the country; the potential of biomass is huge, however, its conversion to modern energy is still low. Thus, the following recommendations are proposed:

- Some difficulties which are still faced in the increase of biomass conversions should be minimized..
- Academic and Research institution should play an important role in accelerating biomass utilization and conversion to modern energy.
- The Research and Development collaboration among researchers in East Africa community (EAC) members and SADC region should be developed and realized.
- A biomass user network among East African community should be established to deal with biomass utilization.

http://dx.doi.org/10.5772/ 52956

11. Glossary

URT-United Republic of Tanzania

REA- Rural Energy Agency

RETs-Renewable Energy Technologies

TANESCO- Tanzania Electric Supply Company Limited

HFO- Heavy fuel Oil

GPD- Gross Domestic Product

C-Carbon

H-Hydrogen

N-Nitrogen

O-Oxygen

Cl-Chlorine

S-Sulphur

CH_4-Methane gas

CO- Carbon monoxide

CO_{2-}Carbon dioxide

NGOs-Non – Government Organizations

CMSW- City and Municipal Solid Waste

MSW-Municipal solid waste

MEM-Ministry of Energy and Minerals

CHP- Combined Heat and Power

Anaerobic- Digestion Combustible gas called biogas produced from biogas through low temperature biological processes

Bagasse-The fibre residue that remain after juice extraction from sugarcane

Bioenergyhe- conversion of biomass into useful forms of energy such as heat, electricity and liquid fuels

Biogas-The common name for a gas produced by the biological process of anaerobic (without air) digestion of organic material

Biomass-Organic, non-fossil material of biological origin constituting an exploitable energy source

Carbon Dioxide (CO_2)-The gas formed in the ordinary combustion of carbon, given out in the breathing of animals, burning of fossil fuel, etc. Human sources are very small in relation to the natural cycle

Commercial Energy-Energy supplied on commercial terms; distinguished from non-commercial energy comprising fuelwood, agricultural waste and animal dung collected usually by the user

Energy crops-Crops designed either exclusively for biomass energy feedstock or for the co-production of energy and other agricultural products

Ethanol-Clean burning high efficiency fuel produced from fermentation of biomass that can substitute for conventional liquid petroleum fuels such as gasoline and kerosene

Fossil Fuel-A device that produces electricity directly from chemical reactions in a galvanic cell wherein the reactants are replenished

Gasification-Combustible gas called producer-gas produced from biomass through a high temperature thermochemical process. Involves burning biomass without sufficient air for full combustion, but with enough air to convert the solid biomass into a gaseous fuel

Methane (CH_4)-Gas emitted from coal seams, natural wetlands, rice paddies, enteric fermentation (gases emitted by ruminant animals), biomass burning, anaerobic decay or organic waste in landfill sites, gas drilling and venting, and the activities of termites

Photovoltaic-The use of lenses or mirrors to concentrate direct solar radiation onto small areas of solar cells, or the use of flat-plate photovoltaic modules using large arrays of solar cells to convert the sun's radiation into electricity

UNIT

MW-MegaWatt

Author details

Mashauri Adam Kusekwa*

Address all correspondence to: kusekwa_adam@yahoo.com

Electrical Engineering Department, Dar es Salaam Institute of Technology, Tanzania

References

[1] Casmiri, Damian. (2009). Energy Systems: Vulnerability-Adaption-Resilience. Regional Focus: Sub-Saharan Africa-Tanzania. *56, rue de Passy-75016 Paris-France.*

[2] URT. (2009). Economic Survey report.

[3] URT. (2002). Population and Housing Sensor.

[4] Kusekwa, M. A. (2011). A Review on the Renewable Energy Resources for Rural Application in Tanzania. Renewable Energy-Trends and Applications 978-9-53307-939-4 10.5772/25715

[5] Mrindoko and Mbise. (2011). Annual Engineers Day. *Dar es Salaam (AED), Tanzania.*

[6] SADC. (2009). Southern African Development Community Electrical Report.

[7] MEM. (2012). Budget report 2012/1013.

[8] Magessa, F. (2005). Contribution of Tanzania GVEP activities in Achieving Millennium Development Goals. Retrieved October 27, 1010 from http://deafrica.net/workshops/Tanzania%201/Tanzania%20GVEP.ppt

[9] MEM. (2011). Final report on joint energy sector review for 2010/2011.

[10] Gwang'ombe, F. (2004). Renewable Energy Technologies in Tanzania-Biomass Based Co-generation. *AFREPREN/FWD, Nairobi, Kenya.*

[11] TPDC. (2012). Press release.

[12] Kassenga, G. R. (1997). Promotion of Renewable Energy Technologies in Tanzania. Resources, Conservation and Recycling. , 19(4), 257-263.

[13] Mnjokava, T. T. (2008). Geothermal Exploration . *in Tanzania-Status Report Lake Naivasha, Kenya.*

[14] Jacob, Mayalla., Kato, Kabaka., Ngereja, Mgejwa., & Gabriel, Mbogoni. (2011). Geothermal Development in Tanzania. *Kenyatta International Conference Centre, Nairobi, Kenya.*

[15] Frey, I. (2009). Business Guide Ermeuerbare Energien Tanzania. *Berlin: GTZ*, 23.

[16] Zafar, S. (2009). Biomass wastes, alternative energy. *eMagazine-Altenergy magazine.*

[17] Gavrilescu, M., & Christi, Y. (2005). Biotechnology-a sustainable alternative for chemical industry". *Biotechnology Advances*, 23, 471-499.

[18] Mckendry, P. (2001). Energy production from biomass part (1): Overview of Biomass. Bioresource Technology , 83, 37-46.

[19] Mckendry, P. (2002). Energy production from biomass part (2): Conversion Technologies. *Bioresource Technology*, 47-54.

[20] Loeser, M., & Redfern, M. A. (2008). Micro-Scale Biomass Generation Plant Technology: Stand-Alone Design for remote customer. *16th European Biomass Conference and Exhibition, 2-6 June, Valencia, Spain.*, 1468-1477.

[21] Chiaramonti, D., Oasamaa, A., & Solantausta, Y. (2007). Power Generation using fast pyrolysis liquid from biomass. *Renewable and Sustainable Energy Review,* 11(6), 1056-1086.

[22] Demirbas, A. (2000). Mechanisms of liquefaction and pyrolysis reaction of biomass. *Energy conversion and Management*, 633-646.

[23] EU. (1999). Biomass Conversion Technologies. *EUR 18029 EN*, 9-28285-368-3.

[24] Chynoweth, D. P., Owens, J. M., & Legrand, R. (2001). Renewable methane from anaerobic digestion of brewery effluent. *Renewable Energy*, 22(1), 1-8.

[25] Dagnall, S., Hill, J., & Pegg, D. (2000). Resource mapping and analysis of farm livestock manures-assessing the opportunities for biomass-energy schemes. *Bioresource Technology*, 71(3), 225-234.

[26] Pangraz, E. (2008). Biomass and Waste-to-Energy Technologies. *Environmental Impact assessment, Report based on the activities of technology Development and Adaption of the Micro Energy to Waste (MicrE) Northern Periphery Programme project.*, http://nortech.oulu.fi.

[27] Jenner, M. (2006). Turning local biomass into new energy options. *BioCycle*, 62-70.

[28] Kilimo Kwanza Initiative. (2012). http://www.tanzania.gov.tz.

Section 5

Ocean Energy

Chapter 12

Ocean's Renewable Power and Review of Technologies: Case Study Waves

Ehsan Enferad and Daryoush Nazarpour

Additional information is available at the end of the chapter

http://dx.doi.org/10.5772/53806

1. Introduction

In last decades, in effect of high price of fossil fuel, environmental pollution due to fossil fuel utilization and greenhouse effect, renewable energy resources are considered as an alternative energy resource to the World's excessive energy demand. Nowadays, different technologies are utilized to energy generation from hydro power, fuel cell and hydrogen, biomass, geothermal, solar thermal, photovoltaic and wind, while the technology for converting ocean powers are still in infancy. The aim of this chapter is to introduce potential renewable power sources of ocean, mostly ocean wave power, as well as available technologies for extracting wave power. Due to high energy amount available in ocean, the issue has a strong importance to investigate. Furthermore there are variety of technologies that are developed for harnessing wave power each of which has an individual mechanism. Harvesting ocean wave power and converting to electrical power is a challenge for marine, mechanical, electrical and control engineers and we hope to give essential information about ocean wave, methods of energy extracting from wave and related electrical equipment.

1.1. Ocean

The oceans contain 97.2% of total world water which are covering 71% of Earth's surface [1]. Also the oceans intrinsically are couple with atmosphere via air-water interface and they exchange heat, moisture, momentum and trace constituents by means of air-water interface [2]. The fundamental processes that transfer energy from atmosphere to ocean are energy input to ocean by wind and net surface heat flux [3]. Furthermore, ocean absorbs heat of geothermal energy via geothermal vent in ocean bed. So that, oceans are vigorous and ubiquitous sources of renewable energy which contain 93100 TWh of energy annually [4]. Energy in oceans comes in various forms such as tides, surface wave, thermal gradient, ocean circulation and salinity gradients. It is apparent that ocean with its high amount of energy and global realm on Earth surface can be appropriately utilized for generating electric power. To date, diverse technologies has been developed for extracting different energy forms of ocean most of which are in infancy stage and there is a challenging road before scientists and engineers to generates electricity from ocean in a cost-effective manner.

1.2. Ocean waves

The most potent form of ocean energy is ocean wave. According to the International Energy Agency report, It is estimated that ocean waves have approximately 10000-15000 TWh of energy annually [4], also last researches clarified that 2.11±0.05TW (to 95% confidence) of ocean waves power facing total coastlines of the world [5].

Ocean wave is created by wind, as a byproduct of the solar energy. As solar energy is converted to wind energy, the time-averaged power flow is spatially concentrated, from an intensity of typically 0.1–0.3 kW/m^2 horizontal surface of the earth to 0.5 kW/m^2 envisaged area perpendicular to wind direction. As wind energy is converted to wave energy, even more spatial concentration takes place. Just below the ocean surface, average power flow intensity is typically 2–3 $\mathrm{kW}/\mathrm{m}^2$of envisaged area perpendicular to direction of wave propagation [6]. Because of increase in power intensity, the wave energy is more persistent than wind energy or solar energy. In addition to wind, passing ships and subsea earthquakes generate waves while their contribution in total generated ocean wave power is negligible in comparison to wind.

When wind generates disturbance on ocean surface, gravity or surface tension act as restoring forces that tend to drive water toward its equilibrium state consequently ocean waves manifest themselves. Ocean waves can travel thousands of kilometers with little energy loss. When a wave is propagating, the water particles are not traveling, in fact they move clockwise around a small ellipse with the same period as the progressive wave which drives the motion [7]. The ellipse has its axes vertical and horizontal. By approaching to the seabed, the ellipses become progressively thinner in the vertical direction. At the sea bed, the water particles slip back and forth horizontally. Since the entire particle paths are closed loops, there is no net mass transport by the wave. Equation of water particles' pathway while taking part in a gravity wave motion is as follow;

$$\left(\frac{X_1}{\cosh(Z_0+h)}-\frac{A_I gk}{\omega^2\cosh(kh)}\sin(kX_0)\right)^2+\left(\frac{Z_1}{\sinh(Z_0+h)}+\frac{A_I gk}{\omega^2\cosh(kh)}\cos(kX_0)\right)^2=\left(\frac{A_I gk}{\omega^2\cosh(kh)}\right)^2 \quad (1)$$

Where A_I is amplitude of wave, X_0 and Z_0 are the initial position of water particle in x (wave propagation direction) and z (gravity acceleration direction) directions in Cartesian coordinates respectively, h is ocean depth, ω is angular velocity of ocean wave harmonics, g is gravity acceleration and k is wave number which is achieved from dispersion relation. According to the Eq. (1), by approaching to the sea bed $Z_0 \rightarrow -h$ so that $\sinh(Z_0 + h)$ and Z_1 tend to zero and water particle displacement abates in vertical direction.

Another important characteristic of gravity waves is propagation velocity of ocean waves. The importance is due to intrinsic relation of ocean wave power with group propagation velocity. Angular velocity for different harmonics of ocean waves is calculated according to the Eq. (2), so called dispersion relation, and phase velocity is achieved from Eq.(3);

$$\omega^2 = gk\tanh(kh) \tag{2}$$

$$v_p = \frac{\omega}{k} = \sqrt{\frac{g\lambda}{2\pi}\tan(\frac{2\pi h}{\lambda})} \tag{3}$$

In above equation, v_p is phase velocity of ocean wave harmonics and λ is wavelength of ocean wave. As Eq. (3), the phase velocity depends on ocean depth so that a distinct wave propagates in different depth of ocean with different velocities. There are two possible values for phase velocity. In shallow water, where ocean depth is significantly less than wavelength ($h \ll \lambda$), the phase velocity is;

$$v_p = \sqrt{gh} \tag{4}$$

And phase velocity in deep water, where ocean depth is more than wave length ($h \gg \lambda$), is as follow;

$$v_p = \sqrt{\frac{g\lambda}{2\pi}} \tag{5}$$

With respect to Eq. (4), in shallow water, wave is not dispersive and various harmonics of wave propagates with same velocity, while in deep water, according to Eq. (5), waves are dispersive. It means wave with different wavelength propagate with different velocities proportional to their wavelength.

Total power that is carried by one harmonic of wave in unit length of wave crest, called wave energy transport, is as below equation;

$$J = \frac{\rho g^2 A_I^2}{4\omega}\left(1 + \frac{2kh}{\sinh(2kh)}\right)\tan(kh) = \frac{\rho g A_I^2}{2} v_g \tag{6}$$

While, $\rho$is density of ocean water and v_g is group velocity of wave that is equal to phase velocity in shallow water$v_g = v_p = \sqrt{gh}$ and in deep water is equal to half of phase velocity

$2v_g = v_p = \sqrt{g\lambda / 2\pi}$. According to different wave propagation velocity in deep and shallow waters and related energy transport by wave in these environments and by considering various methods of extracting wave power by different devices and related commercial, installation and maintaining issues, ocean waves study is divided to three different areas called; shoreline, near-shore and off-shore.

Off-shore is a location of ocean in where depth is more than 40 meters. In this location ocean waves have the most power.

Near-shore is location with ocean depth of 10-30 meters and typically has a distance of 0.5-2Km from coastline. In near-shore, seabed fraction is the major source of incident wave power reduction. For instance, in location with ocean depth of 10 meters, different harmonics of ocean waves losses 2-10% of their total power [8].

Shoreline is the location of ocean where depth is less than 10 m. In this location most of the wave power is declined due to seabed fraction and wave breaking.

Not only amount of wave power is various in off-shore and in-shore (shoreline and near-shore) but also ocean wave power is not uniformly separate in all oceans. Fig. 1 illustrates global distribution of wave power density [5]. The arrows on the plot show the mean best wave propagation direction. This figure represents that most of wave power is concentrated in western part of continents which is due to west to east winds. The highest levels in the Northern Hemisphere are off the west coast of the British Isles, Iceland and Greenland, with somewhat lower energy levels in the Pacific off the western seaboard of the US and Canada in Southern Hemisphere Chile, South Africa and the entire south and south west coasts of Australia and New Zealand.

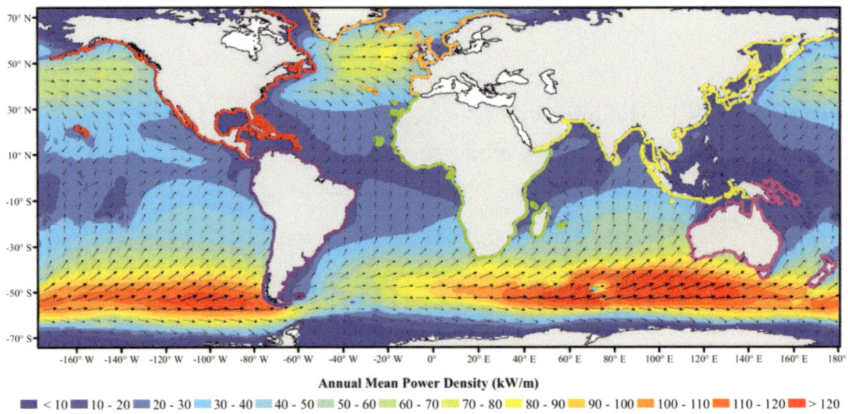

Figure 1. Global annual mean wave power density and annual mean best direction (arrow) [5].

In other point of view, ocean wave power is denser between 40º-60º latitudes in both Northern and Southern hemisphere. Stephen Barstow et al. represented relevance of annual mean ocean wave power density to latitude as Fig. 2. It is shown that ocean wave power is mostly travelling between 40º-60º latitudes [9].

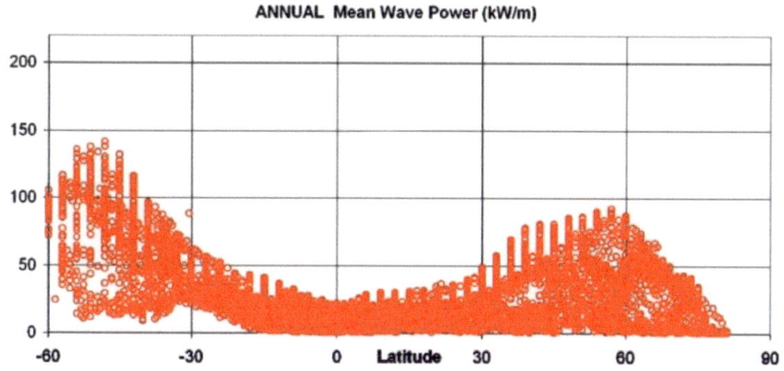

Figure 2. Global annual mean wave power density with respect to latitude [9].

Ocean waves are variable in different time scales. The average wave energy for a winter month can be 5–10 times the mean value for a summer month. The wave energy can vary 10 times from one week to the next. The wave energy during one storm can be five times higher than the mean value for the week the storm occurs. Wave energy in a wave group can be up to 50 times the wave energy between wave groups [10]. In Fig. 3, Monthly mean ocean wave power is plotted for different months of year. According to this figure ocean wave power in both Hemispheres is significantly higher in winter season in comparison to summer. Also seasonal variation of wave power in southern Hemisphere is lower than northern one.

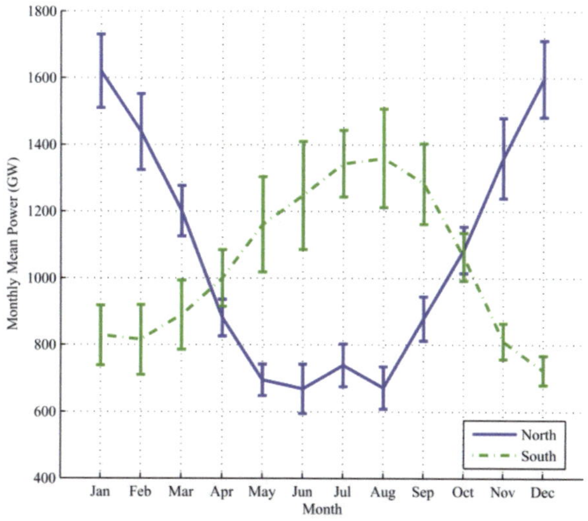

Figure 3. Monthly mean wave power for Northern and southern Hemispheres [5].

1.3. Wave measurement and prediction

Ocean waves are variable in different time scales. Not only wave elevation and wave length (wave power) but also wave propagation direction varies with time. These detailed statistics are important for designing and controlling of particular Wave Energy Convertor (WEC). In most of cases, WECs extract wave power with oscillation in special directions and these WECs absorb maximum power whereby are adjusted in the resonance frequency, a restricted frequency span that WECs oscillation frequency is equal to exciting wave frequency. Implementing upcoming characteristic parameters of ocean wave (amplitude, propagation direction and wave length or period) has a significant impact on capture width of WECs. This information can be used for generator speed control. In all WECs, rotating or linear electric generator is utilized for generating electrical power, implementing information of upcoming wave can be used for speed control of generator and adjusting system in resonance frequency [11]. Briefly, measurement and anticipation of upcoming ocean wave parameters play a significant role in controlling of a WEC. To date, there are various instruments for measuring and predicting ocean wave some of which are described as follow;

1.3.1. Buoy

Buoys are the oldest method for measuring wave parameters. Buoys can measure wave amplitude and wave period but for detecting wave propagation direction at least two buoys are needed. Different sensors have been used on buoys for measuring such as down looking laser profiles, current meter triplets, pressure transducers.

However, when using buoys to measure the waves, these tend to drift with the water particle motion and give imprecise spatial measurement (see Fig.4 left side). In order to solve these problem different methods have been offered as an example see ref. [12]. In Fig. 4 right side a measurement buoy named WaveScan is depicted, this buoy measures heaven and sway acceleration [9].

1.3.2. Synthetic aperture radar

It has been amply demonstrated that synthetic aperture radar (SAR) data can be used to estimate parameters of the two-dimensional (2-D) sea surface elevation field [13, 14]. Due to their high spatial resolution and all-weather and daylight capabilities, spaceborne SAR systems are the only sensors that can provide directional ocean wave information on a continuous and global scale.

The SAR radar sends a pulse down to the ocean surface at nadir. The significant wave height is obtained from the slope of the leading edge of the return pulse, while the total backscatter gives us the wind speed. The major goals of the variable SARs were applications in ocean wave research and wave forecasting. Two-dimensional ocean wave spectra can be derived from SAR images by inversion of the SAR imaging mechanism.

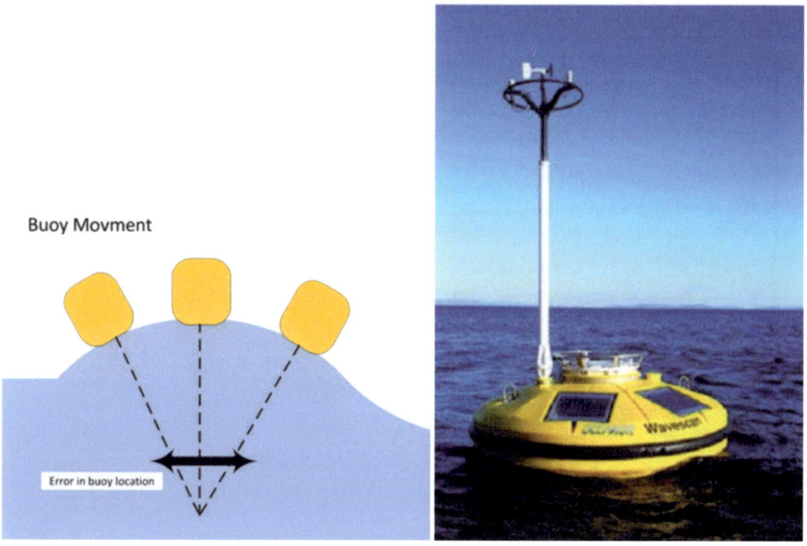

Figure 4. Left side; Error in measurement due to buoy movement. **Right side;** The WaveScan buoy [9].

2. Wave energy convertors

Ocean Waves has been considered as a source of energy since late 18th century and the first patent for capturing wave power was filled in France in 1799 by a father and son named Girard [15]. After World War I when petrol became most important source of energy the interest for harnessing ocean wave power faded. In 1940s, the Japanese wave power pioneer Yoshio Masuda developed an innovative device for absorbing wave power which is known as Oscillating Water Column (OWC) [16]. Oil crisis in 1973 was a great stimulator for governmental funds and researches in wave power extracting but petrol price decline in early 1980s abated interest in wave power [15]. Recently, the Kyoto Protocol on reduction of CO_2 emission, has intensified the interest in this field among researchers. To date, there are various devices which capture power from ocean waves [16, 17]. These devices are distinct in installation location (Off-shore, Near-shore and shoreline) as well as manner of energy harvesting. This section categorise Wave Energy Convertors (WECs) based on manner of interaction with ocean waves.

Excising WECs harvest wave power based on two different principles, Oscillating-WECs and Interface-WECs.

Oscillating-WECs are devices that are in direct contact with ocean waves and oscillate in specific direction related to device design and degree of freedom. J. Falnes, one of the distinguished pioneers in ocean wave power absorption, has been clarified that there are six degree of freedom for a body to oscillate with ocean waves [18]. According to Fig. 5, Oscillation in direction of 1st, 2ed and 3rd arrows are known as surge, sway and heave respectively. Also rotation around x-axis (mode 4), y-axis(mode 5) and z-axis (mode 6) are named respectively roll, pitch and yaw. Oscillating-WECs are displaced in only these six modes, most of these WECs have oscillation in only one of these modes which are known as single-mode oscillators and some others have oscillation in more than one mode that are regarded as multi-mode oscillators. (Note that in most of Oscillating-WECs there is no variation in y direction, in this case there are only three modes for body oscillation; Surge, Have and Pitch).

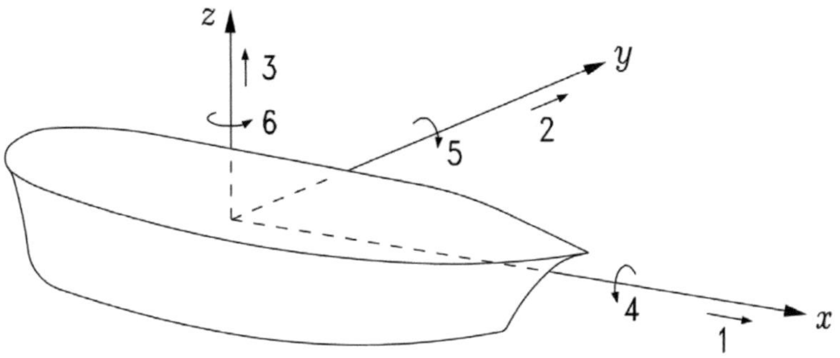

Figure 5. Different oscillation mods of a Wave Energy Convertor: Surge (1), Sway (2), Heave (3), Roll (4), Pitch (5) and Yaw (6) [18].

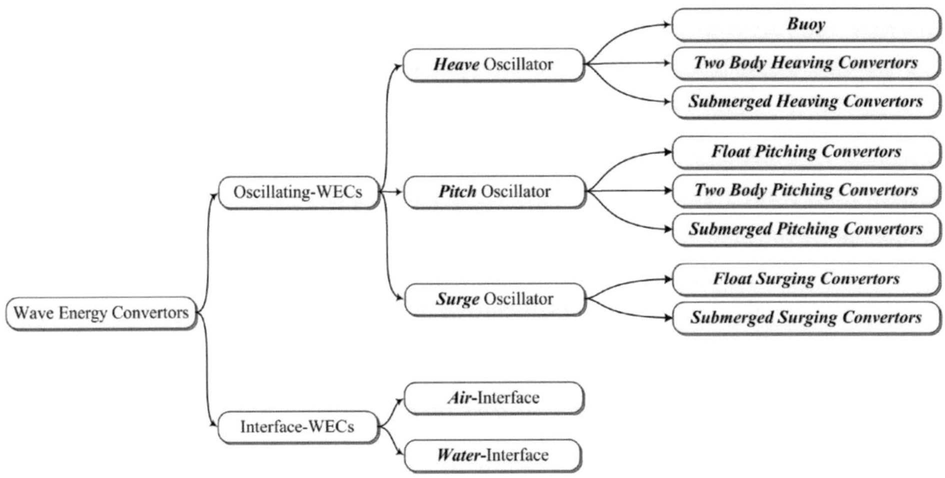

Figure 6. Categories of Wave Energy Convertors.

Interface-WECs are special forms of WECs that are not oscillating in aforementioned modes. In fact these WECs are devices or structures that are fixed in a location and have not interaction with ocean waves. Interface-WECs are used to deliver wave power by an interface (water or air) to the PTO (power take-off). The scheme of classification of WECs is illustrated in Fig. 6.

2.1. Oscillating-WECs

In this subsection, different WECs, which have one or two degrees of freedom, are investigated. Indeed, these WECs are simple in motion due to their restricted motion modes and they can be divided to three distinct categories; Heave Oscillators, Pitch Oscillators and Surge Oscillators.

2.1.1. Heaving oscillators

Heave Oscillators are the simplest oscillators for absorbing wave power. These devices, extract wave power with motion in perpendicular direction to the Sea Water Level and according to the mooring and working principle are divided to three groups; *Buoys, Two Body Heaving Convertors* and *Submerged Heaving Convertors* [16].

2.1.1.1. Buoy

Point absorbers or buoys are convertors with one floating body in sea level which is connected to the PTO (Power Take-Off) via a steel structure or a cable (translator). The body fluctuates with ocean waves in earth gravity direction (heave) which cause the steel structure or cable to oscillate with it hence the bidirectional movement of buoy is transferred to the PTO in the other point of translator and PTO generates electricity. To date various buoys have been designed for wave energy harvesting which are same in principal but different in detail. One of these is a buoy that was designed by Budal et al. in Norway [19]. The device was linked to anchor on the sea bed via universal joint (see Fig. 7). An air turbine was implemented on the device for energy converting and it was controlled by latching control.

Figure 7. Norwegian buoy (courtesy of J. Falnes) [16].

Figure 8. Left side; Taut moored heaving buoy with linear electrical machine PTO [16]. **Right side;** Component of tout moored WEC, Sweden [22].

Another buoy type is floating body connected to bottom fixed structure via a cable (taut moored). Due to cable flexibility, this device is not restricted in heave mode and it also has oscillation in surge direction. PTO of device is fixed in ocean bed and cable transfer WEC motion to PTO. There are two different PTO that has been coupled with this type of WEC. Hydraulic PTO was implemented by this WEC in Denmark in 1990s which by piston pump supplies high pressure water to a hydraulic turbine [20] and another is linear electric generator that is housed in inside a steel hull mounted on a concrete ballast structure and converts linear motion of cable (translator) to electric power [21]. This system is developed in Sweden and the scheme of system is represented in Fig.8. Barbarit et al. have investigated different power capture preformation of various WECs by numerical method [22] and capture width of taut-moored WEC is as Fig.9.

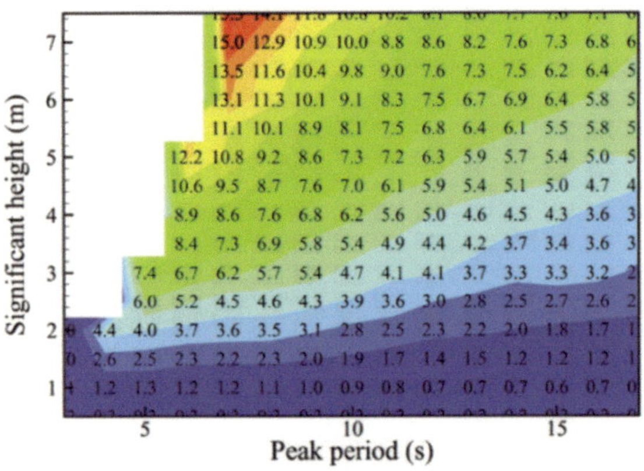

Figure 9. Power capture performance of taut moored WEC for various waves (the vertical axis is wave amplitudes in meter and horizontal axis is wave period in second and the annual men absorbed power amount on the table are in kW) [22].

Figure 10. Wave Star WEC, Denmark [22].

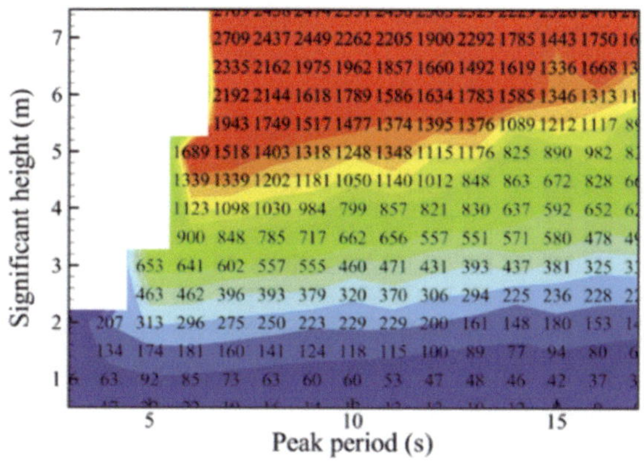

Figure 11. Power capture performance of Wave Star WEC [22].

In the case of heaven buoys, Wave Star Energy company has developed an innovative WEC namely Wave Star WEC. In this device a jack up structure stands on sea bed and provides a reference to the buoys (see Fig. 10). A full sized system is consisting of different buoys in which a hinging arm is utilized to transfer each buoy's motion to the PTO. The buoys have ability to be polled up (survival mode) in case harass ocean condition. Furthermore, hydraulic rams are employed to convert hinging motion of arms. Power capture preformation of 20 buoys Wave Star WEC was investigated by Babarit et el. [22] and the result is as Fig. 11.

2.1.1.2. Two body heaving convertors

Two body heaven convertors are multi-body convertors in which ocean power is extracted from the relevant motion between two bodies. One of these bodies is float on ocean surface and another is completely submerged. Ocean wave interaction with this WEC causes the float body elevates in heave mode as well as it pushes submerged body to the sea bottom due to increase in sea water inertia. One of the important advantages of this WEC is facility in its installation. Because device does not require any sea-bed connection, it can easily be installed in off-shore. Yet, different devices are developed in this principal one of which is ISP buoy developed in Sweden [23], Irish Wavebob [24] and PowerBuoy inspired by Ocean Power Technologies company.

The schematic of Wavebob is depicted in Fig. 12 left side. According this picture wavebob is consist of an inner buoy, submerged body (body2), and floating buoy (body 1) which are axially connected to each other. Additionally a high Pressure oil hydraulic system is implemented to deliver extracted power to electric generator. Fig. 13 illustrates power capture matrix of Wavebob WEC [22].

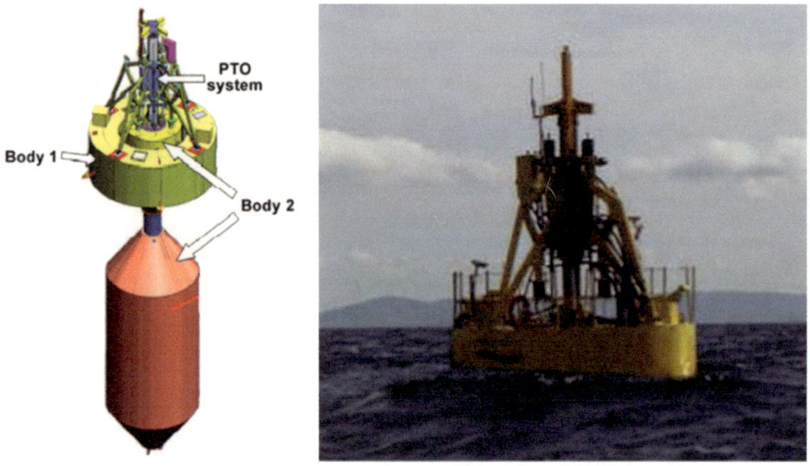

Figure 12. Left side; Principle of Wavebob WEC [16]. **Right side;** ¼ Scale Wavebob WEC, Ireland [22].

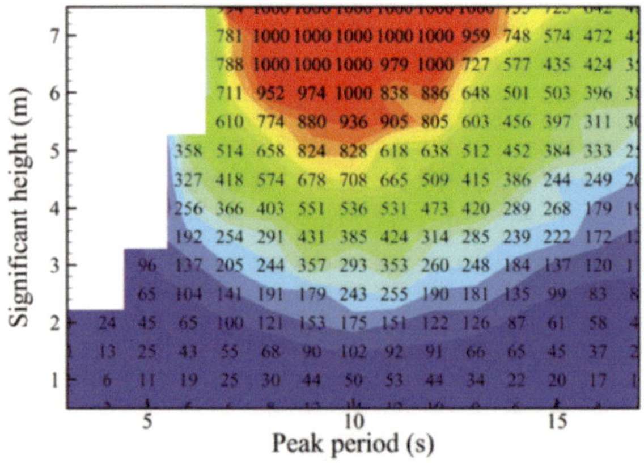

Figure 13. Power capture performance of Wavebob WEC [22].

Figure 14. Archimedes Wave Swing (**Left side:** Scheme [16]. **Right side:** Prototype [37]).

2.1.1.3. Submerged heaving convertors

The Archimedes Wave Swing (AWS) is a fully submerges off-shore WEC. It is constructed from two main parts, Silo or Basement is an air filled cylindrical chamber which is moored on seabed and floater is oscillating upper part as is illustrated in Fig. 14. Floater oscillates by water pressure variation. By crossing wave crest upon AWS the floater moves down compressing the air inside the chamber and by passage of wave trough upon AWS, the air inside the chamber expands and consequently push the floater up. Beside its unique design, AWS is the first WEC that a linear electric machine is implemented as PTO. One side of linear machine is fixed to the basement and the other side is connected to the floater via a translator, hence oscillation of floater excites linear machine. The AWS was successfully tested in 2004 [25].

Due to their mooring structure and the environment they work in (off-shore), almost all of the heaving oscillators requires highly maintenance. Meanwhile delivering generated electric power to the electric network or consumer is another problem intrinsically related to these off-shore WECs. High cost and Long distance underwater cabling is a commercial problem that heaving oscillators face with.

2.1.2. Pitching oscillators

Pitching oscillators are WECs which extract wave power by hinging motion in wave propagation direction. As it is demonstrated in Fig. 5, the hinging motion of pitching oscillator occurs in the axis in which WEC is installed on. According to device mooring and body design theses WECs are categorized to three types; *Float Pitching Convertors*, *Two Body Pitching Convertors* and *Submerged Pitching Convertors*.

2.1.2.1. Float pitching convertors

Developed in Lancaster University and off-shore WEC, *PS Frog Mk 5* is the best example of Floating Pitching Convertors [26]. As it is illustrated in fig. 15, PS Frog Mk 5 is composed of a large buoyant paddle with an internal ballasted handle below it. It oscillates in pitching mode and is float on sea level. When wave acts on paddle the ballast provides necessary reaction for pitch motion, consequently the wave power is absorbed by partially resisting the sliding of a PTO mass, which moves in guides above sea level. The sliding mass (PTO) converts the wave motion into differential mechanical motion within device then the mechanical motion is transferred via hydraulic circuit to an electrical generator.

The main advantages of PS Frog Mk 5 is its self-orienting capability with which the device spontaneously adjusts to face incident waves. Meanwhile, by moving ballast in the hall or by controlling sliding mass, it is viable to control PS Frog Mk 5 in resonance frequency in which device has the maximum capture width.

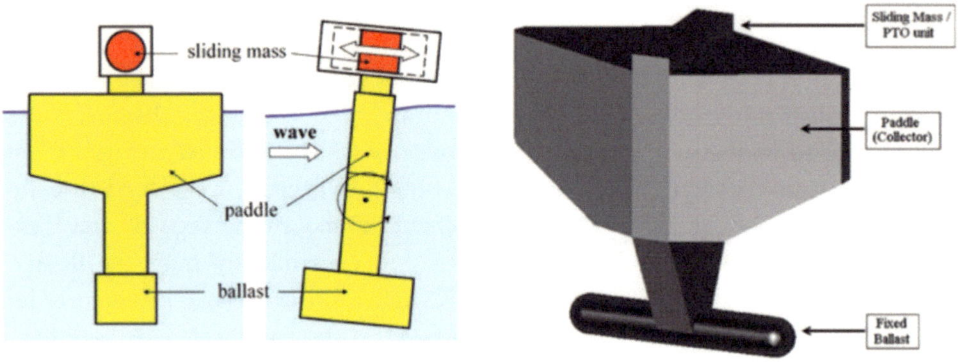

Figure 15. Left side: 2D illustration of PS Frog Mk 5 [16]. **Right side:** Perspective view of PS Frog Mk 5 [26].

2.1.2.2. Two body pitching convertors

Pelamis is a multi-body, floating, off-shore WEC. This device consists of several slender semi-submerged cylinders linked by hinged joins [27, 28]. When wave acts on Pelamis, adjacent cylinders start to fluctuating by angular motion in the joins in which wave power is absorbed in Pitch and Yaw mods. The scheme of Pelamis is presented in Fig. 16. In this figure, the left side represents the working concept of Pelamis and the right side shows a full scale Pelamis. In Pelamis the motion of cylinders is used to move hydraulic cylinders which pump fluid to high pressure fluid accumulators for short term energy storage. Furthermore, the smooth supply of high pressure fluid in accumulators drives hydraulic motors which are coupled with grid-connected electric generators. About device mooring, because of the self-referencing no rigid connection to the sea-bed is required and a slack mooring is sufficient to hold the device on station.

http://dx.doi.org/10.5772/53806

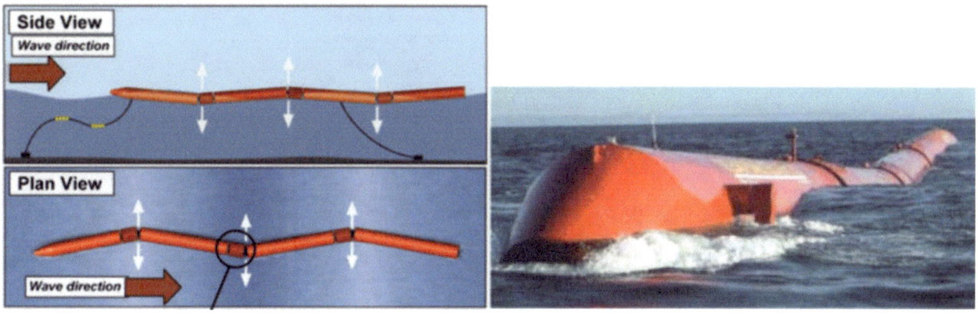

Figure 16. Left side: Scheme of Pelamis and principle of working [27]. **Right side:** Full scale prototype (By Ocean Power Delivery Ltd).

The PTO section of Pelamis is divided to two main parts which are called primary and secondary transmissions. The primary transmission, which stores wave power in hydraulic accumulator, consists of the hydraulic cylinders and their controllers. The secondary transmission, consisting of hydraulic motors coupled to electric generators, converts the energy stored in the hydraulic accumulators into electricity transmitted to shore. The separation provided by high pressure accumulator and controlling of electronically controlled valves, which controls input and output fluid of accumulators, makes it feasible efficient power absorption from ocean waves.

Full scale prototype Pelamis WEC, 120 m long and 3.5 m diameter, has been constructed and has been successfully connected to local electrical power network.

2.1.2.3. Submerged pitching convertors

Yet, there are different devices that fall in this category [29, 30] one of which is Oyster [31], Aquamarine Power Ltd developed near-shore WEC. The Oyster is a bottom hinged rigid flap which completely penetrates the water column from above the surface to the sea bed. When wave attach the Oyster, WEC starts to oscillate in pitch mode, rotational motion around hinging axis, and this motion moves a double acting high pressure sea water pump. A set of non-return valves rectify the flow from the double acting pump consequently the flow is regulated by a gas accumulator. The flow (water) is transferred to the shore through pipeline. In the onshore hydraulic plant, hydraulic pressure is converted into electric power via a Pelton wheel. Finally the water passes back to device in a close lop via a second low pressure return pipeline. The schematic of Oyster is presented in fig. 17. Another WEC that work in the same principle is WaveRoller WEC. In spite of Oyster, the rigid flaps of WaveRoller are short and it harvests wave power only near seabed [30].

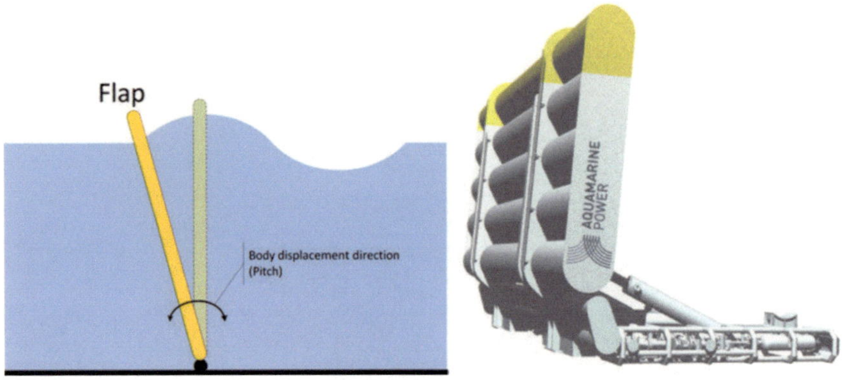

Figure 17. Left Side: Schematic of pitching flap WEC. **Right Side:** The Oyster WEC by Aquamarine Power Ltd [32].

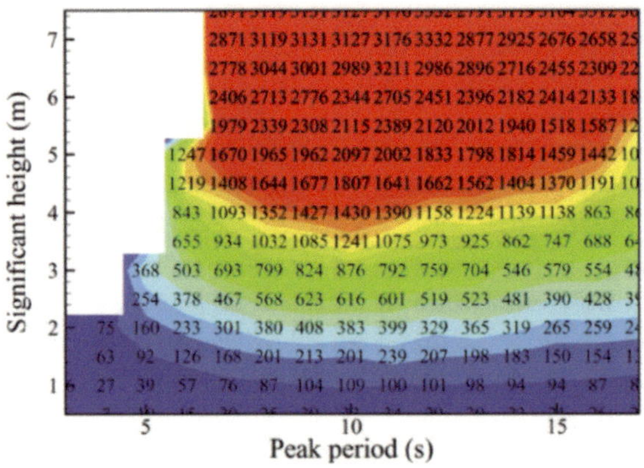

Figure 18. The power matrix of bottom hinged pitching flap [22].

The power capture performance of submerged pitching convertor is presented in Fig. 18.

2.1.3. Surging oscillators

However, waves' force in near-shores is concentrated in surge direction [8], surge wave energy convertors are less developed WECs in comparison to other types.

Developed in Western Ontario University namely "Surfing Wave Energy Convertor" is one of the surge WECs [33]. The Surfing Wave Energy Convertor is comprised of several paddles connected to a common drive train, which mechanically links them to an electric machine. The operating cycle begins with a paddle suspended in the path of the incoming wave. As the waves impact the paddle it is driven horizontally in the direction of wave travel in a 'Surfing' like motion. The horizontal motion of the paddle in-turn drives the electric machine, generating power. Once the paddle reaches the downstream end of the system, the electric machine is switched to motor operation and drives the system, lifting the first paddle out of the path of the waves and lowering the next paddle into the wave path.

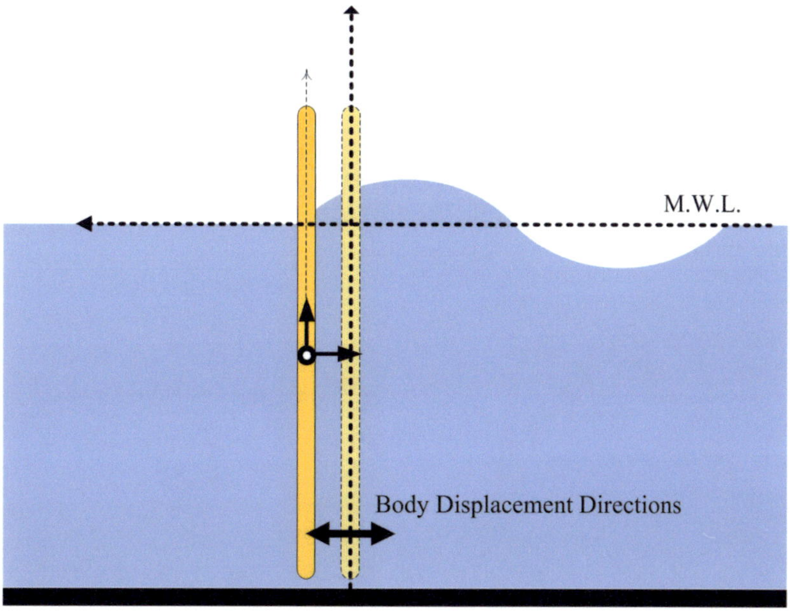

Figure 19. Schematic of Surge WEC.

Another surge WEC is designed by these authors [34, 35]. It consists of a rigid plate which is installed in near-shore vertical to the sea bed. The plate has been connected to a linear generator. When waves attack the plate it fluctuate in horizontal direction, parallel to the wave propagation direction (See Fig. 19). Various power electronic instruments has been implemented to deliver ocean wave power to electric network. The device is still under development in Altin Tara Electric Ltd. in Iran.

2.2. Interface-WECs

Interface-WECs are device that have no oscillation in interacting ocean wave. In fact these devices are designed to deliver wave energy via an interface (Air or Water) to the PTO. In comparison to Oscillation-WECs they require less maintain.

2.2.1. Air-interface

Oscillating water column (OWC) WEC is an Air-interface WEC in which wave power is converted to the electrical power without direct oscillation of WEC body by water particle. OWC is consists of a floating or bottom fixed structure whose upper part encloses a column of air and whose immersed part is open to the wave action [36]. The scheme of OWC is presented in Fig. 20.

When waves attack the OWC, the water pressure below the chamber compresses the trapped air in the chamber and the air is guided to a turbine hence the flow of air to outside of chamber turns the turbine and the coupled generator. Meanwhile, by decline of water pressure below the chamber, the pressure of trapped air reduces and consequently air flows into chamber and drives the turbine again. To date, two type of self-rectifying turbine has been used in OWC; the Well Turbine and Impulse Turbine, both of which have the capacity to rectify air flow that makes it feasible to have unidirectional motion in electrical generator.

The early inventor and developer of OWC is Yoshio Masuda who had commercialized the floating OWC in Japan since 1965 [16]. The OWC has been used for energy converting in both shoreline and near-shore.

Figure 20. Left side: Schematic of OWC. **Right side:** Back view of an OWC [37].

2.2.2. Water-interface

Overtopping wave energy convertors are the main members of water-interface WEC. To date, three overtopping wave energy convertor has been developed; The Tapchan (Tapered Channel Wave Power Device) [38], Wave Dragon [39] and SSG (Seawave Slot-Cone Generator) [40]. One of the well-known overtopping WECs is Wave Dragon. The Wave Dragon is an off-shore WEC, installed in location with depth of 25-40m, which is moored like a ship and consists of three main sections. The main platform, a reservoir with a double curved ramp, is float on the ocean surface and other parts are mounted on it. Two wave reflectors are installed on both sides of platform and intensify wave amplitude approaching to the platform. Finally a Hydro turbine, a set of low head Kaplan turbine, converts the hydraulic head in the reservoir. The Wave Dragon is represented in Fig. 21.

When waves approach to Wave Dragon, the reflectors focus them and guide the water toward the ramp, water overtop the ramp and fill in the reservoir then the turbine generates electricity from water motion inside the device. In Wave Dragon the wave energy convertor has not oscillation with wave, indeed wave power is transferred to PTO by water. In spite of air-interface WECs, water-interface WECs are not capable to be analyzed by Linearized Wave Theories.

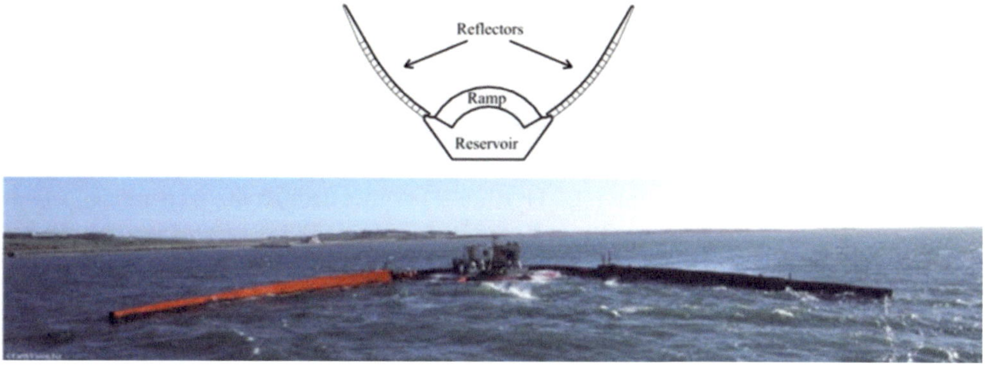

Figure 21. Wave Dragon. (**Above:** Scheme [39] **Below:** Prototype [39])

3. Power take-off

In process of wave energy convertion, after power extraction in WEC, another energy conversion happens in Power Take-Off (PTO). Power take-Off is (mostly mechanical or hydraulic) device in which the absorbed power is transferred to an electric generator. To date, three type of PTO is more commonly used.

- *Air/Water Turbine* forms part of an integral system that consists of a capture device, which also includes an electric generator. This PTO system is used in Interface-WECs (Oscillating wave columns and Overtopping devices). The air turbines which are used in OWC are mostly self-rectifying turbines which convert reciprocating air flow to unidirectional torque. The two implemented turbine in OWC are Well turbine [41] and Impulse turbine [42]. Probably one of the advantages of turbines in PTO is facility of using flywheel for energy storage.

- *Hydraulics PTO* system is consisting of a hydraulic circuit which transfers absorbed power of wave to a hydraulic motor which drives an electrical generator [43]. This kind of PTO is used in Pelamis and bottom hinged pitching flap. Implementation of accumulator is an effective method for energy storing in this PTO.

- *Direct Drive* is consisting of a moving part named translator on which linear generators is mounted. Direct drive is the simplest and probably the most efficient Power Take-Off system. PTO system of heaving oscillators and submerge heaving oscillators are from this kind.

Regard less to the type of PTO, all of the PTOs are connected to an electrical generator which generates useful energy from waves (except especial uses of WEC that is not intended to generate electricity [32]). The two type of electrical generators implemented in WECs are Linear Generators and Rotational Generator.

3.1. Linear generators

Linear generator are the kind of electrical machine in which the rotor (translator) and stator are linear and translator displace in straight path inside stator. Since most of the WECs harness wave power in reciprocating directions, Linear Generators are the most appropriate machines for generating electricity by WECs. It is due to, there is not require for any other interface or other power conversion (e.g. Hydraulic) which decline the transferred power in transmission process. Linear generator is connected to the WEC via translator and displace with it. Yet, three topology of linear generators are considered in wave energy industry.

3.1.1. Linear Permanent Magnet Synchronous Machine (LPMSM)

LPMSAs are constructed in three different configuration; Single-Side, Double–Side and Tubular. In all of these structures the translator (actuator) have moving magnets and primary constitute mover. The core of the primary of a flat LPMSA is made of longitudinal laminations with uniformly distributed slots, which house the windings. Since the windings are located in open slots, the effective air gap is greater than the actual air gap.

In tubular structures the laminations of the shape primary core is longitudinal or disk-shaped. Stacking of the disk laminations increases the effective air gap. The core of the secondary of a tubular LPMSA is generally made of solid magnetic steel. In LPMSM the flux density is supplied by rare earth permanent magnet. Due to ease in assembly, disk shaped laminated tubular LPMSM are preferred to flat types also slitting technique is use on disk lamination to reduce eddy current [44]. The schematic of single side LPMSM is illustrated in Fig. 22.

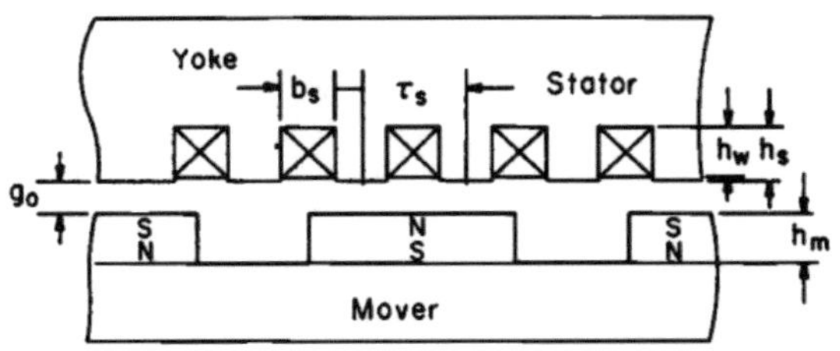

Figure 22. Schematic of single-side LPMSM [44].

3.1.2. Variable Reluctance Permanent Magnet Machine (VRM)

A family of permanent magnet machine, known as variable reluctance permanent magnet machine (VRM), has been developed with high force density. Despite its high force density, low power factor is a characteristic of VRM machines. The machine has a very high inductance, such that the current can be out of phase for almost 90 degree. The variable reluctance permanent magnet family is divided for two subcategory; Transverse Flux Permanent Magnet (TFM) and Vernier Hybrid Machine (VHM).

3.1.2.1. *Transverse Flux Permanent Magnet Machine (TFM)*

Transverse Flux Permanent Magnet Machines (TFM) has higher shear stresses than other machine topologies, This implies that TFM machines may be more suitable for wave energy application than the synchronous machines. There are two topology of TFM. In the first topology, the coils are in the stator and the magnets are on the translator. The translator is longer than the stator, which means that a part of the (rather expensive) magnets is not used (see Fig. 22. Left). This machine was used in AWS.

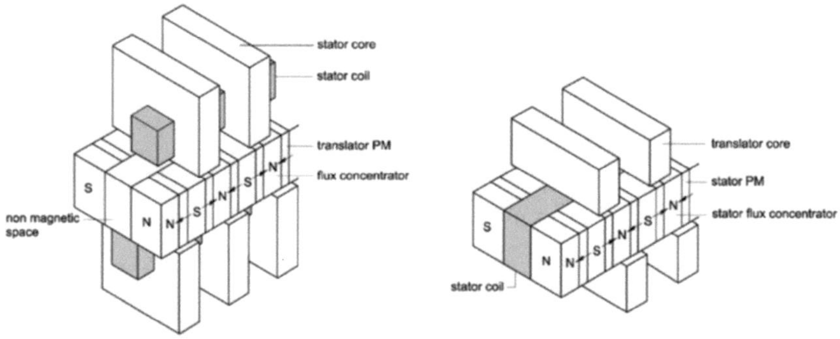

Figure 23. Left side: TFM machine with flux concentration and moving magnets [45]. **Right side:** TFM machine with flux concentration and stationary magnets [45].

As it is illustrated in Fig. 22 Right, the second topology, namely the double-sided moving-iron TFM, is a double-sided machine in which the translator consists only of iron. "In this topology, the stator consists of coils and U-cores on both sides of the translator, which consists of two rows of magnets and flux concentrators with space for construction material in between. Both the conductors and the magnets are kept stationary. The U-cores are now simpler in shape because space for coils is no longer required, and these cores or yokes form the translator."[45]

3.1.2.2. *Vernier Hybrid Machine (VHM)*

The Vernier Hybrid Machine is constructed from a linear toothed translator constructed from iron laminates which move inside two C-cores. The C-core has coil wounded on each pole also magnets are mounted on pole face. The translator tooth and slots width are similar in dimension to the magnet pitch, hence rapid flus reversal happens over short distance as the translator teeth move from the aligned to the unaligned position. Due to rapid flow change in a short distance, the electric frequency of the flux pulsation is greater than the translator's frequency [46, 47]. The schematic of one pole VHM is presented in Fig. 24. In spite of TFM, the VHM can be constructed from lamination which makes construction easy and small. However power factor of VHM is low, the high shear stress, small sizes and facility of constructing are main advantages of this machine.

3.1.3. Tubular air cored permanent magnet generator (TAPM)

The TAPM machine is proposed for wave application by Baker and Muller [48]. The principle of TAPM is shown in Fig. 25. In this topology, the magnets are magnetized in the axial directions with altering direction of flux [37]. Flux concentrators are placed between the magnets and a varying flux wave is created outside the translator. The large constructional advantage of TAPM is elimination of normal force by removing the steel in stator. Significant support structures are needed to overcome these forces and maintain a constant air gap width. The main problem in this topology is that the magnetic reluctance in the magnetic circuit is increased considerably, since the distance the flux travels in air now is in the range of the pole pitch and not, as in steel stator machines, just over the air gap [37]. The flux of an air cored machine is thus much smaller and the power per air gap area is considerably lower.

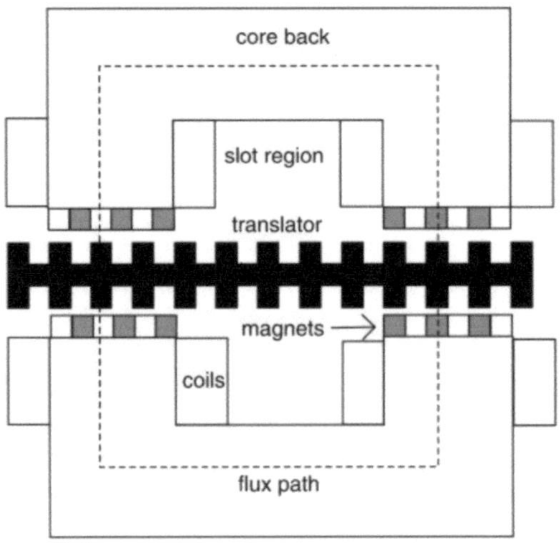

Figure 24. One phase of Linear Vernier Hybrid Machine [46].

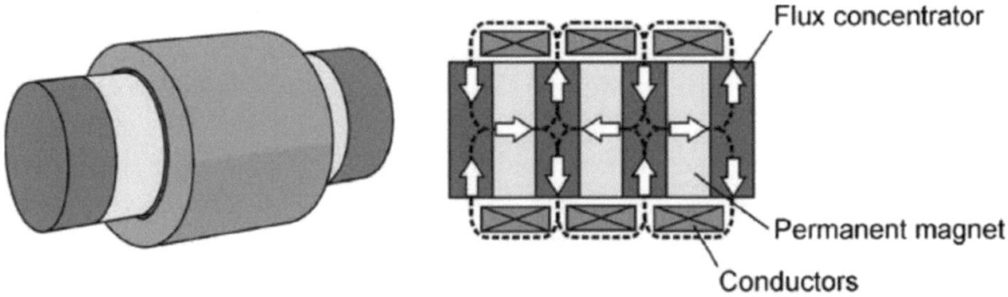

Figure 25. Tubular Air Cored Permanent Magnet Machine [37]. (**Left side:** Three dimensional view. **Right Side:** Cross section view)

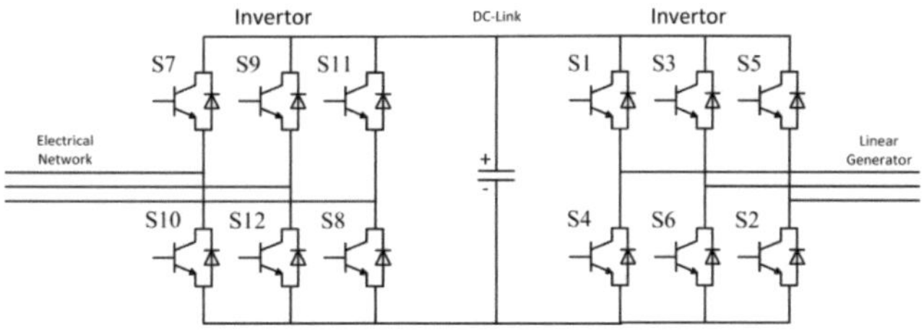

Figure 26. Common used power electronic circuit for connecting linear generator to the electrical network.

Regardless to the type of the linear machine, a linear machine generates fluctuating voltage from reciprocating motion of WEC, which is varying in both amplitude and frequency. Thus, it is not possible to connect linear generator directly to electric grid. For solving this problem, effective power electronic instrument is implemented [49]. One of the common power electronic circuits, implemented with linear generator, has a diode bridge rectifier or invertor rectifier in generator side, in which the generated voltage is delivered to a battery or capacitor after rectifying. Also another invertor is used in grid side to deliver power in DC-Link (battery or capacitor) to the electric. This power electronic circuit is depicted in Fig. 26.

3.2. Rotational generators

However utilization of rotational generator in wave industry is a questionable issue, conventional generators have been used with different wave energy convertors. Pelamis, Oscillating Water Column and overtopping devices are the main WECs which use rotational generator for mechanical electrical conversion. In these cases, the Asynchronous Generator or Induction machine is a preferred machine. It is due to the capacity of this type of machine to act in variable speed condition. Meanwhile, utilization of wound rotor induction machine makes it feasible to control rotor current and frequency which in consequence control the velocity and extracted power by mechanical part (This use of induction machine is so called Double Fed Induction Generator (DFIG)). Amundarain et al. considered the DFIG on OWC for controlling the electrical machine in a proper manner [50]. In comparison to the linear machine, it is much more simple to connect an induction machine to the electrical grid. While, since these machines was designed to act in higher speed, in most of the cases the utilization of rotational machine for convert low speed WEC motion is a low effective manner.

4. Environmental consequences

Renewable sources have been considered as new source of energy to decline our dependence on fossil fuel and be a solution to the global warming and CO_2 emission. Considering, each technology (WEC) beside its advantage might have some side effect and by taking into account the importance of environment in which a wave energy convertor acts (Ocean), finding out and preventing possible negative consequence of WEC carries a huge importance. Thus, before any decision to build a large WEC farm this issue should be considered accurately to prevent occurring of any other disaster like CO_2 emission and global warming. There are some probable problems that may afflict the ecosystem as a consequence of WEC application, some of which is as follow;

Constructing Wave Farm in a location might cause military importance of location, avoidance of shipping and fishing which not only might change the economy of area but also cause a biological change in the area [51]. The WECs are attractive to the fish and it is due to fish use these devices as protection from predator, availability of food near them, spewing substrates and so on [52]. Hence, restriction for fishing and popularity of WECs by fish may enhance fish population in Wave Farm and this has a negative impact on local species diversity and population density [53].

WECs alter the currents and waves in implemented area and this might change sediment size distribution in area which may favour the accumulation of organic material in area [54].

It was indicated that some species of shark are sensitive to the electromagnetic field of 25-100 uT [55] and also it was shown that migrating European eels can detect the magnetic field of underwater cables [56]. It is worthwhile to find out possible consequence of external electromagnetic field on marine species and reduce the emitted magnetic field from under water cables.

Noises generation during site construction, maintenance and even normal working of WEC can affect dolphin, whales, seals and other fish species, which use underwater sound for communication and finding mat and so on [57, 58]. However little is known about long-term effects of noise on these species, the possible preventing technics should be developed for less interacting of WECs in ocean environment.

5. Conclusion

This chapter presents essential information about generation and potential of ocean wave power. Also the available Wave Energy Convertors has been categorized according to the principle of interaction with ocean waves. The manners of acting and mooring and in some case the average power output of WECs are presented. Meanwhile, the implemented electrical generator by WECs has been described in section three. Finally the possible environmental consequence of WECs has been investigated at the end of this chapter. This chapter reviews the available technologies of harnessing wave power.

Author details

Ehsan Enferad[1,2*] and Daryoush Nazarpour[2]

*Address all correspondence to: st_e.enferad@urmia.ac.ir

1 Altin Tara Electric Ltd, Urmia, Iran

2 Department of Electrical Engineering, Urmia University, Urmia, Iran

References

[1] A. H. Strahler and A. N. Strahler. Modern Physical Geography. New York : John Wiley and Sons, 1992.

[2] Air-sea interaction: connecting the ocean and atmosphere. Rogers, D.P. 1995, Review of Geophysics, Vol. 33.

[3] Massel, Stanis aw R. Ocean Waves Breaking and Marine Aerosol Fluxes. New York : Springer, 2007.

[4] Annual Report of International Energy Agency. s.l. : Implementing agreement on ocean energy systems (IEA-OES), 2007.

[5] Quantifying the global wave power resource. Kester Gunn, Clym Stock-Williams. s.l. : Elsevier Ltd, 2012, Renewable Energy, pp. 1-9.

[6] A review of wave-energy extraction. Johannes, Falnes. 20, 2007, Marine Structures, Vol. 4, pp. 185-201.

[7] J. Billingham & A. C. King. Wave Motion. Cambridge : Cambridge University Press, 2000.

[8] The effect of water depth on the performance of a small surging wave energy converter. M. Folley, T.J.T. Whittaker and A. Henry. 34, 2007, Ocean Engineering, pp. 1265–1274.

[9] Stephen Barstow, Gunnar Mørk, Denis Mollison and João Cruz. The Wave Energy Resource. [book auth.] João Cruz. OceanWave Energy Current Status and Future Prepectives. Berlin : Springer-Verlag, 2008.

[10] The Norwegian "wave climate mapping" programme. Torsethaugen, K. Tapir, Trondheim, Norway : s.n., 1982. Proceedings of the second international symposium on wave energy utilization. pp. 81–97.

[11] Wave prediction based on a modified grey model MGM(1,1) for real-time control of wave energy converters in irregular waves. D.Q. Truong, K.K. Ahn. 2012, Renewable Energy.

[12] An application of the Fast Fourier Transform to the short-term prediction of sea wave behaviour. J. Ross Halliday a, David G. Dorrell, Alan R. Wood. 36, 2011, Renewable Energy, p. 685e1692.

[13] Measurement of 2-D Sea Surface Elevation Fields Using Complex Synthetic Aperture Radar Data. Johannes Schulz-Stellenfleth and Susanne Lehner. 6, 2004, IEEE Transactions on Geoscience and Remote Sensing, Vol. 42, pp. 1149-1160.

[14] Ocean wave extraction from RADARSAT synthetic aperture radar inter-look image cross spectra,. M. Dowd and P.W. Vachon. 2001, IEEE Transactions on Geoscience and Remote Sensing, Vol. 39, pp. 21-37.

[15] Ross, D. Power from sea waves. Oxford : Oxford University Press, 1995.

[16] Wave energy utilization: A review of the technologies. Falcao, Antonio F. de O. 3, 2010, Renewable and Sustainable Energy Reviews, Vol. 14, pp. 899–918.

[17] A review of wave-energy extraction. Falnes, Johannes. 4, 2007, Marine Structures, Vol. 20, pp. 185-201.

[18] Falnes, Johannes. Ocean Waves and Oscillating systems. Cambridge : Cambridge University Press, 2004.

[19] The Norwegian wave-power buoy project. Budal K, Falnes J, Iversen LC, Lillebekken PM, Oltedal G, Hals, et al. Trondheim, Norway : Proceedings of 2nd International Symposium on Wave Energy Utilization, 1982. pp. 323–44.

[20] Point absorber—optimization and survival testing. Nielsen K, Smed PF,. s.l. : Proceedings of 3rd European Wave Energy Conference, 1998. 207–14.

[21] Experimental results from sea trials of an offshore wave energy system. Waters R, Stalberg M, Danielsson O, Svensson O, Gustafsson S, Stromstedt E, et al. 90, s.l. : Applied Physics Letters, 2007, Vol. 3.

[22] Numerical benchmarking study of a selection of wave energy converters. A. Babarit, J. Hals, M.J. Muliawan, A. Kurniawan, T. Moan, J. Krokstad. s.l. : Renewable Energy, 2012, Vol. 41, pp. 44-63.

[23] Contribution to the theory andexperience of energy production and transmission from the buoy-concept. Cleason L, Forsberg J, Rylander A, Sjostro mBO,. s.l. : Proceedings of 2nd International Symposium on Wave Energy Utilization, 1982. pp. 345–70.

[24] Wavebob—research & development network and tools in the context of systems engineering. Weber J, Mouwen F, Parrish A, Robertson D. s.l. : Proceedings of 8th European Wave Tidal Energy Conference, 2009. pp. 416–20.

[25] Learning experience of AWS pilot plant test offshore Portugal. FE, Gardner. s.l. : Proceedings of 6th European Wave Energy Conference, 2005. pp. 149–54.

[26] Developments in the design of the PS Frog Mk 5 wave energy converter. McCabe AP, Bradshaw A, Meadowcroft JAC, Aggidis G. 2006, Renewable Energy, Vol. 31, pp. 141-51.

[27] Design, simulation, and testing of a novel hydraulic power take-off system for the Pelamis wave energy converter. Henderson, Ross. 2006, Renewable Energy, Vol. 31, pp. 271–283.

[28] Measurements of the slow drift dynamics of a model Pelamis wave energy converter. Retzler, Chris. 2006, Renewable Energy, Vol. 31, pp. 257–269.

[29] WaveNet. European Community. 2003. ERK5-CT-1999-20001.

[30] Company, AW-Energy. http://www.aw-energy.com/concept.html. [Online] AW-Energy Company.

[31] Design of the Next Generation of the Oyster Wave Energy Converter. L. Cameron, R. Doherty, A. Henry, K. Doherty, J. Van 't Hoff, D. Kaye and D. Naylor, S. Bourdier, T. Whittaker. Bilbao : s.n., 2010 October. 3rd International Conference on Ocean Energy.

[32] The cost of water from an autonomous wave-powered desalination plant. Matt Folley, Trevor Whittaker. 2009, Renewable Energy, Vol. 34, pp. 75–81.

[33] Electric power generation by 'Surfing' water waves. Ben D. Hazlett, Ion I. Inculet a, Diana R. Inculet. 2009, Renewable Energy, Vol. 34, pp. 2510–2514.

[34] Enferad, Ehsan. Engine for Producing Energy from Sea Waves. 47700 Iran, April 2008.

[35] Application of PMSM for Electric Generation from Ocean Waves by EPEW. Ehsan Enferad, Murtaza Farsadi and Shirin Enferad. Bursa, Turkey : s.n., 2011. nternational Conference of ELECO. pp. 258-262.

[36] OWC wave energy devices with air flow control. A.F. de O. Falcao, P.A.P. Justinob. 1999, Ocean Engineering, Vol. 26, pp. 1275–1295.

[37] Cruz, João. Ocean Wave Energy Current Status and Future Prepectives. Berlin : Springer-Verlag, 2008.

[38] Evans DV, Falcao A.F. de O. Mehlum E Tapchan. Hydrodynamics of ocean wave energy utilization. Berlin : Springer, 1989.

[39] Prototype testing of the wave energy converter Wave Dragon. Kofoed JP, Frigaard P, Friis-Madsen E, Sørensen HC. 2006, Renewable Energy, Vol. 31, pp. 181-189.

[40] SSG wave energy converter: Design, reliability and hydraulic performance of an innovative overtopping device. L. Margheritini, D. Vicinanza b, P. Frigaard. 2009, Renewable Energy, Vol. 34, pp. 1371-1380.

[41] AA, Wells. Fluid driven rotary transducer. Spec No. 1595700 British Patent, 1976.

[42] IA, Babinsten. Apparatus for converting sea wave energy into electrical energy. No. 3922739 U.S. patent, 1975.

[43] Power conversion mechanisms for wave energy. Salter SH, Taylor JRM, Caldwell NJ. 2002, Proc Inst Mech Eng Part M J Eng Maritime Environ, Vol. 216, pp. 1-27.

[44] I. Boldea, Syed A. Nasar. Linear Electric Actuators and Generators. s.l. : Cambridge University Press, 1997.

[45] Conventional and TFPM Linear Generators for Direct-Drive Wave Energy Conversion. Henk Polinder, Barrie C. Mecrow, Alan G. Jack, Phillip G. Dickinson, and Markus A. Mueller. 2, 2005, IEEE TRANSACTIONS ON ENERGY CONVERSION, Vol. 20.

[46] Power conditioning of the output from a linear vernier hybrid permanent magnet generator for use in direct drive wave energy converters. P.R.M. Brooking and M.A. Mueller. 2, 2005, IEE Proc.-Gener. Transm. Distrib, Vol. 152.

[47] Modelling the performance of the vernier hybrid machine. M.A. Mueller and N.J. Baker. 6, 2003, IEE Proc.-Electr. Power Appl, Vol. 150.

[48] Permanent magnet air-cored tubular linear generator for marine energy converters. Baker NJ, Mueller MA. Edinburgh : s.n., 2004. IEE Power Electronics and Electrical Machines & Drives Conference.

[49] Experimental verification of linear generator control for direct drive wave energy conversion. J.K.H. Shek D.E. Macpherson M.A. Mueller. 5, IET Renewable Power Generation, Vol. 4, pp. 395–403.

[50] Wave energy plants: Control strategies for avoiding the stalling behaviour in the Wells turbine. Modesto Amundarain, Mikel Alberdi, Aitor J. Garrido, Izaskun Garrido, Javier Maseda. 2010, Renewable Energy, Vol. 35, pp. 2639-2648.

[51] The impact of marine reserves: do reserves work and does reserve size matter? BS, Halpern. 2003, Ecological Applications, Vol. 13.

[52] Urban structures as marine habitats: an experimental comparison of the composition and abundance of subtidal epibiota among pilings, pontoons and rocky reefs. SD, Connell. 2001, Marine Environmental Research, Vol. 53, pp. 115–125.

[53] Temporal changes in the composition and abundance of the macro-benthic invertebrate communities at dredged material disposal sites in the Anse a Beaufils, baie des Chaleurs, eastern Canada. Harvey M, Gauthier D, Munro J. 1998, Marine Pollution Bulletin, Vol. 36, pp. 41–55.

[54] Effects of an artificial reef on the surrounding soft-bottom community (central Adriatic Sea). Fabi G, Luccarini F, Panfili M, Solustri C, Spagnolo A. 2002, ICES Journal of Marine Science, Vol. 59, pp. 343–349.

[55] Sharks can detect changes in the geomagnetic field. Meyer CG, Holland KN, Papastamatiou YP. 2004, J R Soc Interface, Vol. 2, pp. 129–30.

[56] Orientation of silver eel (Anguilla anguilla) in a disturbed geomagnetic field. Westerberg H, Begout-Anras ML. 2000. Proceedings of the 3rd Conference on Fish Telemetry. pp. 149–58.

[57] Sound detection and processing by fish: critical review and major research questions. Popper AN, Fay RR. 1993, Brain Behavior and Evolution, Vol. 41, pp. 14–38.

[58] Attraction of settlementstage coral reef fishes to reef noise. Simpson SD, Meekan MG, McCauley RD, Jeffs A. 2004, Marine Ecology Progress Series, Vol. 276, pp. 263–8.

Section 6

Geothermal

Chapter 13

ORC-Based Geothermal Power Generation and CO_2-Based EGS for Combined Green Power Generation and CO_2 Sequestration

Basel I. Ismail

Additional information is available at the end of the chapter

http://dx.doi.org/10.5772/52063

1. Introduction

Electrical power generation using innovative renewable and alternative geothermal energy technologies have shown merits and received renewed interest in recent years due to an increasing concern of greenhouse gas (GHG) emissions, being responsible for global warming & climate change, environmental pollution, and the limitations and conservation of natural energy resources. Organic Rankine Cycle (ORC) power generation using low-temperature geothermal resources is one of these innovative geothermal power generation technologies. The vast low-temperature geothermal resources found widely in most continental regions have not received much attention for electricity generation. Continuous development of ORC power generation and state-of-the-art drilling technologies and other factors make this renewable and nonconventional energy source one of the best future viable, alternate and available source to meet the required future electricity demand worldwide, significantly reducing GHG emissions and mitigating global warming effect. The first part of this chapter will introduce the ORC-based geothermal power generation technology. It will also present its fundamental concept for power generation and discusses its limitations, environmental & economic considerations, and energy conversion performance concept. Another novel "double-benefit" technology is enhanced (engineered) geothermal systems (EGS) using CO_2 as the working fluid for combined renewable power generation and CO_2 sequestration. CO_2 is of interest as a geothermal working fluid mainly because it transfers geothermal heat more efficiently than water. While power can be produced more efficiently using this technology, there is an additional benefit for carbon capture and sequestration (CCS) for reducing GHG emissions. Using CO_2 as the working fluid in geothermal power systems may permit utilization of lower-temperature geologic formations than those that are currently deemed economically viable,leading to more widespread utilization of geothermal energy. The second part of this chapter will present and discuss the merits, limitations, environmental, economic and fundamental aspects of CO_2-based EGS technology.

2. ORC-based geothermal power generation

2.1. Developments & utilization of low-temperature geothermal energy resources for power generation

The geothermal resources of the Earth are vast and abundant. For example, the part of geothermal energy stored at a depth of 3 km is estimated to be 43,000,000 EJ (equivalent to 1,194,444,444 TWh) which is much larger compared to all fossil fuel resources, whose energy equivalent is 36,373 EJ, combined (Chandrasekharam & Bundschuh, 2008). Conventional energy resources, such as oil, natural gas, coal, and uranium, being widely consumed in the world, originate from finite energy sources embedded in the crust of the Earth. Only one energy resource of the crust is renewable, namely geothermal energy. The word "geothermal" is originated from Greek words; "geo" meaning the Earth and "therme" meaning heat, so geothermal energy means the natural heat energy from the Earth. The source of geothermal energy is the continuous energy flux flowing from the interior of the Earth towards its surface. Unlike other conventional and renewable energy sources, geothermal energy has unique characteristics, namely it is abundantly available, stable at all times throughout the year, independent of weather conditions, and has an inherent storage capability (Hammons, 2004). Distinct from fossil-fuelled power generation, geothermal power generation is also considered to be a clean technology and environmentally friendly power source which could significantly contribute to the reduction of GHG emissions by replacing fossil fuels and other non-clean energy sourcesused for power generation (Chandrasekharam& Bundschuh, 2008).

Depending on the temperature and depth of the resource, the rock chemical composition and the abundance of ground water, geothermal heat energy resources vary widely from one location to another (Gupta & Roy, 2007). Geothermal heat sources are typically classified based on their available temperature, thus enthalpy energy level, from about 50 °C to 350 °C. The high-temperature (high-enthalpy) geothermal resources (with temperature > 200 °C) are typically found in volcanic regions and island chains, whereas the moderate-temperature (150-200 °C) and low-temperature (low-enthalpy) geothermal resources (<150 °C) are usually found broadly in most continental regions and by far the most commonly available heat resource (Chandrasekharam& Bundschuh, 2008; Gupta & Roy, 2007). The increase in temperature with depth in the Earth's crust can be expressed in terms of what is known as the geothermal temperature gradient. Down to the depths accessible by drilling with modern technology (e.g. over 10 km), the average geothermal gradient is about 2.5-3.0 °C/100 m (Dickson& Fanelli, 2005). For example, at depth around 3 km below ground level, the temperature is about 90 °C. There are, however, areas in which the geothermal gradient is far from the average value (e.g. in some geothermal areas the gradient is ten times the average value) due to geothermal structure and composition of these areas (Dickson& Fanelli, 2005). The type of geothermal resource determines the type of system and method of its harvesting and utilization for electrical power generation. For example, high-temperature geothermal resources (vapour- and liquid-dominated) can be harvested and utilized to generate electricity using one of the following methods depending on the compositional and thermal characteristics of the resource: (1) single-flash steam power systems, (2) double-flash steam power systems, and (3) dry-steam power systems. Generating electricity from medium- and low-temperature geothermal resources (i.e. water-dominated resources) can be efficiently accomplished using a Binary-cycle technique, such as, ORC (Ismail, 2011a; Chandrasekharam& Bundschuh, 2008; Dickson & Fanelli, 2005; DiPippo, 2008).

Generating electricity from geothermal steam resources using an experimental 10 kW-electrical generator was made at Larderello of Italy in 1904 (Dickson& Fanelli, 2005; Panea et al., 2010). The commercial success of this attempt indicated the industrial value of geothermal energy and marked the beginning of a form of exploitation that was to develop significantly from the on. By 1942, the installed geothermal-electric capacity had reached approximately 128 MW_e (Dickson& Fanelli, 2005). In the early 1950's, many countries were attracted by geothermal energy, considering it to be economically competitive with other forms of energy. It was estimated (Dickson& Fanelli, 2005; Ruggero, 2007) that the worldwide installed geothermal-electric capacity reached 1.300 GW_e (in 1975), 4.764 GW_e (in 1985), 6.833 GW_e (in 1995), 7.974 GW_e (in 2000), 8.806 GW_e (in 2004), 8.933 GW_e (in 2005), 9.732 GW_e (in 2007). In 2010, it was reported (Holm et al., 2010) that 10.715 GW_e is online generating 67,246 GWh which represents a 20% increase in geothermal power online between 2005 and 2010. While power on-line grew 20% between 2005 and 2010, countries with projects under development grew at a much faster pace. In 2007, Geothermal Energy Association (GEA) reported that there were 46 countries considering geothermal power development. In 2010, this report identified 70 countries with projects under development or active consideration, a 52% increase since 2007. It should be noted that projects under development grew the most intensely in two regions of the world; namely, Europe and Africa (Holm et al., 2010).Very recently, it was reported (GEA, 2012) that as of May 2012, approximately 11.224 GW_e of installed geothermal-electric power capacity was online globally, and is increasingly contributing to the electric power supply worldwide. It was estimated (Ruggero Bertani, 2007) that geothermal energy provides approximately 0.4% of the world global power generation, with a stable long term growth rate of 5%; the largest markets being in USA, Mexico, Indonesia, Philippines, Iceland, and Italy. Security for long-term electricity supply and GHG emission from fossil fuelled power plants is becoming a cause of concern for the entire world today. It was estimated (Chandrasekharam& Bundschuh, 2008) that the world net electricity demand is going to increase by approximately 85% from 2004 to 2030, rising from 16,424 TWh (in 2004) to 30,364 TWh in the year 2030. It was also reported (Chandrasekharam& Bundschuh, 2008; Dickson & Fanelli, 2005) that the emissions of GHG from geothermal power plants constitute less than 2% of the emission of these gases by fossil-fuelled power plants. To meet future energy demands renewable energy sources should meet the following criteria (Chandrasekharam& Bundschuh, 2008): (1) the sources should be large enough to sustain a long-lasting energy supply to generate the required electricity for the country, (2) the sources should be economically and technically accessible, (3) the sources should have a wide geographic distribution, and (4) the sources should be environmentally friendly and thus should be low GHG emitters in order to make significant contribution to global warming mitigation. Low-temperature (low-enthalpy) geothermal energy resources meet all the above criteria. It was reported in (Chandrasekharam& Bundschuh, 2008; Cui et al., 2009) that this huge low-temperature geothermal energy resource has already been used for power generation by typical countries, such as USA, Philippines, Mexico, Indonesia, Iceland, Germany, and Austria. The installations of several commercial low-temperature geothermal power systems in these countries have substantially proved the ability of low-temperature geothermal fluids to generate green electricity (Chandrasekharam& Bundschuh, 2008).

In most developing countries, low-temperature geothermal resources have not received much attention for electricity generation. The main reason for not utilizing these resources by most developing countries (and several industrialized countries) for commercial exploitation is that they are not considered as economically feasible for generating electricity (Chandrasekharam& Bundschuh, 2008). In contrast, in some industrialized countries, especially USA and in Europe, increasing energy demand and environmental awareness related to climate change have urged these countries to develop technologies which utilize low-temperature geothermal resources economically for power generation (Chandrasekharam& Bundschuh, 2008; Dickson & Fanelli, 2005). It was reported (Chandrasekharam& Bundschuh, 2008; Galanis et al., 2009) that developing countries, in general, need to benefit from these new and continually improving technologies for using low-temperature geothermal resources for generating electricity. It should be noted that for many developing countries, the use of low-temperature geothermal resources is not new. Many of developing countries have been using these resources for the past centuries for direct heating (but not power generation) applications (Chandrasekharam& Bundschuh, 2008). Recent increases in the cost and uncertainty of future conventional energy supplies for power generation are improving the attractiveness of low-temperature geothermal resources. Continuous development of innovative drilling and power generation technologies makes this nonconventional, renewable and clean energy source the best future viable, alternate and available source to meet the required future electricity demand worldwide, significantly reducing GHG emissions and mitigating global climate change (Chandrasekharam& Bundschuh, 2008).

As mentioned earlier, generating electricity from low-temperature geothermal resources (water-dominated resources) can be effectively achieved using a binary ORC technology. Low-temperature geothermal ORC technology has virtually no GHG emissions to the atmosphere (DiPippo, 2008; Hettiarachchi et al., 2007) and is an attractive energy-conversion technology due to its simplicity and its limited number of components, all of them being very common and commercially available. Nowadays, the ORC can be considered asthe only proved technology that is commonly used in ranges of afew kW up to 1 MW (Schuster et al., 2009). Despite the fact that ORC technology is currently associated with low conversion efficiencies,new applications of this technology are commonly examined and implementeddue to its possibility to utilize the low-grade heat from sources, such as low-temperature geothermal resources, for power generation (Ismail, 2011a). A number of successful &

innovative ORC binary power plants were installed in different locations (e.g. remote and rural sites) worldwide which demonstrate the ability of this promising alternative technology to utilize renewable low-temperature geothermal energy sources for generating electricity. For example, two plants were installed in Nevada, USA in 1984 and 1987 with electric power generation capacity of 750 and 800 kW_e, respectively (Chandrasekharam& Bundschuh, 2008). The production wells supply geo-fluid (water) temperature at 104 °C with a flow rate of 60 l/s to these plants. The ORC binary fluid used was initially R-114 but due to non-availability of this working fluid the plant switched to iso-pentane in 1998. In another location near Empire, Nevada, approximately four 1 MW_e units were installed and commissioned in 1987. Two geothermal production wells with geo-fluids temperature of 137 °C were used (Chandrasekharam& Bundschuh, 2008). In 1998, a third well with geo-fluid temperature of 152 °C was drilled to maintain the capacity of the plant at approximately 4 MW_e. The modular approach was used so that high plant availability factors of 98% and more were achievable (Hammons, 2004). In 1987, another plant was installed and commissioned in Taiwan with an electric power generation of 300 kW_e. The plant draws geo-fluids from a 500 m deep well at a temperature of 130 °C. It was reported that the power generated from this facility was sold to the national power grid at 0.04 US$/kWh (Chandrasekharam& Bundschuh, 2008). In 1986, a low-temperature geothermal ORC unit (Mulka plant) with a power capacity of 15 kW_e was commissioned in Australia. The unit was coupled to a geothermal production well which was drilled down toa depth of 1,300 m, and supplying geo-fluid at 86 °C. The unit was operated non-stop for about three and a half years, showing frequency stability and response to load changes (Rosca et al., 2010).

In 1992, a binary ORC power generation unit which utilized a low-temperature geothermal water resource with a temperature ranging from 90 to 115 °C was tested at a location near arderello, Italy. The geothermal power plant generated between 800 and 1,300 kW_e of electricity (Rosca et al., 2010). In Germany, the first low-temperature geothermal power plant using ORC technology was installed at Neustadt-Glewe, with a power capacity of approximately 230 kW_e using a geo-fluid temperature of 98 °C (RuggeroBertani, 2007). Another plant was commissioned in Thailand in 1989, with an installed capacity of 300 kW_e. The actual production was reported to vary from 150 to 250 kW_e and the geo-fluid temperature is 116 °C with a flow rate of approximately 8 l/s (Chandrasekharam & Bundschuh, 2008). In Japan, binary ORC technology was experimentally operated for 5 years starting in 1993 by NEDO (Yamada & Oyama, 2004). More recently, in 2006, the first binary ORC plant which utilizes a low-temperature geothermal resource at a temperature of 74°C reported by (RuggeroBertani, 2007) to be the lowest low-temperature geothermal energy resource worldwide) was installed at Chena Hot Springs, Alaska, with a power generation capacity of 200 kW_e. A photograph of Chena ORC-based geothermal power plant is shown in Figure 1. A second ORC unit was added, reaching the total installed capacity of 400 kW_e net. The total project cost of this binary geothermal plant was $2.2 million with a simple payback period of 4 years (Holdmann, 2007). In Altheim, Austria, a geo-fluid of temperature 106 °C is utilized both for district heating and electric power generation using a binary plant technology. The net electric output of this plant is 500 kW_e, selling to the electric grid 1.1 GWh in 2006 (RuggeroBertani, 2007).

Figure 1. Photograph of Chena ORC-based geothermal power plantat Chena, Alaska, USA (Holdmann, 2007).

2.2. Energy conversion and performance aspects of ORC-based low-temperature geothermal power generation

The ORC is a thermodynamic Rankine cycle that uses an organic working fluid instead of steam (water). A schematic diagram showing a low-temperature geothermal ORC binary-fluid system used for electric power generation is shown in Figure 2. In this system, the first (primary) fluid being the geo-fluid (brine) is extracted from the low-temperature geothermal resource through the production well. The geo-fluid carries the heat from the liquid-dominated resource (thus called the geo-fluid heat carrier) and efficiently transfers this heat to the low-boiling point (BP) organic working fluid (the secondary fluid) using an effective heat exchanger; shell-and-tube heat exchangers arewidely used (Chandrasekharam& Bundschuh, 2008). In this binary-fluid system, the low-boiling point organic liquid absorbs the heat which is transferred by the geothermal fluid and boils at a relatively much lower temperature (compared to water) and as a result develops significant vapor pressure sufficient to drive the axial flow or radial inflow turbine. The turbine is coupled to an electric generator which converts the turbinemechanical shaft power into electrical power. The organic working fluid expands across theturbine and then is cooled and condensed in thecondenser before it is pumped back as a liquid to the heat exchanger using a condensate

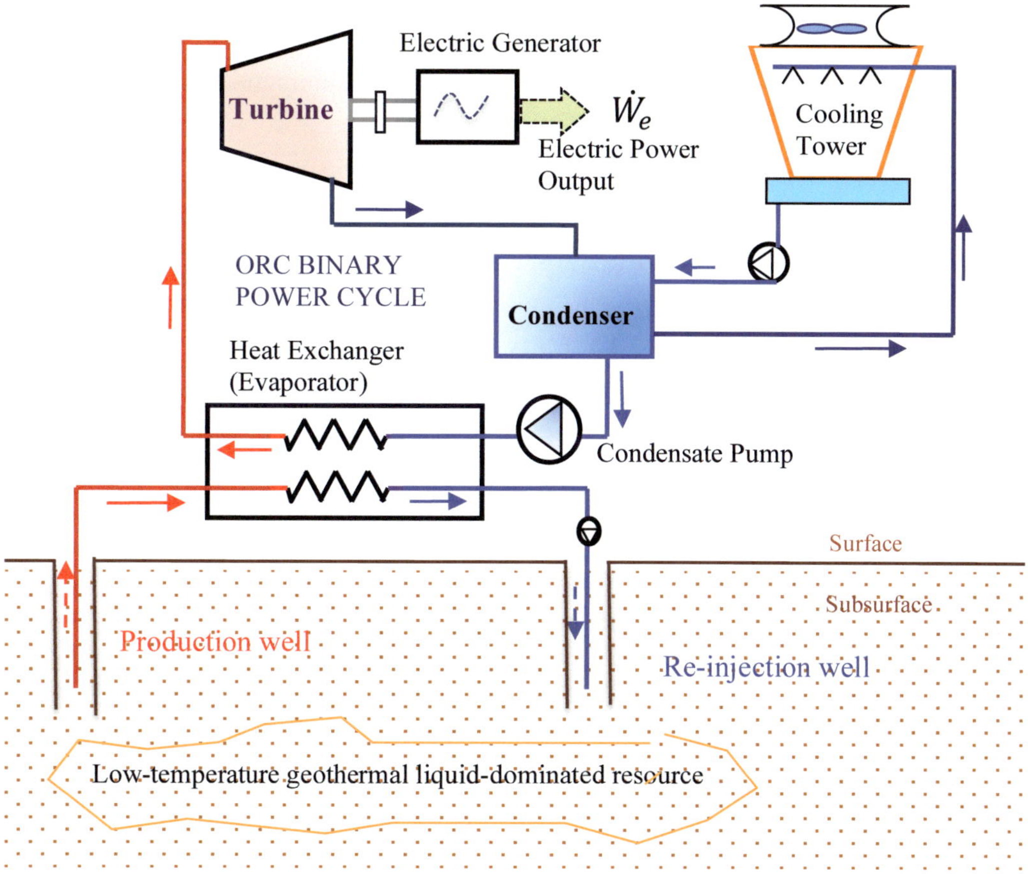

Figure 2. A schematic diagram showing the basic concept of a low-temperature geothermal binary ORC system for electrical power generation.

pump to be re-evaporated, and the power cycle repeats itself. One of the most important performance criteria in low-temperature geothermal ORC power generation technology requires the optimal selection of the ORC organic working fluid. Organic fluids used in binary ORC technology have inherent feature (compared to water) and that is they have low boiling temperature and high vapor pressure at relatively low temperatures, compared with steam (water) (Dickson & Fanelli, 2005).

Typical ORC organic fluids may include pure hydrocarbons (e.g. pentane, butane, propane, etc), refrigerants (e.g. R134a, R218, R123, R113, R125, etc), or organic mixtures (Panea et al., 2010; Saleh et al., 2007; Hung, 2001; Wei et al., 2007). The optimal energy conversion performance of a low-temperature geothermal ORC power generation system depends mainly on the type of organic fluid being used in the system (Ismail, 2011a). The selection of the type of organic fluid is normally based on the following criteria (Hettiarachchi et al.,2007; Saleh et al., 2007; Chandrasekharam& Bundschuh, 2008; Ismail, 2011b):

- The ORC organic fluid should be environmentally friendly; less in ozone depletion potential (ODP) and global warming potential (GWP).
- It should result in high thermal efficiency by allowing maximum utilization of the available low-temperature geothermal heat source.
- It should be safe (non-flammable and no-toxic) and non-corrosive.
- It should have a low-boiling temperature and should evaporate at atmospheric pressure.
- It should lead to optimum design and cost effectiveness of the ORC system.
- It should not react or disassociate at the pressures and temperatures at which it is used.
- It should have suitable thermal stability and high thermal conductivity.
- It should have appropriate low critical temperature and pressure.
- It should have small specific volume, low viscosity and surface tension.
- It should result in low maintenance.

It should be noted that many binary ORC fluids may not meet all these criteria (Chandrasekharam& Bundschuh, 2008) but the selection of the organic fluid should be optimized, in terms of the above requirements, while meeting the demanded power generation. In general, binary ORC systems exhibit great flexibility, high safety (installations are perfectly tight), and low maintenance (Wei et al., 2007). It was reported that the selection of suitable organic fluids for application in binary ORC systems for generating electricity still deserves extensive thermodynamic and technical studies (Maizza, V., & Maizza, A., 2001).

The quality of heat energy which can be supplied by any heat source depends on its temperature level. For ORC-based geothermal power system, this is the temperature of the produced geo-fluid from the geothermal production well. The theoretical overall performance of low-temperature geothermal binary systems can be evaluated using the thermal efficiency of a heat engine, given by (Cengel& Boles, 2008)

$$\eta_{th} = \frac{\dot{W}_{out}}{\dot{Q}_{geo}} \tag{1}$$

In Eq. (1), $\dot{W}_{out}$ is the net power output produced by the geothermal power system (in kW_e); and $\dot{Q}_{geo}$ is the thermal heat supplied by the geo-fluid from the available geothermal resource (in kW_t). A correlation is proposed (Dickson & Fanelli, 2005) to calculate the actual net power output (used for a quick estimate with rough accuracy) as a function of the available thermal power from the geo-fluid flowand inlet temperature of the geo-fluid, given by

$$\dot{W}_{out} = 0.0036\ \dot{Q}_{geo}(0.18\ T_{geo,in} - 10) \tag{2}$$

Substituting Eq. (2) in Eq. (1), the estimated thermal efficiency of the low-temperature based geothermal power generation system, as a function of geo-fluid inlet temperature (in °C) available at the production well, is given by

http://dx.doi.org/10.5772/52063

$$\eta_{th} = 0.000648\ T_{geo,in} - 0.036 \tag{3}$$

For example, using Eq. (3) it can be estimated that a thermal efficiency of approximately 4.8% could be achieved for power generation with a geo-fluid extracted from a low-temperature geothermal resource available at 130 °C. The thermal efficiency as a function of the geothermal heat resource temperature, $T_{geo,in}$ (in K), and ambient temperature, T_o (in K) is given by (DiPippo, 2007)

$$\eta_{th} \cong 0.58 \left(\frac{T_{geo,in} - T_o}{T_{geo,in} + T_o} \right) \tag{4}$$

So for example, with a geothermal heat resource temperature of 130°C and ambient temperature of 25°C, the thermal efficiency is estimated to be 8.7%, using Eq. (4). It should be noted that Eq. (4) is valid for resource temperatures between 100 and 140 °C. The estimated net power output produced by the geothermal power system can also be determined using (DiPippo, 2007)

$$\dot{W}_{out} \cong 2.47\ \dot{m}_{geo} \left(\frac{T_{geo,in} - T_o}{T_{geo,in} + T_o} \right) (T_{geo,in} - T_{sink}) \tag{5}$$

In Eq. (5), $\dot{m}_{geo}$ is the geo-fluid mass flow rate; and T_{sink} is the heat sink temperature. It should be noted that the above correlations given by Eqs. (2) through (5) provide quick estimate of the thermal efficiency and net power output. However, for more accurate system performance predictions, a detailed energy analysis should be performed to predict the net power, the available geothermal heat, and overall thermal efficiency using Eq. (1). Since the geothermal energy is produced at low enthalpy levels,ORC-based low-temperature geothermal power generation plants tend to have low thermal efficiencies: 10-13% reported by (DiPippo, 2007), 2.8-5.5% reported by (Gupta & Roy, 2007), and 5-9% reported by (Hettiarachchi et al., 2007). Maximizing generating power capacity is normally sought from these power plants by maximizing the geo-fluid flow rate (depending on the capability of the production well) with a limited geo-fluid temperature available from the geothermal resource. It was reported (Chandrasekharam& Bundschuh, 2008) that low-temperature geothermal production wells with geo-fluid temperature < 150 °C and geo-fluid flow rate > 900 l/min could generate electric power ranging from 50 to 700 kW$_e$. When appropriate, multiple production wells could be installed using the same low-temperature geothermal energy reservoir so that a number of ORC power generation units could be cascaded to obtain larger power production rates from the plant (Gupta & Roy, 2007). Limited by the second-law of thermodynamics, the ideal (absolute maximum) efficiency of a thermoelectric power cycle, such as the low-temperature geothermal ORC power cycle, operating as a reversible heat engine between a heat source at a temperature T_H and a heat sink at a temperature T_L is Carnot efficiency, given as (Cengel& Boles, 2008)

$$\eta_{ideal} = \eta_{Carnot} = 1 - \frac{T_L}{T_H} \tag{6}$$

For example, for an ORC power system using a geo-fluid extracted from a low-temperature geothermal heat source at 130 °C (403 K) and a heat sink (condenser) at 40 °C (313 K), the maximum ideal Carnot efficiency can be calculated using Eq. (6) to be approximately 22.3%. For an actual (irreversible) ORC-based geothermalsystem operating between the same temperature limits would have lower efficiency. Another measure of the performance of the low-temperature geothermal ORC power plant can be obtained using the Second-Law of thermodynamics in the form of exergetic efficiency, η_{ex} , given as

$$\eta_{ex} = \frac{\dot{W}_{out}}{\dot{Ex}_{geo}} \tag{7}$$

The exergetic efficiency in Eq. (7) is defined as the ratio of the actual net power output from the power generation system to the maximum theoretical power that could be extracted from the geo-fluid at the geothermal resource state relative to the thermodynamic dead-state. This involves determining the rate of exergy carried by the geo-fluid to the ORC power system. Typically, the design and operation of geothermal binary power generation systems should be optimized in order to increase their thermal and exergetic efficiencies guided by the Carnot efficiency (Ismail, 2011b).

2.3. ORC-based low-temperature geothermal power generation: Environmental & economic aspects

Geothermal power generation is relatively pollution-free and considered to be a clean technology for power generation (Dickson & Fanelli, 2005) and it tends to have the largest technological potential compared to other renewable energy sources used for power generation (Hammons, 2004). Once up and running, GHG emissions are typically zero when low-temperature geothermal energy reservoirs are utilized using ORC power systems, since all of the produced geo-fluid is injected back into the reservoir (Hammons, 2004). In this case, one of the effective ways of getting rid of hazardous chemical constituents of geothermal water (e.g. trace metals) is re-injection. ORC-based low-temperature geothermal power generation systems are far less environmentally intrusive than alternative power generation systems in several respects, e.g. they are essentially zero-GHG emission systems and have low land usage per installed megawatt (DiPippo, 2008). As far as physical environmental effects, geothermal projects may cause some kind of disruption activities as other same size and complexity of civil engineering projects. Also, the locations of excavations and sitting of boreholes and roads will have to be taken into account, soil and vegetation erosion, which may cause changes in ecosystems, has to be watched. It should be noted that many geothermal installations are in remote areas where the natural level of noise is low and any additional noise is very noticeable (Dickson & Fanelli, 2005). There is a relatively larger production of waste-heat energy in geothermal systems, and this needs to be dissipated in an environmentally acceptable way. In ORC-based low-temperature geothermal power systems, the thermal impact is much reduced by disposing of waste geothermal water using deep re-injection approach so that the thermal impact of the waste heat becomes insignificant (Dickson & Fanelli, 2005). Appropriate measures should be applied to prevent leakage of the binary working fluid from ORC power generation units to the environment (Yamada & Oyama, 2004); normally the installations of these units are made perfectly tight to meet high safety standards.

In theory, geothermal energy potential is present below the entire surface of the Earth. In practice however, special geologic settings are required for geothermal energy to be economically exploited (Grasby et al., 2011). Generating electricity using ORC-based geothermal technology is very cost-effective and reliable (Chandrasekharam& Bundschuh, 2008; Dickson & Fanelli, 2005). Table 1 compares electrical energy costs produced by various renewable energy technologies.The cost of geothermal energy for generating electricity is favourable compared to other energy sources. The reported costs of low-temperature based small geothermal power plants vary from 0.05 to 0.07 US$/kWh for units generating < 5 MW_e (Chandrasekharam& Bundschuh, 2008).

Renewable Energy Source	Current Energy Cost (US cents/kWh)	Turnkey Investment Cost (US$/kW$_e$)	Potential Future Energy Cost (US cents/kWh)
Geothermal	2-10	800 – 3,000	1-8
Wind	5-13	1,100 – 1,700	3-10
Solar photovoltaic	25-125	5,000 – 10,000	5-25
Solar thermal	12-18	3,000 – 4,000	4-10
Biomass	5-15	900 – 3,000	4-10
Tidal	8-15	1,700 – 2,500	8-15
Hydro	2-10	1,000 – 3,000	NA

Table 1. Energy and investment costs for electric power production from different renewable energy sources (Hammons, 2004; Dickson & Fanelli, 2005).

The unit cost of electricity generated from low-temperature geothermal based small power plants is compared in Table 2. Moreover, the unit cost of electricity from small-scale geothermal plants (<5 MW_e) is much lower than the average cost of 0.25 US$/kWh supplied through diesel generators (Chandrasekharam& Bundschuh, 2008). The total investment for a geothermal power plant mainly includes the following types of costs: (1) cost of exploitation, (2) cost of drilling, (3) cost of power plant (capital cost of design and construction), and (4) operating & maintenance costs (Chandrasekharam& Bundschuh, 2008). The first two types are referred to as subsurface costs whereas the other two are referred to surface costs. The high initial investments incurred through the exploration, drilling and development of wells and the production field is an important constraint on future geothermal power development. Despite low maintenance and operational costs, high initial investments are often a strong restrictive(Grasby et al., 2011). For small-scale geothermal power plants (<5 MW_e) utilizing low-temperature resources, the subsurface cost typically accounts for approximately 30% of the total investment costs whereas the surface cost accounts for the remaining 70%.

Net Power (kW_e)	Capital Cost (US$/net kW_e) Geothermal Resource Temperature (°C)			O&M Cost (US$/year)
	100	120	140	
100	2,786	2,429	2,215	21,010
200	2,572	2,242	2,044	27,115
500	2,357	2,055	1,874	33,446
1000	2,143	1,868	1,704	48,400

Table 2. Unit cost of electricity generated from low-temperature based small power plants (Chandrasekharam& Bundschuh, 2008; DiPippo, 2008).

Generating electricity using low-temperature geothermal ORC technology is very reliable due to its advanced technological aspects. However, the maintenance costs and shutdowns could be reduced when the technical complexity of the plant is on a level that is accessible to local technical personnel or to experts who are readily available (Dickson & Fanelli, 2005). As mentioned before, geothermal ORC power generation plants are normally constructed and installed in small modular power generation units. These units can then be linked up to create power plants with larger power production rates. Their cost depends on a number of factors, but mainly on the temperature of the geothermal fluid produced, which influences the size of the ORC turbine, heat exchangers and cooling system. It was reported (Dickson & Fanelli, 2005) that the total size of the plant has little effect on the specific cost, as a series of standard modular units is linked together to obtain larger power capacities. It was also reported (Panea et al., 2010) that the modular units have a satisfying economic efficiency, because modular construction reduces installation time and costs. Ultimately, the economic viability of the geothermal power plant depends on its ability to generate revenue in the long-term.

3. CO_2 – based EGS for combined power generation& CO_2 sequestration

3.1. Enhanced geothermal systems (EGS) – developments & utilization

EGS, also known as engineered geothermal systems, are reservoirs that have been stimulated(e.g. hydraulic stimulation) and engineered to extracteconomical amounts of heat from unproductive geothermal resources that lack heat-carrier fluid circulation, permeability and/or porosity. EGS is a new type of geothermal power technologyand has the potential to become a significant sustainable and renewable power source for the future (Grasby et al., 2011, Kalra et al., 2012). The EGS concept is currently the subject of several international research investigations and once brought to successful production, will significantly expand the regions where geothermal powergeneration is feasible. In an EGS, a fluid (typically cold water or brine) is injected and fractures are induced to form subsurface heat exchange systems. The heated fluid is then produced from a parallel well where heat can be used at surface to generate electricity (Huenges, 2010; Majorowicz& Grasby, 2010). Additional production-injection wells are drilled to extract more heat from large volumes of rock mass to meet power generation requirement (Azim et al., 2010). This technology does not require conventional natural convective hydrothermal resources located at depth, nor an initial high permeability of the reservoir, for power generation(once linked with ORC power technology). A schematic diagram showing a typical EGS concept is shown in Figure 3. EGS has the potential for accessing the Earth's vast resources of heat located at depth to help meet future increasing energy demands. The EGS concept has driven increased interest in widelydevelopment of this geothermal energy potential by orders of magnitude (Huenges, 2010; Majorowicz& Grasby, 2010; Azim et al., 2010). It was reported (Chandrasekharam& Bundschuh, 2008) that developing countries can access all low-temperature geothermal and EGS sources for green electricity generation immediately.

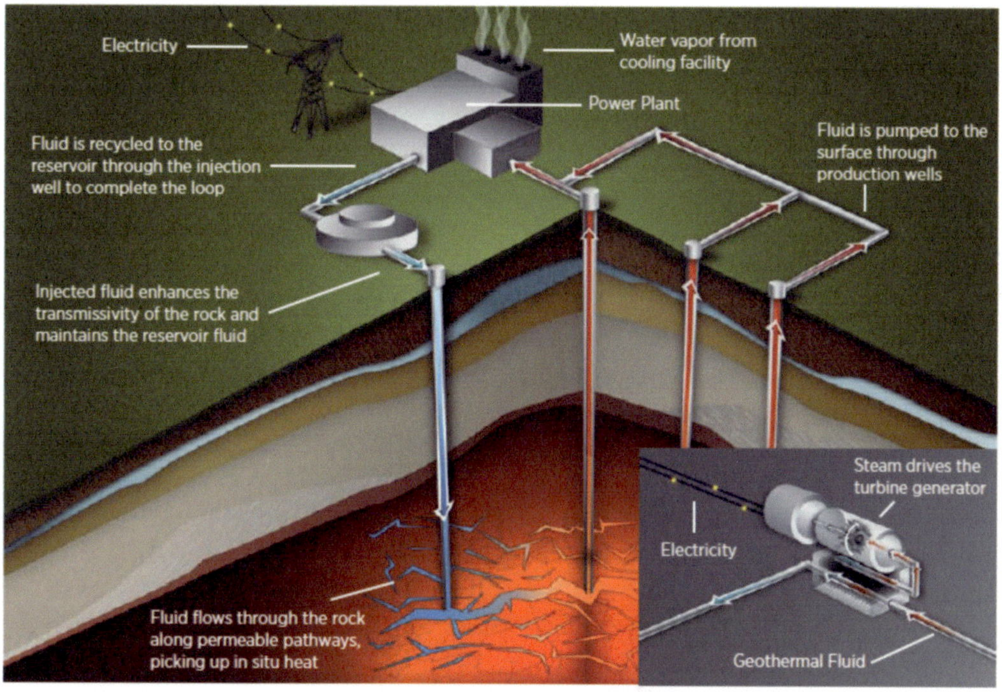

Figure 3. A conceptual model showing how EGS works (Source: http://energyinformative.org)

The basis, on which today's EGS projects are developed were laid out in the early 1970s, when an EGS (hot dry rock (HDR)) development concept was implemented at Los Alamos National lab, involving drilling a well into hot crystalline rock, using pressurized water to create a large vertical fracture, and ultimately to drill a second well to access that fracture at some distance above the first wellbore (Huenges, 2010; Stephens & Jiusto, 2010). It was the first attempt anywhere to make a deep, full-scale HDR reservoir. Significant research and production plants are under construction or operation to take advantage of this abundant renewable energy opportunity (Kalra et al., 2012). For example, power plants driven by EGS are currently being developed and tested in Australia, Germany, Japan, France, USA and Switzerland (DiPippo, 2008; Azim et al., 2010).The largest EGS project in the world is a 25 MW_e demonstration plant currently being developed in Cooper Basin, Australia.There are several EGSprojects that are already or will soon produce power. Someof them are just for-research and development (R& D) and some are for commercial purpose. Examples of theEGS projects around the world with their location, capacity, welldepth, plant type, and the project type are summarized in Table 3.

EGS Project	Type	Country	Capacity (MW_e)	Plant Type	Depth (km)
Soultz (EU)	R&D	France	1.5	Binary (ORC)	4.2
Desert Peak	R&D	USA	11-50	Binary (ORC)	(Unknown)
Landau	Commercial	Germany (EU)	3	Binary (ORC)	3.3
Paralana (phase I)	Commercial	Australia	7-30	Binary (ORC)	4.1
Cooper Basin	Commercial	Australia	250-500	Binary (Kalina)	4.3
The Geysers	Demonstration	USA	(Unknown)	Flash	3.5-3.8
Ogachi	R&D	Japan	(Unknown)	Flash	1.0-1.1
United Downs, Redruth	Commercial	United Kingdom	10	Binary (ORC)	4.5
Eden Project	Commercial	United Kingdom	3	Binary (ORC)	3-4

Table 3. EGS-based projects around the world (Azim et al., 2010).

An EGS project has several stages (Majorowicz& Grasby, 2010; Huenges, 2010; DiPippo, 2008); namely: (1) Identifying a potential site possessing the necessary characteristics through surface exploration, (2) Drilling an injection well to the depth required to reach the desired temperature, (3) Fracturing the rock in the subsurface by hydraulic stimulation (i.e. injecting a fluid at sufficient pressure to propagate fracture), (4) Creating and testing of the EGS reservoir storage capacity, (5) Drilling a production well for a doublet system or two-production wells for a triplet system (one injection & two production wells), (6) Creating fracture connectivity between the injection andthe production wells, (7) Extracting heat en-

http://dx.doi.org/10.5772/52063

ergy from the rock by injecting cool fluid (typically water) through the injection well and producing hot fluid (in some cases steam)from the production wells, and (8) Operating the ORC power plant and maintain the EGS reservoir.

EGS for power generation is still relatively novel technology and remains to be proved on a large scale. Engineers & researchers in several countries throughout the world have been working on advancing EGS technology for few decades, but the technology has received limited attention and minimal financial support from either the public or private sector, with the exception of Australia's significant market investments. The high cost of drilling, which is estimated to account for a third to a half of EGS projected costs, is a major challenge to the technology (Stephens & Jiusto, 2010). The risks and uncertainties associated with EGS technology are other barriers as well.

3.2. Environmental and economic aspectsof EGS for power generation

It was reported (Grasby et al., 2011) that impacts of geothermal development are relatively minor compared to many other energy developments, however there are still importantchallenges to be addressed. More particularly, it was reported (Azim et al., 2010) that the overall impact of EGS power generation plants on environment is remarkably lower than other conventional fossil fuel-fired and nuclear power plants.In addition, EGS plants may have lower impacts in comparison with other renewable energy sources such as solar, wind, hydroelectric, and biomass on an equivalent energy-output basis. This is primarily because a geothermal energy source is contained subsurface and need not to be exposed in the atmosphere and the surface energy conversion system (e.g. ORC unit) is relatively compact, thus making the overall size of the entire system attractively small. EGS power plants also provide environmental benefits by having minimal GHGand other emissions (zero emissions for the case of using ORC technology). Distinct from the conventional fossil fuels, there are minimal discharges of CO_2, nitrogen or sulphur oxides or particulate matter resulting from its use, and there is no need to dispose radioactive materials. However, still there are impacts that must be considered and managed if enhanced geothermal energy resource is to be developed as part of a more environmentally sound, sustainable energy source for the future. The major environmental challenges for EGS are associated with ground water use and contamination, with related concerns about induced seismicity or subsidence as a result of water injection and production. Induced seismicity is a phenomenon in which a change in fluid pressure within a stressed rock formation leads to movement of the fractured rocks; the energy released is transmitted through the subsurface rock and may reach the surface with enough intensity to be heard or felt by people in the surrounding region (DiPippo, 2008; Majer et al., 2007). To mitigate risks related to induced seismicity, strategies and procedures are needed to set requirements for seismic monitoring and for prolonged EGS field operation. Technologies for imaging fluid pathways induced/injected by hydraulic stimulation in EGS fields would constitute a key improvement of the EGS concept. Issues of noise, safety and land use associated with drilling and production operations are also important but can be fully manageable (Huenges, 2010; Majorowicz& Grasby, 2010; Azim et al., 2010).

EGS technology has impacts on the global and local environment (Huenges, 2010; DiPippo, 2008). Therefore, it is important to identify and evaluate any impact which results from the implementation of an EGS plant at the beginning of a project. The goal must be to avoid or minimize adverse impacts on the environment during all stages of an EGS project (e.g., construction, operation, and deconstruction) and to meet the objectives and requirements of environment protection and preservation of finite resources (Huenges, 2010; Stephens & Jiusto, 2010). Potential impactsneed to be addressedduring planning environmentally sound EGS plants. The relevance and extent of the addressed impacts can vary from location to location. Even if EGS plants are not related to (continuous) gaseous emissions during operation (due to the carrying of the geothermal fluid in a closed loop system on the surface), environmental impacts such as airborne emissions or the consumptions of the finite energy resources (such as steel used for well completion or fuel for drilling rig operation) occur during other life cycle stages, therefore, all life cycle stages need to be considered in order to analyze the environmental performance of an EGS plant. In this regard, life cycle analysis (LCA) is a widely applied approach to evaluate and compare specific environmental impacts of different products or technologies (Huenges, 2010; Frick et al., 2010). The idea is to carry out a detailed analysis of the life cycle of the product or a duty emerged in response to increased environmental awareness of the public, the governments, and the industries. An LCA involves two main stages: the collection of a data, related to the product or duty and relevant for the environment, and interpretation of the collected information. For transparency and traceability of LCA results, standards, such as ISO 14040, ISO 14044 have been developed (Huenges, 2010; Frick et al., 2010). Based on this approach, aspects, which influence the environmental impact during the life cycle, and parameters, which need to be considered in the planning of the environmentally sound EGS plants, can be identified. According to given standards, the LCA is carried out in four steps: (1) Goal and scope definition to assess selected environmental effects (e.g., global warming potential, cumulated demand of finite energy resources, etc.) in the different life cycle stages and throughout the whole life cycle of EGS plants, (2) Inventory analysis, (3) Impact analysis to quantify the environmental effects (all inventoried material and energy flows are transformed to different impact indicators based on certain conversion factors), and (4) Interpretation of the results from the impact analysis.

EGS plants are related to different impacts on diverse parts of the local environment and different characteristics regarding their duration (continuous or temporary), reversibility, and their degree of probability. Many of the impacts on the local environment are related to assessing the EGS reservoir. Most of these impacts are known and technologicallymanageable based on the oil and gas exploration experiences. Adverse effects due to reservoir exploitation can be avoided with proper reservoir management and monitoring. Environmental impacts from the construction and operating the surface facilities are comparatively low (Dickson & Fanelli, 2005; DiPippo, 2008). In an EGS project, some environmental impacts and risks, however,need to be considered and evaluatedin connection withthe following EGS phases and activities (Huenges, 2010; Majer et al., 2007; DiPippo, 2008; Stephens & Jiusto, 2010):

http://dx.doi.org/10.5772/52063

1. Drilling operations: Drilling operations have a large impact on the surrounding environment and are associated withvarious risks. They are normally limited to the period of drilling operations (i.e., last only a couple of weeks or months). Environmental effects related to drilling operations may include: noise emissions, subsurface emissions, site preparations and alterations, airborne emissions, water usage (usually taken from nearby surface or groundwater bodies), waste disposal, and visual impact (due to night-time lighting, etc., from drilling rig).

2. Reservoir stimulation: Enhancing a geothermal reservoir for power generation includes injection of stimulation fluids under high pressure which could produce geo-mechanical changes in the subsurface (rocks may slip along pre-existing fractures. Depending on the surrounding rock formations and their magnitude, geo-mechanical alterations can be followed by build-up and dispersion of microseismic activities up to the surface. Induced seismicity, which can result from stimulation, helps to identify the extent of the fracture network in the reservoir. Despite the fact that geo-mechanical changes can lead to damage of buildings and even be hazardous for human beings and animals, the earth tremors caused by EGS reported so far can be categorized a sensible but not as adverse impacts on the environment.In almost all cases, these events in the deep reservoir are of such low magnitude that they are not felt at the surface. However, based on the present state of knowledge, larger impacts cannot be totally excluded since the knowledge about the stress situation and the development of larger microseismic events in the subsurface is still insufficient.

3. Reservoir exploitation: The exploitation of an EGS reservoir can lead to different alterations in the reservoir and the surrounding subsurface. Impacts on the local subsurface environment such as hydraulic and thermal alterations as well as circulation losses need to be considered for sustainable reservoir management, but are not considered as adverse environmental impacts.

4. Installation and operation of surface facilities: The installation of the subsurface part of an EGS power plant, such as a binary ORC power unit is related to general environmental impacts which come with construction work such as noise emissions and dust creation.

5. Dismantling of well infrastructure: In EGS plants, the decommissioning of the deep wells need be managed since abandoned wells are a potential source for material emissions to the subsurface. With proper closing and filling of the wells, this can be eradicated.

For the development of EGS projects, different environmental regulations, standards, and permission procedures are binding depending on the country's legislations. A widely used tool in this context is the environmental impact assessment (EIA); first established in USA in the 1970s and used today in many countries to ensure that all possible environmental impacts of planned EGS projects are identified and assessed before a decision is made for getting permission for an EGS project for power generation (Huenges, 2010). The most important steps of atypical EIA process are briefly outlined in Figure 4.

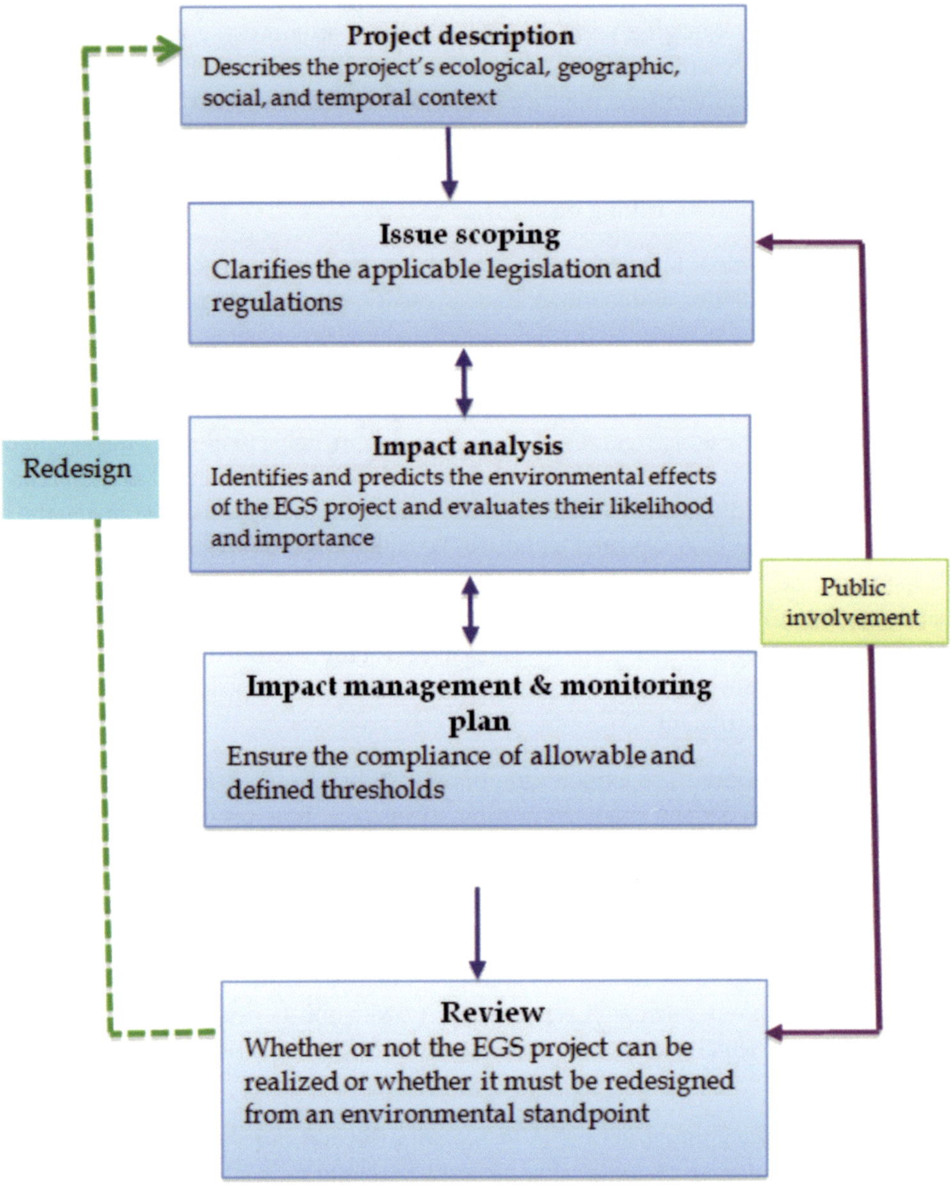

Figure 4. A schematic diagram showing the important steps in an EIAprocessfor an EGS project (Huenges, 2010).

Costs for geothermal plants have been dropping and are becoming more competitive (Pruess, 2006). EGS-based ORC power technology has the potential to replace other more costly and environmentally destructive technologies.It also has the potential to provide greenpower generation at affordable prices, thereby improving the standard of living andsocio-economic potential through creation of jobs in many regions (Grasby et al., 2011). The-planning of EGS projects and especially the decision to realize a project is based on the estimation of the costs and revenues, which are related to a project. Since EGS projects are characterized by a long planning period, large initial investments and a long technical life-time, estimating prospective costs and revenues involves uncertainties and risks (Huenges, 2010). This is true because no reliable statements on markets development, detailed geologic site conditions, or technological problems can be made at the start of a project. In order to minimize existing risks, cost influences must be known and risks must be analyzed. The total costs of an ORC-based EGS project are dominated by the investments at the start of the project (Huenges, 2010; Stephens & Jiusto, 2010; DiPippo, 2008). These investments mainly consist of costs for the subsurface components, including: (1) reservoir exploration, (2) well drilling and completion - the most significant cost factor in all geothermal operations is associated with drilling and well completion (Grasby et al., 2011), (3) reservoir engineering measures, (4) installation of the geothermal fluid loop, and (5) construction of the ORC-based power unit.Typical EGS drilling costs as a function of well depth are shown in Table 4. As mentioned before, drilling costs estimated to account for a third to a half of EGS projected costs; a major challenge to the EGS technology (Stephens & Jiusto, 2010). As shown in Table 4, EGS well costs are not a linear function of depth, but additionally reflect, temperature, extent of casing employed, difficulty in drilling, and lithologic characteristics(Grasby et al., 2011). In addition, surface related costs, such as operation & maintenance O& M costs (i.e., annual operating costs for personnel, material consumption, overhaul and maintenance) are considered. The operating and maintenance cost of an EGS power plant has two important components: (a) the O& M for the ORC power plant and (b) the well field maintenance cost. The ORC power plant O& M costs were estimated based on experience in similar power plant configurations and ORC installations. This is usually a percentage of the installed cost of the power plant on a yearly basis (Kalra et al., 2012). It was reported (Grasby et al., 2011) that the techniques and technologies related to EGS projects are evolving rapidly so as its estimated costs and that EGS technology may not be commercially viable at this time. However, it was suggested (Grasby et al., 2011) that renewable resources such as EGS offer attractive power market contributions beyond power generation.

Well Depth (km)	EGS Well Category	Estimated EGS Drilling Cost ($ millions)
1.5	Shallow well	0.90
2.5		1.81
3.0		2.55
4.0	Mid range well	5.10
5.0		6.45
6.0	Deep well	8.92
7.5		13.83

Table 4. Typical EGS drilling costs as a function of well depth (Azim et al., 2010).

3.3. EGS using CO_2 as the working fluid for green power generation and simultaneous carbon sequestration

It was reported (Pruess, 2006) that previous attempts to develop EGS in Japan, USA, Europe and Australia have all employedwater as a heat transmission fluid. Although, water has many properties that make it a favorable medium for this purpose, it also has serious shortcomings. An unfavorable property of water is that it is astrong solvent for many rock minerals, especially at elevated temperatures. In this case, injecting water at high pressure intohot rock fractures, as part of an EGS resource operation & utilization, results in strong dissolution and precipitation effects that change fracture permeabilityand make it very difficult to operate an EGS reservoir in a stable manner. In 2000, Brown, D. (Pruess, 2006) proposed a novel EGS concept that would utilize supercritical CO_2 instead of water as heat exchange (carrier) fluid, and would simultaneously achieve CO_2 geologic sequestration as an additional benefit. There are only very few investigations that characterized the performance of CO_2 as working fluid in EGS applications. For example, Pruess (Pruess, 2006) performed numerical simulations and evaluated thermophysical properties in order to explore the heat transfer and fluid dynamics characteristics in an EGS reservoir that would be operated with CO_2. It was found that CO_2 is superior to water in its ability to exchange heat from hot fractured rock. Carbon dioxide also offers certain advantages with respect to wellbore hydraulics, in that its larger compressibility and expansivity as compared to water would increase buoyancy forces and would decrease the parasitic power consumption (thus reduce pumping cost) of the EGS fluid circulation system. This is because the larger expansivity of CO_2 would generate large density differences between the cold CO_2 in the injection well and the hot CO_2 in the production well, and therefore provide buoyancy force that would reduce the power consumption of the fluid circulation system. Another interesting feature of CO_2 is that its lower viscosity, tend to yield larger flow velocities for a given pressure gradient. In addition, CO_2 would be much less effective as a solvent for rock minerals, which would reduce or eliminate scaling problems, such as silica dissolution and precipitation in water-based systems (Pruess, 2006). It was also reported (Pruess, 2006) that while the thermal and hydraulic aspects of aCO_2-based EGS system look promising, major uncertainties remain with regard to geochemical interactions betweenfluids and rocks. It was concluded in (Pruess, 2006) that an EGS system running on CO_2 has sufficiently attractive features to warrant furtherinvestigation. It was suggested that an EGS using CO_2 as heat transport and exchange fluid could have favorable geochemical properties, as CO_2 uptake and sequestration by rock minerals would be quite rapid.

Supercritical CO_2 can also be used as the working fluid of the power cycle before it is sent back to the EGS reservoir. For example, ina study by (Gurgenic et al., 2008), it was reported that there is a significant potential to use supercritical CO_2 as working fluid in the power loop as illustrated (Gurgenic et al., 2008) in Figure 5. Significantly higher energy conversion efficiencies were predicted using a single-loop system with the CO_2 being both the heat exchange and the power cycle working fluid. It was reported (Gurgenic et al., 2008; Atrens et al., 2011) that the loops in either of the two cycles (i.e. subsurface loop and surface power loop) do not have to be closed. For example, if there is ready access to CO_2 (e.g., at a geothermal installation situated close to a coal-fired power plant), the captured CO_2 from the plant can be run through the geothermal reservoir first and then sequestered in a geologic sequestration site of choice.

CO_2-based EGS has been examined in (Atrens et al., 2011) from a reservoir oriented perspective, and as a result thermodynamic performance was investigated. It was reported (Atrens et al., 2011) that economics of the system are still not well understood, however. In their study, the economics of the CO_2–based EGS technology was explored for an optimized power plant design and best-available cost estimation data. It was demonstrated in (Atrens et al., 2011) that near-optimum turbine exhaust pressure can be estimated from surface temperature. It was found that achievable cooling temperature is an important economic site consideration alongside EGS resource temperature. The role of sequestration as part of CO_2–based EGS was also examined in (Atrens et al., 2011), and it was concluded that if fluid losses occur, the economic viability of the concept depends strongly on the price associated with CO_2 (Atrens et al., 2011). Potential barriers to implementation of CO_2–based EGS technology include access to CO_2 at an acceptable cost, proximity of the EGS to the electricity grid, and access to cooling water. Similar issues related to long-term responsibility for the resultant reservoir, including the liability for future CO_2 leakage from the geologic sequestration site.In another study by (Randolph & Saar, 2011), it was suggested that using CO_2 as the working fluid in geothermal power systems may permit utilization of lower temperature geologic formations than those that are currently deemed economically viable,l eading to more widespread utilization of geothermal energy. However, additional exploration of economics regarding the opportunities and issues for CO_2–based EGS technology for combined carbon sequestration and power generation is needed.

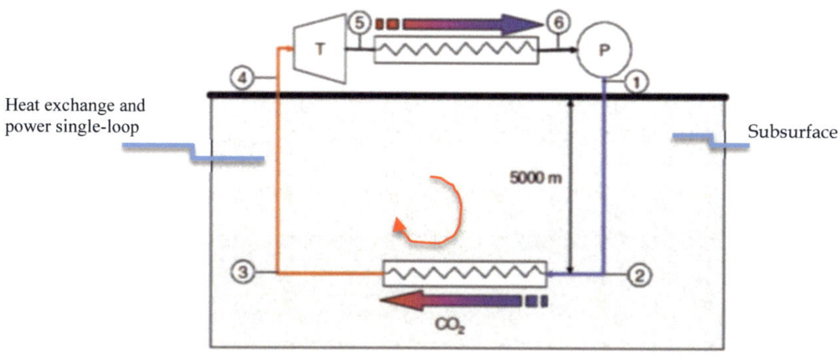

Figure 5. A conceptual model showing a single-loop system with CO_2 used for combined heat exchange and power cycle (Gurgenic et al., 2008).

4. Conclusion

An increasing concern of environmental issues of emissions & pollution, in particular global warmingand the constraints on consuming conventional energy sources has recently resulted in extensive research into innovative renewable and green technologies of generating electrical power. One of these innovative emerging technologies includes renewable low-temperature (low-enthalpy) geothermal energy source for clean electrical power generation. This promising technology offers potential applications in generation of electric power which can be produced using the vast renewable low-temperature geothermal energy resources available worldwide.In this chapter, the concept of ORC binary technologyfor power generation using low-temperature geothermal heat source was introduced and its potential applications and limitations for small-scale geothermal power generation and its relevant environmental and economic considerations were presented and discussed. Also, recent developments of ORC-based low-temperature geothermal power generation with their significant and relevant applications were presented and discussed. A number of successful ORC binary plants were installed in different locations (e.g. remote and rural sites) worldwide which demonstrated the ability of this promising alternative and green technology to utilize renewable low-temperature geothermal energy sources for generating electricity. Also, several patents were reported on the application of this innovative technology. Geothermal ORC power generation plants are normally constructed and installed in small modular power generation units. These units can then be linked up to create power plants with larger power production rates. Their cost depends on a number of factors, but mainly on the temperature of the geothermal fluid produced, which influences the size of the ORC turbine, heat exchangers and cooling system. Currently, ORC power cycles exhibit great flexibility, high safety (installations are perfectly tight), and low maintenance when coupled with low-enthalpy geothermal heat sources. The future use of low-temperature geothermal energy resources for generating electricity would very much depend on further overcoming technical barriers both in utilization and production, and its economic viability compared to other conventional and renewable energy sources used for power production. Another emerging "dual-benefit" technology is EGS using CO_2 as the working fluid for combined clean power generation and geologic CO_2 sequestration. CO_2 is of interest as a geothermal working fluid mainly because it transfers geothermal heat more efficiently than water. While power can be produced more efficiently using this technology, there is an additional benefit CCS for reducing GHG emissions. The second part of the chapter presented the merits and fundamental aspects of CO_2-based EGS technology.In 2000, Brown, D. (Pruess, 2006) proposed a novel EGS concept that would utilize supercritical CO_2 instead of water as a more efficient heat exchange (carrier) fluid (due to its favorable properties over water), and would simultaneously achieve CO_2 geologic sequestration as an additional benefit.It was found that CO_2 is superior to water in its ability to exchange heat from EGS hot fractured rock and reduce hydraulic power consumption for fluid injection and circulation in the EGS reservoir. It was concluded that an EGS system running on CO_2 has sufficiently attractive features to warrant furtherinvestigation.It was also concluded that EGS for power generation is still relatively a novel technology and remains to be proved on a large scale and that further research is needed for additional exploration oftechnological and economic aspects regarding the opportunities and challenges for CO_2–based EGS technology for combined carbon sequestration and power generation.

Acknowledgements

The author of this chapter would like to acknowledge the funding contribution by Goldcorp Canada Ltd.-Musselwhite Gold Mine that mainly supported the collaborative geothermal energy & heat pump (GHP) technology research project (author was the PI of the project) at their site in Northern Ontario; a contracted research project with Lakehead University (2007-09). Acknowledgement also goes to Natural Sciences and Engineering Research Council of Canada (NSERC) for the Discovery Grant (Individual) funding that was provided to the author's research in the area of clean energy technologies related to CO_2 membrane gas separation from industrial flue gases for GHG emissions reduction.

Author details

Basel I. Ismail*

Address all correspondence to:

Department of Mechanical Engineering, Lakehead University, Thunder Bay, Ontario, Canada

References

[1] Atrens, A. D., Gurgenic, H., Rudolph, V., & (2011, . (2011). Economic optimization of a CO_2-based EGS plant. *Energy & Fuels,*, 25, 3765-3775, ACS Publications.

[2] Azim, M. R., Amin, M. S., & Shoeb, A. (2010). Prospects of enhanced geothermal system in baseload power generation. *IEEE,*, 10, 176-180, 0000-9781.

[3] Cengel, Y. A., & Boles, M. A. (2008). Thermodynamics: an engineering approach (6th ed.),. McGraw-Hill press,, 978-0-07352-921-4, New York.

[4] Chandrasekharam, D., & Bundschuh, J. (2008). Low-enthalpy geothermal resources for power generation. CRC Press Taylor & Francis Group,, 978-0-41540-168-5, New York., 10.1201/9780203894552.

[5] Cui, J., Zhao, J., Dai, C., & Yang, B. (2009). Exergetic performance investigation of medium-low enthalpy geothermal power generation. *IEEE Computer Society,*, 636-639, 0000-9780.

[6] Dickson, M. H., Fanelli, M., & (2005, . (2005). Geothermal energy: utilization and technology,. Earthscan, an imprint of James & James (Science Publishers) Ltd. in Association with the International Institute for Environment and Development,, 1-84407-184-7.

[7] Di Pippo, R. (2008). Geothermal power plants: principles, applications, case studies and environmental impact. 2nd ed, Elsevier,, 978-0-75068-620-4, New York.

[8] Di Pippo, R. (2007). Ideal thermal efficiency for geothermal binary plants. *Geothermics,*, 36, 276-285, 0375-6505.

[9] Energyinformative. How enhanced geothermal systems (EGS) work. Available from:, http://energyinformative.org.

[10] Frick, S., Kaltschmitt, M., Schröder, G., & (2010, . (2010). Life cycle assessment of geothermal binary power plants using enhanced low-temperature reservoirs. *Energy*, 35, 2281-2294, 0360-5442.

[11] Galanis, N., Cayer, E., Roy, P., Denis, E. S., & Desilets, M. (2009). Electricity generation from low temperature sources. *Journal of Applied Fluid Mechanics,*, 2(2), 55-67, 1735-3645.

[12] Geothermal Energy Association, GEA,. (2012). Geothermal: International market overview report. Available from:, http://www.geo-energy.org.

[13] Grasby, S. E., Allen, D. M., Bell, S., Chen, Z., Ferguson, G., Jessop, A., Kelman, M., Ko, M., Majorowicz, J., Moore, M., Raymond, J., & Therrien, R. (2011). Geothermal energy resource potential of Canada. Geological Survey of Canada,. Open File 6914. Available from:, http://geoscan.ess.nrcan.gc.ca, 10.4095/288745.

[14] Gupta, H., & Roy, S. (2007). Geothermal energy: an alternative resource for the 21st century,. Elsevier,, 978-0-44452-875-9, New York.

[15] Gurgenic, H., Rudolph, V., Saha, T., & Lu, M. (2008). Challenges for geothermal energy utilisation. *Proceedings, 33rd Workshop on Geothermal Reservoir Engineering,*, Stanford University, Stanford, California,.

[16] Hammons, T. J. (2004). Geothermal power generation worldwide: global perspective, technology, field experience, and research and development. *Electric Power Components and Systems,*, 32, 529-553, 1532-5008.

[17] Hettiarachchi, H. D. M., Golubovic, M., Worek, W. M., & Ikegami, Y. (2007). Optimum design criteria for an organic Rankine cycle using low-temperature geothermal heat sources. *Energy,*, 32, 1698-1706, 0360-5442.

[18] Holdmann, G. (2007). ORC technology for waste heat applications. Presented at the Diesel Heat Recovery and Efficiency Workshop, Chena Power, Chena, Alaska, December 2007. Available from:, http://www.akenergyauthority.org.

[19] Holm, A., Blodgett, L., Jennejohn, D., Gawell, K., Geothermal, Energy., & Association, G. E. A. (2010). Geothermal energy: International market update. Available from:, http://www.geo-energy.org.

[20] Huenges, E., & (2010, . (2010). Geothermal energy systems, Wiley-VCH Verlag GmbH & Co. 978-3-52740-831-3, KGaA, Weinheim, Germany., 10.1002/9783527630479.

[21] Hung T.-C. (2001). Waste heat recovery of organic Rankine cycle using dry fluids. *Energy Conversion and Management*, 42, 539-553, 0196-8904.

[22] Ismail, B. I. (2011a). Advanced electrical power generation technology using renewable & clean low-enthalpy geothermal energy sources. *Recent Patents on Mechanical Engineering,*, 4(2), 168-179, 0187-4477 X.

[23] Ismail, B. I. (2011b). Power generation using nonconventional renewable geothermal and alternative clean energy technologies,. Chap. 18, in:, *Planet earth 2011 Global warming challenge and opportunities for policy and practice,*, Edited by Carayannis, E. G., INTECH Open Access Publisher,, 978-9-53307-733-8, Croatia.

[24] Kalra, C., Becquin, G., Jackson, J., Laursen, A. L., Chen, H., Hardy, A., Klockow, H., & Zia, J. (2012). High-potential working fluids and cycle concepts for next generation binary Organic Rankine Cycle for enhanced geothermal systems. *Proceedings, 37th Workshop on Geothermal Reservoir Engineering,*, Stanford University, Stanford, California,.

[25] Maizza, V., & Maizza, A. (2001). Unconventional working fluids in organic Rankine-cycles for waste energy recovery systems. *Applied Thermal Engineering*, 21, 381-390, 1359-4311.

[26] Majer, E., Baria, R., Stark, M., Oates, S., Bommer, J., Smith, B., Asanuma, H., & (2007, . (2007). Induced seismicity associated with Enhanced Geothermal Systems. *Geothermics,*, 36, 185-222, 0375-6505.

[27] Majorowicz, J., & Grasby, S. E. (2010). Heat flow, depth-temperature variations and stored thermal energy for enhanced geothermal systems in Canada. *J. Geophys. Eng.,*, 7, 232-241, 1742-2132.

[28] Panea, C., Rosca, G. M., & Blaga, C. A. (2010). Power generation from low-enthalpy geothermal resources. *Journal of Sustainable Energy,*, 1(2), 1-5, 2067-5538.

[29] Pruess, K. (2006). Enhanced geothermal systems (EGS) using CO_2 as working fluid-A novel approach for generating renewable energy with simultaneous sequestration of carbon. *Geothermics,*, 35, 351-367, 0375-6505.

[30] Randolph, J. B., & Saar, M. O. (2011). Coupling carbon dioxide sequestration with geothermal energy capture in naturally permeable, porous geologic formations: Implications for CO_2 sequestration. *Energy Procedia,*, 4, 2206-2213.

[31] Rosca, M. G., Karytsas, K., & Mendrinos, D. (2010). Low enthalpy geothermal power generation in Romania. *Proceedings World Geothermal Congress,*, Bali, Indonesia, April 2010.

[32] Ruggero-Bertani, E. (2007). World geothermal generation in 2007,. *GHC Bulletin,*, 1-12.

[33] Saleh, B., Koglbauer, G., Wendland, M., & Fischer, J. (2007). Working fluids for low-temperature organic Rankine cycles. *Energy,*, 32, 1210-1221, 0360-5442.

[34] Schuster, A., Karellas, S., Kakaras, E., & Spliethoff, H. (2009). Energetic and economic investigation of Organic Rankine Cycle applications. *Applied Thermal Engineering,*, 29, 1809-1817, 1359-4311.

[35] Stephens, J. C., Jiusto, ., & , S. (2010). Assessing innovation in emerging energy technologies: Socio-technical dynamics of carbon capture and storage (CCS) and enhanced geothermal systems (EGS) in the USA. *Energy Policy,*, 38, 2020-2031, 0301-4215.

[36] Wei, D., Lu, X., Lu, Z., & Gu, J. (2007). Performance analysis and optimization of organic Rankine cycle (ORC) for waste heat recovery. *Energy Conversion and Management*, 48, 1113-1119, 0196-8904.

[37] Yamada, S., & Oyama, H. (2004). Small capacity geothermal binary power generation system. *Fuji Electric Review,*, 51, 86-89.

Section 7

Fuel Cell

Chapter 14

Methodology of Designing Power Converters for Fuel Cell Based Systems: A Resonant Approach

Maria Teresa Outeiro and Adriano Carvalho

Additional information is available at the end of the chapter

http://dx.doi.org/10.5772/54674

1. Introduction

There has been an increased interest in the use of renewable energy sources, which is due to the limitations in fossil fuel reserves and to planet pollution. The research in the new sources of energy, such as photovoltaic, wind and fuel cells can be used to enhance the safety, reliability and sustainability of the planet, because they are environmentally friendly, highly efficient and renewable cycles. Particularly fuel cells appear in this context as an attractive power source because they generate electricity from hydrogen through an electrochemical process, which is virtually free of emissions and noise and only water and heat are the by-products. In addition, they present several advantages such as: silent, high potential for cogeneration applications, adaptable to a wide range of power and applications.

In this context, the chapter main goal is the analysis design and implementation of power generation systems based on fuel cells, which demands to careful selection of both; i) the fuel cell model and ii) the power electronic converter.

Then, the first part of the chapter presents and characterizes a semi-empirical model of the PEM fuel cell, including its static and dynamic behaviours and thorough some experimental tests made with the commercial system Mark 1020 constraints imposed by the PEM to the converter are established, regardless the topology of the power converter.

The second part of the chapter make considerations on the most suitable topologies of converter for this application type, and considering the requirements imposed by the PEM Mark 1020, a new efficient high power converter topology is selected, designed and implemented experimentally.

The converter follows a resonant approach that provides low component stresses, high frequency operation, soft-switching commutation, and operation under a wide range of input and output conditions.

The control of the converter is divided into two parts, namely: i) the voltage controller, which is responsible for keeping constant the output voltage of the converter under loading variations and ii) the PEM controller, which is responsible for improving the performance by keeping the PEM fuel cell in its optimal operating point.

The results are firstly presented for the PEM fuel cell and then for the whole system with load disturbance.

The results demonstrate that the proposed converter is a good selection to improve the efficiency of PEM fuel cells because it allows an adequate control of the power delivered by the fuel cell while maintaining the requirements imposed by the load and minimizing the losses by using soft-switching control.

2. PEM fuel cell

A PEM fuel cell can be described as a static device that converts the chemical energy of a fuel directly, isothermally, and continuously into electrical energy. In this process, only the reaction between hydrogen and oxygen occur. The only by-products are water and heat. Similarly to a battery, a fuel cell consists of two electrodes (anode and cathode) and an electrolyte. Whereas a storage battery contains all the substances in the electrochemical oxidation-reduction reactions involved and has therefore a limited capacity, the fuel cell is supplied with its reactants externally and operates continuously as long as it is supplied with fuel. The basic scheme for a single cell is represented in Figure 1 and the reactions involved in the anode side, the cathode side and the overall reaction of the process are described by the equations 1 to 3 follows.

$$H_2 \rightarrow 2H^{+} + 2e^{-} \tag{1}$$

$$2H^{+} + \frac{1}{2}O_2 + 2e \rightarrow H_2O \tag{2}$$

$$H_2 + \frac{1}{2}O_2 \rightarrow H_2O \tag{3}$$

2.1. Modelling of the PEM fuel cell

Many proton exchange membrane (PEM) fuel cell models have been investigated and presented in the literature [1-12]. The process of selecting the fuel cell model needs to clarify what are the necessary features to take into account in the model [7].

http://dx.doi.org/10.5772/54674

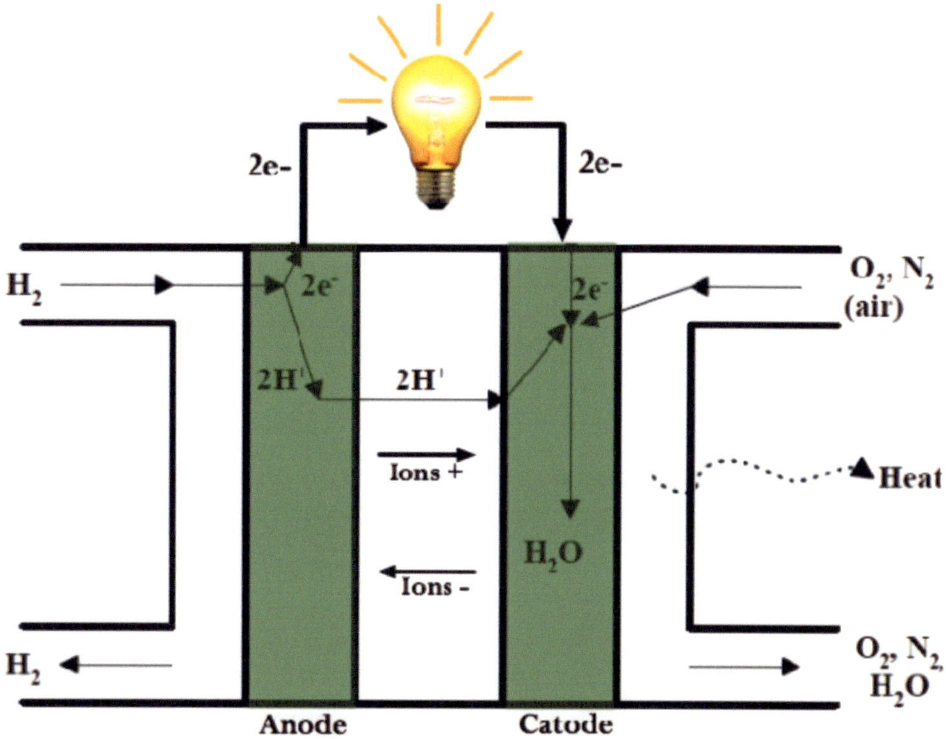

Figure 1. Scheme of a single cell.

The model selection differs for each application and user and the initial decisions are important to avoid changes later in the model evaluation process. The theoretical models are normally detailed, complex and usually require large computation time [8,9,11-12]. The semi-empirical models give a general voltage-current relationship without examining in depth the physical and electrochemical phenomena involved in the operation [1,5,6,10]. These models are usually characterized by simple implementation and faster simulation.

The electrical equivalent circuit represented Figure 2 corresponds to the semi-empirical model adopted for this study. This circuit is the electrical equivalent of the static and dynamic behaviour of the PEM fuel cell and includes the effects of the thermodynamic potential of the fuel cell and the losses. The equations 4 to 9 represent the static behaviour of the PEM while the dynamics is represented by equations 10 and 11. The capacitor C corresponds to the fuel cell phenomenon known as "charge double layer" on which the interface electrode/electrolyte acts as storage of energy element. The electrical power and efficiency are represented by equations 12 and 13 respectively.

Output voltage of one cell:

$$V_{FC} = E_{Nernst} - V_{act} - V_{Ohmic} - V_{con} \tag{4}$$

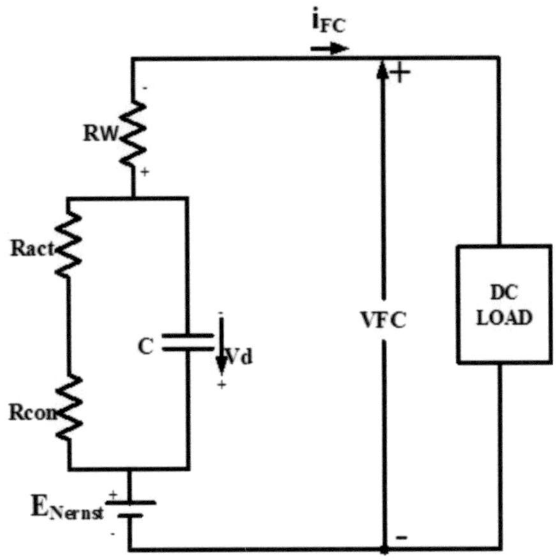

Figure 2. Electrical equivalent circuit of the PEM fuel cell.

Thermodynamic potential:

$$E_{Nernst} = 1.229 - 0.85 \times 10^{-3} \times (T - 298.15) + +4.31 \times 10^{-5} \times T \times \left[\ln(P_{H2}) + \frac{1}{2}\ln(P_{O2}) \right] \tag{5}$$

Activation over-potential:

$$V_{act} = -\left[\xi 1 + \xi 2 \times T + \xi 3 \times T \times \ln(CO_2) + \xi 4 \times T \times \ln(i_{FC}) \right] \tag{6}$$

Ohmic over-potential:

$$V_{ohmic} = i_{FC} \ (R_M + R_C) \tag{7}$$

Concentration over-potential:

$$V_{con} = -B \times \ln\left(1 - \frac{J}{J_{max}}\right) \tag{8}$$

Output voltage of the stack:

$$V_s = n \times V_{FC} \tag{9}$$

Voltage across capacitor:

$$\frac{dV_d}{dt} = \left(\frac{1}{C} \times i_{FC}\right) - \left(\frac{1}{\tau} \times V_d\right) \quad (10)$$

Electrical Time-constant:

$$\tau = C \times Ra = C \times \left(R_{act} + R_{con}\right) = C \times \left(\frac{V_{act} + V_{con}}{i_{FC}}\right) \quad (11)$$

Electrical power:

$$P_{FC} = i_{FC} \times V_{FC} \quad (12)$$

Efficiency:

$$\eta = \mu_f \times \frac{V_{FC}}{1{,}48} \times 100\% \quad (13)$$

2.2. Experimental tests made with the PEM Mark 1020

The circuit of Figure 3 corresponds to the experimental setup performed to obtain the electrical PEM characteristics. For each step of load the data is logged when the fuel cell achieved the steady-state operation.

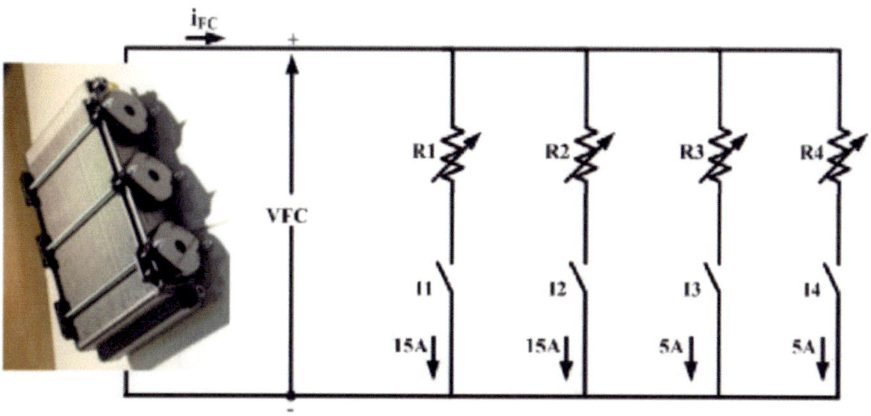

Figure 3. Electrical circuit to test the PEM fuel cell Mark 1020.

2.2.1. Output voltage and power

The output voltage is measured by directly connecting the digital multimetter in parallel with the fuel cell. This is an uncontrolled DC voltage, which fluctuates with the load as well as with changes in the fuel input to the system. Figure 4 shows the stack voltage while Figure 5 corresponds to the stack power of the PEM Mark 1020. The results obtained for the both electrical variables are in accordance with the information provided by the manufacturer of the stack.

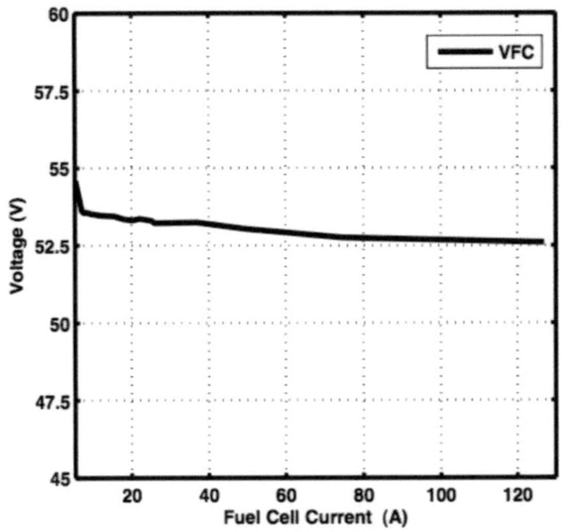

Figure 4. Fuel cell voltage of PEM Mark 1020.

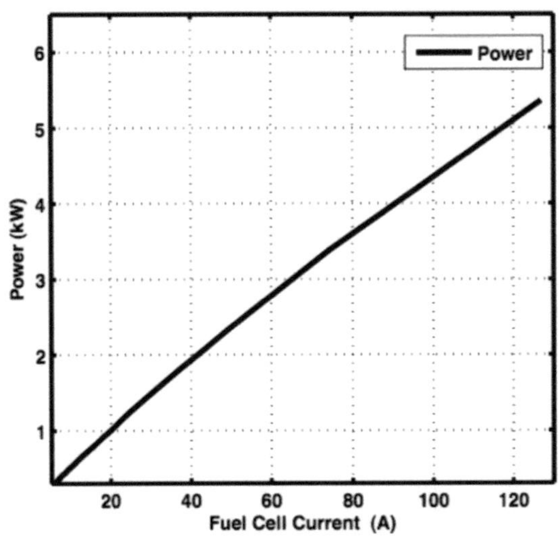

Figure 5. Electrical power of PEM Mark 1020.

http://dx.doi.org/10.5772/54674

2.2.2. *Efficiency and hydrogen consumption*

The efficiency of PEM Mark 1020 is in the range of 40 % - 55 %, which minimum and maximum values are 45.15 % and 55.49 % respectively. The efficiency decreases slightly with the increase of the current density as is shown in Figure 6. The hydrogen consumed by the stack Mark 1020 is represented in Figure 7 below and is proportional to the power delivered by the fuel cell.

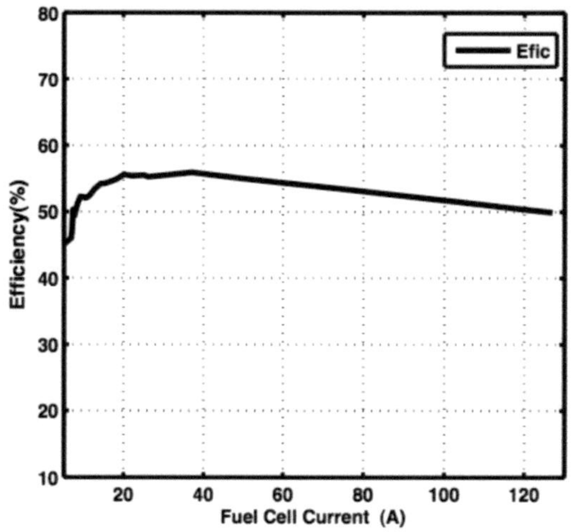

Figure 6. Figure 6. Efficiency of the PEM Mark 1020.

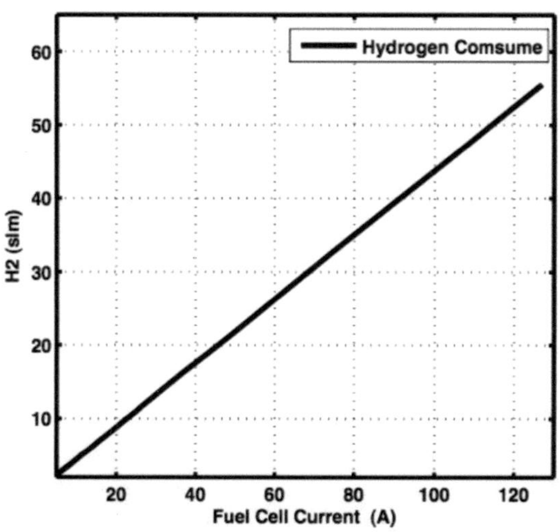

Figure 7. Hydrogen consumed by the PEM Mark 1020.

2.2.3. Constraints imposed to the converter

Regardless the topology of the converter selected, the constraints imposed by the PEM fuel cell should be respected that is, the minimum and maximum values of the voltage, the current and the power, which for the PEM Mark 1020 are listed in the Table 1 below.

Ifc (A)	Vfc (V)	Power (W)
2.8	23.71	66
24	19.11	492

Table 1. Constraints imposed to the power system by PEM MARK 1020.

3. Power electronic converters for PEM fuel cells

Power electronic converters are used in fuel cell systems to convert the DC electrical power generated by the fuel cell into usable AC or DC power through power electronic circuits. The power electronic converter plays an important role on the interface of the fuel cell system as power generating system and many solutions are already presented in the literature [14-30]. The output voltage of the fuel cell varies normally in the range of 20 V to 50 V and the possible converter topologies that can be used are such as; DC-DC together with DC-AC, DC-AC interfacing directly the fuel cell to the grid, or DC-DC together with AC-AC isolated by a transformer.

Figure 8 shows that the DC-DC power converters can be divided according to the operation mode into tree types: 1) the linear mode, 2) the switching mode and 3) the soft switching or resonant mode. The main difference between them is caused by efficiency. The soft-switching or resonant has some advantages compared to the linear like; the high switching frequency, which enables the use of a small ferrite transformer core, it may operate in a much larger DC input voltage range than the linear regulators, and it often has a higher efficiency. However, there are some drawbacks associated too, the noise at the supply may be increased according to different power switching techniques, and the control circuitry is more complicated compared to the linear one. Figure 8 also shows that the switching-mode topologies are divided into two types, the isolated and the non-isolated. Non-isolated DC-DC converter topologies are the Buck, Boost and Buck-Boost converters; and further, the Cuk converter. For many applications, isolation between the input and the output is a necessary requirement within the converter. By inserting isolation transformers into the four basic non-isolated switching topologies presented above, four single-ended isolated switching DC-DC converters can be obtained, namely; Forward, Boost, Flyback and Cuk converters. Nonetheless, the single switch topology is not an ideal solution for higher power converters, since these converters need a higher power transformer. Therefore, another group of DC -DC isolated converters utilizing more than one switch are identified: Push-pull, Half-bridge and Full-bridge converters.

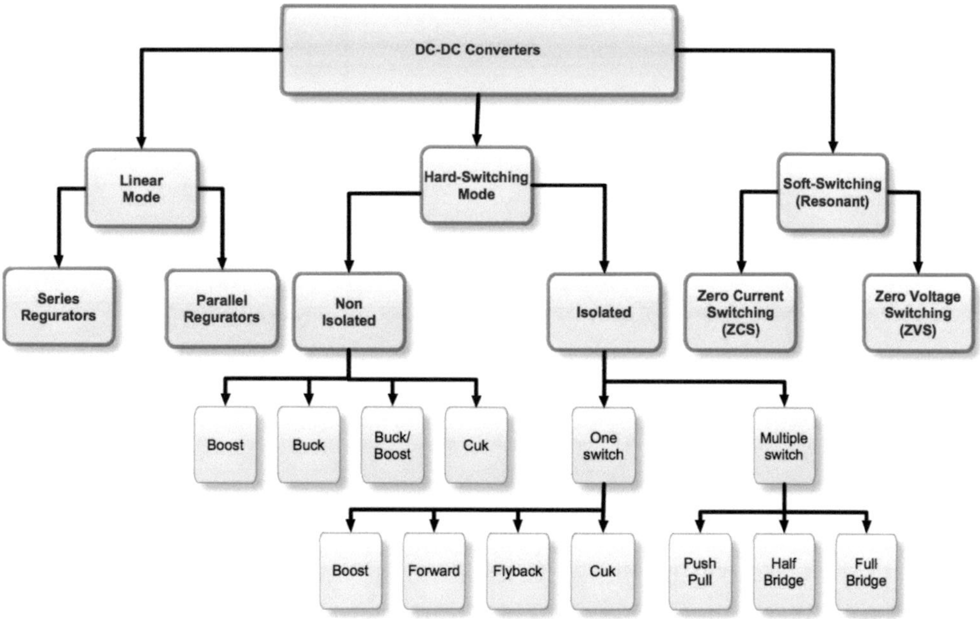

Figure 8. DC-DC power converter family tree.

In switched-mode topologies, finite duration of the switching transitions will cause high peak pulse power dissipation in the devices, degradation of the converter efficiency and, also can lead to transistor damage during the turn-off transition. Employing load-line snubbers can reduce this problem. When using snubbers the stress of the switches are minimised, as shown in Figure 9. However, with the appearance of new power electronic converters based on soft-switching technologies, [15,17,19,20-22,30], the reduction of switching losses and the continual improvement of power switches allow at being able to increase the switching frequency. In this type of converter the turning on and turning off of the converter switches appears when the switch voltage or the switch current is zero, as shown in Figure 9 [23].

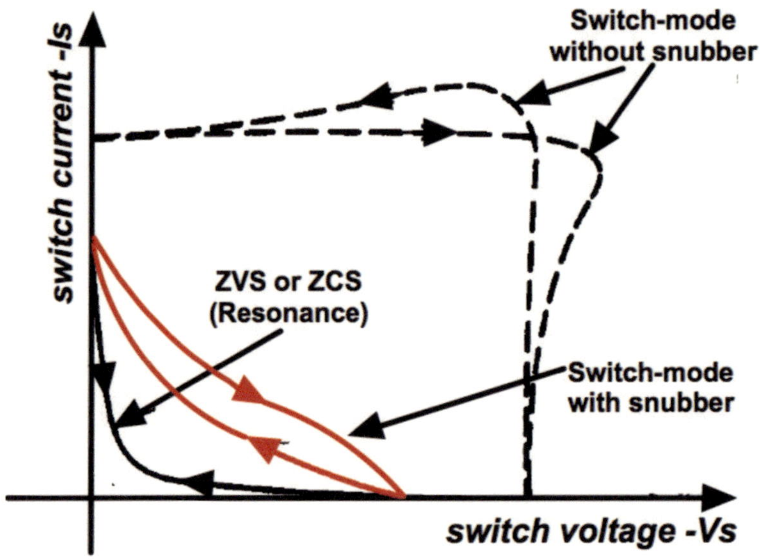

Figure 9. Switching loci trajectories of the different converter types.

Figure 10 provides an arrangement for a soft-switching resonant converter. An inductor -Ls and a capacitor-Cs have been added to help the switch action. A similar *Ld*Cd pair is added to the diode. In any of these soft switching cases, switch action at a zero crossing cuts off the ringing resonant waveform. This technique is often called quasi-resonance.

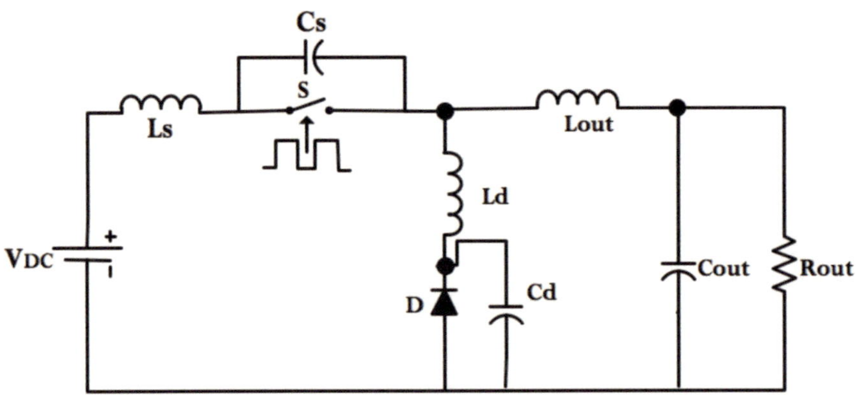

Figure 10. General structure of a resonant converter, where ZVS or ZCS can be obtained.

To create conditions for the ZCS or ZVS in DC-DC converters, the resonance or soft switching approach can be used. The ZVS or ZCS can be obtained by re-arranging the resonant component Figure 10, whose combinations offer several possibilities for resonant action as follows:

1. If the parts are chosen so that Cs and Ld are very small and have minimal effect on the circuit action. With Ls and Cd forming an LC series combination, the transistor operation can take advantages of current zero crossing for ZCS.

2. If the values of Cs and Ld are small, then the transistor supports ZVS [24].

3. It is also possible to use all four parts to support ZVS and ZCS action together, called multi-resonance, but this is not a common technique.

ZCS topologies can eliminate the switching losses at turn-off and reduce the switching losses at turn-on. If a relatively large capacitor is connected across the output diode during resonance, the converter operation becomes insensitive to the diode's junction capacitance. The major limitations associated with ZCS when Mosfet's are used are the capacitive turn-on losses. Thus, the switching loss is proportional to the switching frequency, during turn-on, considerable rate of change of voltage can be coupled to the gate drive circuit through the Miller capacitor, thus increasing switching loss and noise. Another limitation is that the switches are under high current stress, resulting in high conduction loss. ZVS eliminates the capacitive turn-on loss. It is suitable for high-frequency operation. For single-ended configuration, the switches could suffer from excessive voltage stress, which is proportional to the load. The output regulation of the ZCS and ZVS resonant converters can be achieved using variable frequency control. The ZCS [20-22] operates with constant on-time control, while ZVS [24] operates with constant off-time control.

3.1. Requirements for selecting the converter topology

For the selection of the converter topology the following requirements are considered in order to ensure the maximum efficiency and minimum cost of the power generation system.

1. Control of output voltage according to a given reference;

2. Deliver current with little ripple and harmonic contend

3. High efficiency in the whole operating range

4. Properly operation in all conditions

5. Incorporated filtering and storing possibilities

A comparative analysis of the major topologies of DC-DC converters described above is presented below.

3.2. DC-DC converter topologies

Power electronic converters in general and DC-DC converters in particular have a great importance on the performance and efficiency of energy production process based on fuel cells. The control of the operation point of the fuel cell requires appropriate use of static power converters, capable of providing accurate support to the control methods. The main objective to be achieved when applying the converters to fuel cells is obtaining the maximum efficiency using the most appropriate control strategies, taking into account requirements described above. As described above, different converter topologies can be used as represented in figures 11 to 13 below. Usually the DC-DC converter is put between the fuel cell and the inverter, which performs two functions, namely; 1) acts as DC isolation for the inverter; and 2) produces sufficient voltage for the inverter input so that the required magnitude of the AC voltage can be produced. The inverter can be single-phase or three-phase depending on the utility connection.

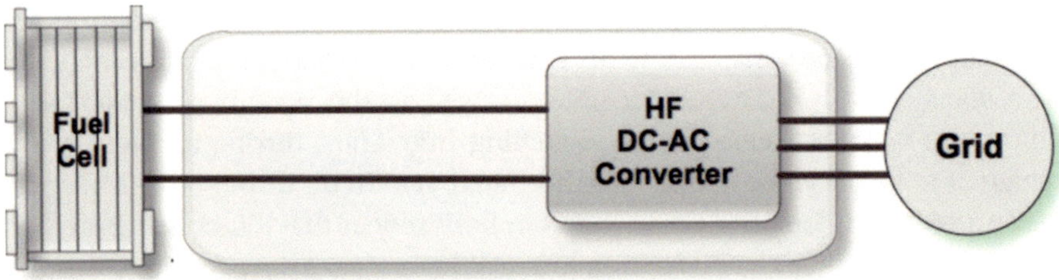

Figure 11. Configuration of a DC-AC converter interfaced directly to the grid.

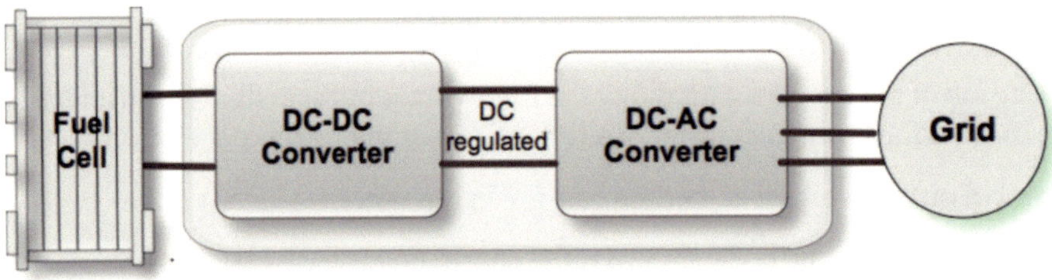

Figure 12. Configuration of a DC-DC followed by a DC-AC converter interfacing the grid.

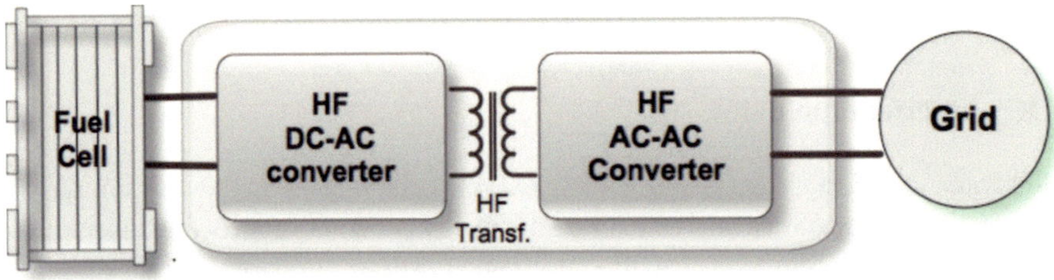

Figure 13. Configuration of a DC-AC followed by a AC-AC converter interfacing the grid.

http://dx.doi.org/10.5772/54674

3.2.1. Series resonant converter with capacitive output filter

The converter topology shown in Figure 14 is in conformity with the considerations presented above. This is a modified series resonant converter (SRC), which uses an (L-C)||L resonant tank for soft switching of HF switches. The main characteristics of such type of converter can be summarized as follows:

1. This configuration gives high efficiency at all varying load and line conditions but it decreases with increase in input voltage.
2. The full range ZVS (full load to light load) is achieved.
3. Switch peak current reduces with load but increases significantly with change in input voltage.

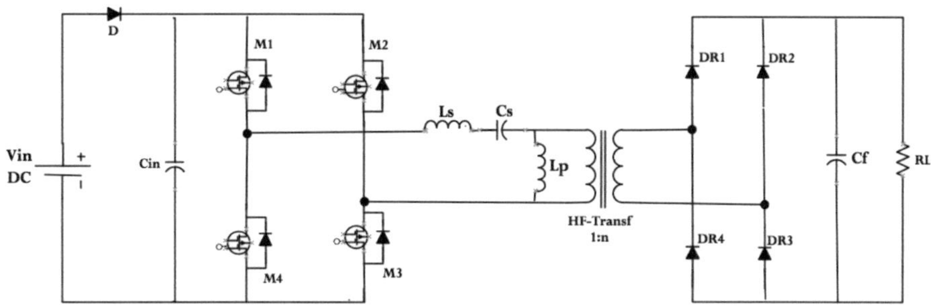

Figure 14. Series Resonant converter with capacitive output filter.

3.2.2. Series resonant converter with inductive output filter

The converter topology shown in Figure 15 is an LCL type SRC with inductive output filter. This uses an LCL resonant tank for soft switching of high frequency switches and an inductive output filter. Its main characteristics can be summarized as follows:

1. The full range ZVS (full load to light load) is achieved.
2. Rectifier diode voltage rating is higher, but with application of ultra fast recovery diodes of high voltage rating with low forward voltage drop does not affect the efficiency much.

Note: This configuration can also be modified to a series resonant converter (SRC), which uses an (L-C)||C resonant tank for soft switching of high frequency switches, which, in this case is also called series-parallel resonant converter (SPRC).

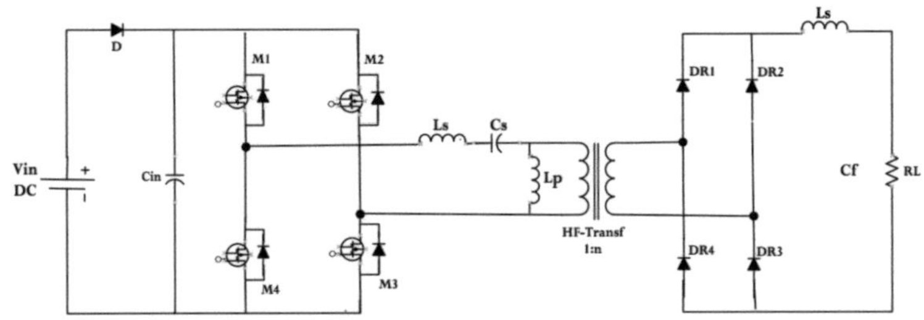

Figure 15. Series Resonant converter with inductive output filter.

3.2.3. PWM full-bridge converter with inductive output filter

The configuration shown in Figure 16 is Phase-shifted full bridge converter with inductive output filter. This is most widely used soft-switched configuration for high power applications. This constant frequency converter features ZVS of the primary switches with relatively small circulating circuit. The ZVS is achieved by filter inductance, transformer leakage inductance, snubber capacitance and parasitic junction capacitances of switches. The control of the output voltage at constant frequency is achieved by phase shift technique. Its main characteristics can be summarized as follows:

1. The major limitation of this configuration is loss of duty cycle on the secondary side.
2. High voltage stress on rectifier diodes on secondary side.
3. Parasitic ringing at the secondary side of transformer.
4. Large inductor may increase ZVS range but needs a transformer turns-ratio (Np/Ns) to be decreased at the same time, which increases primary side current, causing large conduction losses, which decreases converter efficiency.
5. A good compromise between the transformer turns-ratio and leakage inductance is required to keep conduction losses lower.

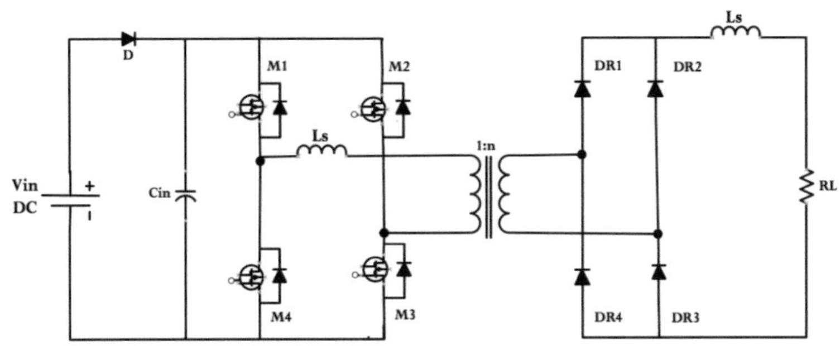

Figure 16. PWM full-bridge converter.

http://dx.doi.org/10.5772/54674

3.2.4. Full-bridge converter with controlled at secondary side

The configuration shown in Figure 17 corresponds to a secondary side controlled converter with a full-bridge inverter on primary side and phase-controlled rectifier on secondary side. The switches on primary side are operated by complementary gating scheme with fixed duty ratio. The switches on secondary side are controlled to produce phase difference between primary and secondary side voltages of HF transformer to control the output voltage with load and input line voltage variations.

The main characteristics of this configuration can be summarized as follows:

1. Control is easy and simple.
2. The switches on secondary side show ZVS or ZCS depending upon the line and load, which represents a drawback of the converter.
3. The switches on secondary side show ZVS for lower input voltage and ZCS for higher input voltage conditions.
4. The efficiency of this converter is near to 92%, obtained at full load and varying line voltage.

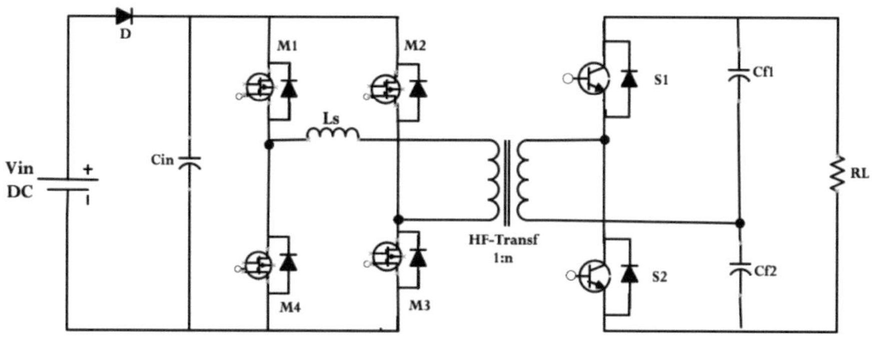

Figure 17. PWM full-bridge converter.

3.2.5. Current-fed two-inductor boost converter

Figure 18 shows the current-fed two-inductor boost converter, which is a dual of voltage-fed half-bridge converter configuration. This topology requires a very small turns ratio transformer with only two switches used on the primary side. The turn off of the primary switches is smooth since voltage across the switches is sinusoidal. Also the same for turn on and turn off of rectifier diodes. However, the peak and average current through rectifier diodes is high and this is very difficult to achieve ZVS at variable line input and load condition. This topology is a good selection for constant-input and constant-output voltage applications.

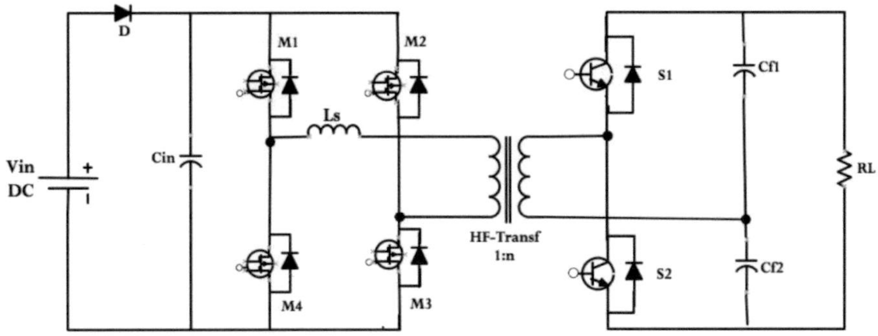

Figure 18. Current-fed two-inductor boost converter.

3.2.6. Bidirectional current–fed converter

The configuration of Figure 19 corresponds to a soft-switched bidirectional current-fed boost converter. The active switches in the both sides of the transformer make this converter bidirectional. Normally the Mosfets are used in the low voltage side and the IGBTs are used in the high voltage side. This configuration can be in push-pull, half-bridge or full-bridge, with the full-bridge considered one of the best topology choices for fuel cell applications.

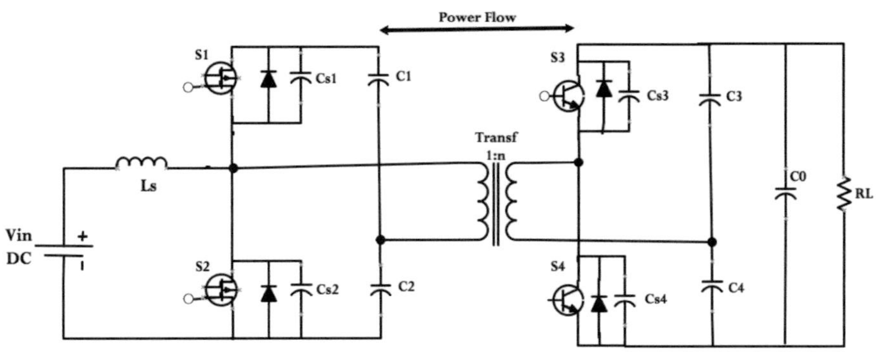

Figure 19. Soft-switched bidirectional current-fed boost converter.

3.3. Selection of converter topology and operation

Considering the analysis made previously a full-bridge series-resonant inverter, followed by a high frequency transformer and a rectifier, composes the topology selected. Additionally, two low-pass filters are included, one in the primary side which is used to protect the PEM of high ripple-current and another in the secondary side, to improve the quality of the energy supplied by the power system to the load or the grid. Figure 20 below represents electrical scheme of its topology, which is also called a series resonant converter (SRC).

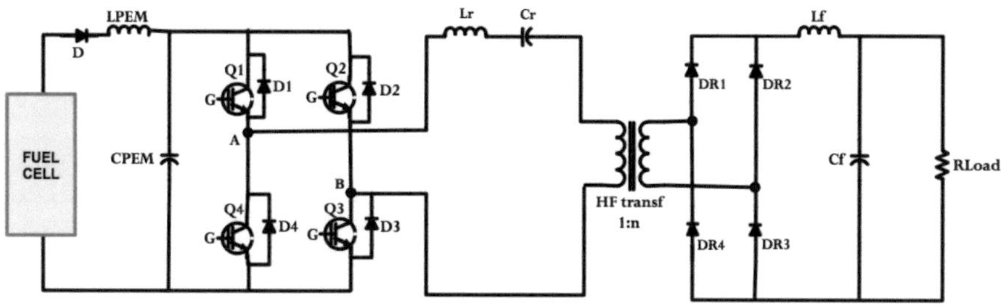

Figure 20. Electrical circuit of the SRC.

The operation of the converter can be described as follows: the voltage supplied by the fuel cell stack, which is typically low must be converted to a high and constant level, for example; 48 V or 400 V_{DC} in order to be able to feed an electric vehicle or to be sent to the grid through an inverter. The HF transformer is a step-up voltage transformer, which also serves as galvanic isolation between the high and low voltage levels of the circuits. The waveforms of the voltage and current in the LC series resonant circuit in the primary side of the transformer are sinusoidal. Selecting appropriate values for the Lr and Cr components, the resonant frequency of the circuit is established. Then, the DC voltage of the fuel cell is firstly inverted in the primary side of the HF transformer, being rectified on the secondary side. The low pass filter in the primary side (L_{PEM}, C_{PEM}) allows at protecting the PEM fuel cell from the ripple current and voltage produced by the converter, and also allows the storage of energy in the DC bus. The low pass filter in the secondary (L_f, C_f) allows at reducing the ripples of current and voltage to the load, respectively.

3.4. Mathematical analysis of the series resonant converter (SRC)

The series-resonant inverter as part of the topology of Figure 21 is constitute by a controlled full-bridge and a series resonant circuit with natural frequency f_r and impedance Z_r.

The analysis of this circuit by sinusoidal approximation neglects the harmonics of the switching frequency and the tank waveforms, namely the capacitor voltage, V_C and the inductance current, i_L are assumed to be purely sinusoidal.

3.4.1. Voltage

In Figure 21 the output voltage of the full-bridge inverter, which in turn is the input voltage of the series resonant circuit, is a square-waveform, wherein in the first half period [0-π] the IGBTs Q1 and Q3 conduct, while in the second half period conduct IGBTs Q2 and Q4 [π - 2π] as represented by Figure 22 and equation 14.

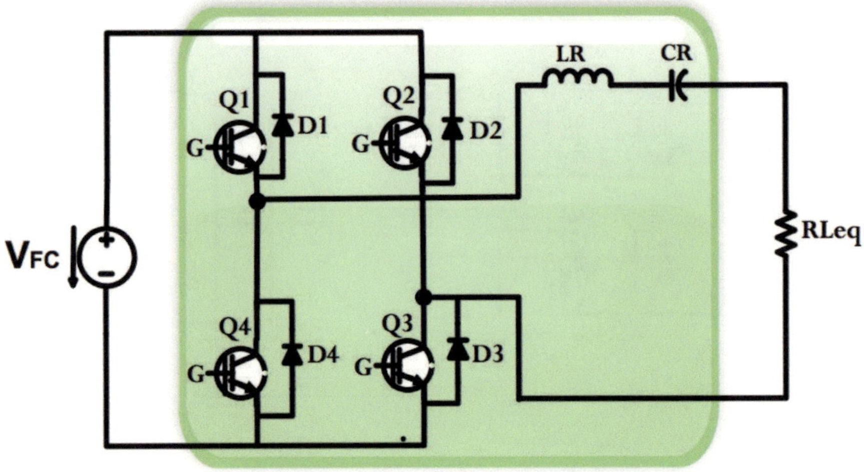

Figure 21. Electrical equivalent circuit of the SRC.

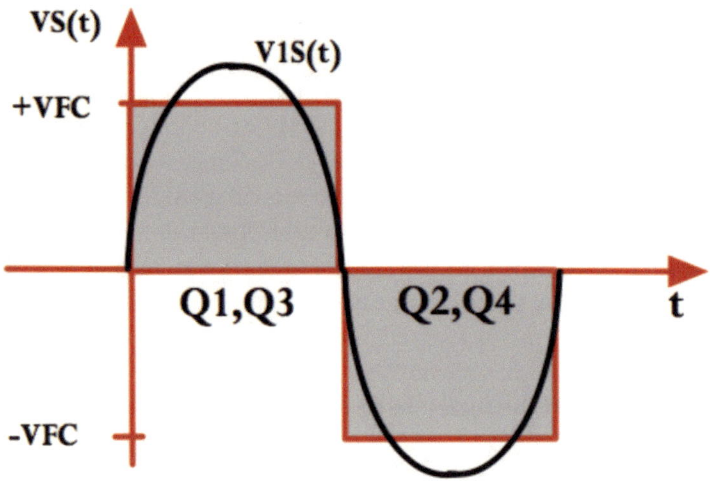

Figure 22. Output voltage in a full-bridge inverter.

$$\begin{cases} +V_{Fc}, \text{ for } 0 < \text{wt} \leq \pi \\ -V_{Fc}, \text{ for } \pi < \text{wt} \leq 2\pi \end{cases} \tag{14}$$

3.4.2. Current

The current is (t) is equal to the output current +is (t) in the first half period [0-π], and is its inverse -is (t), in the second half period [π - 2π]. Under the conditions described above, the tank rings sinusoidal and is(t) is well approximated by a sinusoid waveform as represented by the equations 15 and 16.

$$iQ1 = iQ3 = \begin{cases} +\operatorname{Im}ax \times \sin(wt - \Phi), & \text{for } 0 < wt < \pi \\ 0, & \text{for } \pi < wt < 2\pi \end{cases} \quad (15)$$

$$iQ2 = iQ4 = \begin{cases} 0, & \text{for } 0 < wt < \pi \\ -\operatorname{Im}ax \times \sin(wt - \Phi), & \text{for } \pi < wt < 2\pi \end{cases} \quad (16)$$

A simple equivalent circuit similar to that represented in Figure 23, whose operation can be analyzed by a second-order differential equation as follows, can perform the analysis of the SRC.

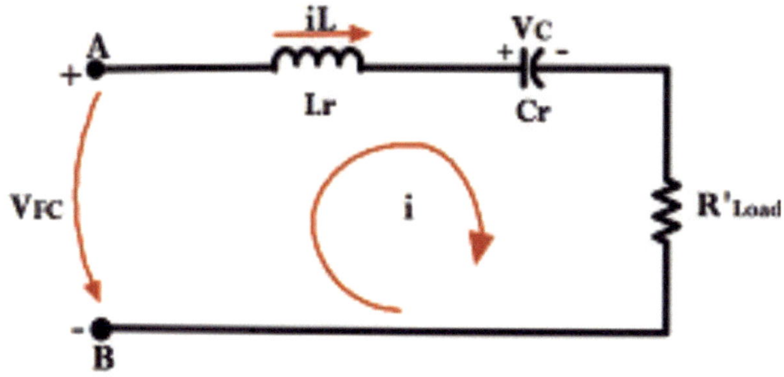

Figure 23. Series resonant circuit with a resistive load.

Applying the Kirchoff loop equation (sum of voltages around loop equals zero), to the RLC circuit of Figure 23 the equation of the series resonant circuit is the follow.

$$Lr\frac{diLr}{dt} + VCr + RI = \bar{V} \quad (17)$$

And the input impedance Z of this circuit is expressed by equation 18 with$R=Z\cos(\theta)$ and $X=Z\sin(\theta)$.

$$\bar{Z}=R+j\left(wL_r-\frac{1}{wCr}\right) \tag{18}$$

3.4.3. Angular frequency and reactance

If the reactance of the resonant circuit becomes zero with $XL_r = XC_r$ then Z=R and the circuit operates at resonant conditions that is, for $f_s=f_r$. This is characterized by the angular frequency w_r and characteristic impedance Z_r as defined by equations 19 and 20 follow.

$$w_r=2\pi f_r=\frac{1}{L_rC_r} \tag{19}$$

$$Z_r=\sqrt{\frac{L_r}{C_r}} \tag{20}$$

3.4.4. Quality factor

The quality factor Q measures the "goodness" or the quality of the circuit. This is the ratio between the power stored in the reactance and the power dissipated in the resistance of the resonant circuit. The quality factor Q, can be defined expressed by the equation 21 or equivalently by equation 22.

$$Q=\frac{w_r\times Lr}{R}=\frac{1}{w_r\times C_r\times R}=\frac{Z_r}{R} \tag{21}$$

$$Q=\frac{1}{R}\times\sqrt{\frac{L_r}{C_r}} \tag{22}$$

3.4.5. Normalized amplitudes of the current and power

Normalized amplitudes of the current and output power in the series resonant circuit as a function of fs/fr and R/Zr=1/Q are represented in figures 24 and 25, respectively. The maximum current value in the circuit occurs at the resonance frequency and with low load resistance. It is also observed that at resonant frequency fr if the load $R\rightarrow 0\Omega$ then the peak current $I\rightarrow\propto$ and the system can de destroyed. This analysis leads to the conclusion that it can't ever work with a short circuit in the load side. The maximum output power occurs at the resonance frequency and with low load resistance.

http://dx.doi.org/10.5772/54674

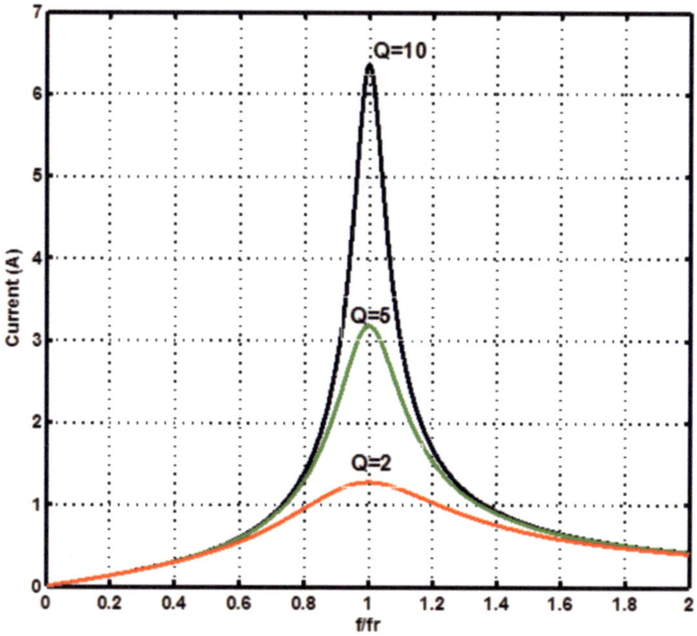

Figure 24. Amplitude of the current in the series resonant circuit

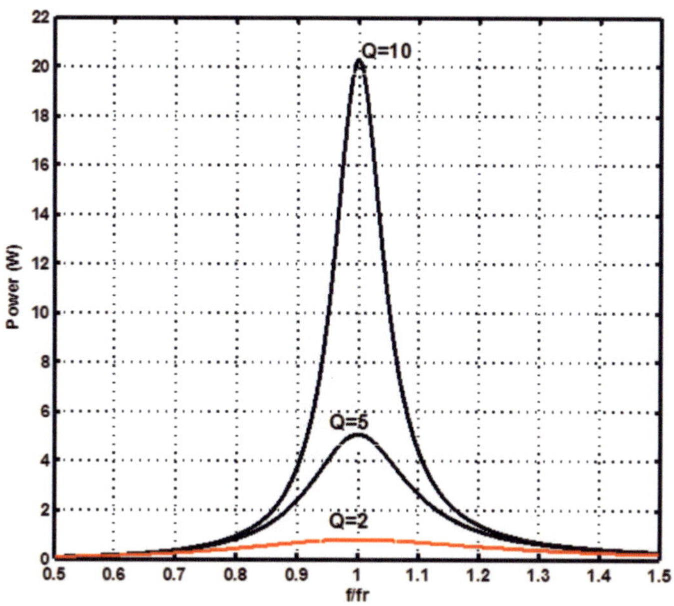

Figure 25. Amplitude of the power in the series resonant circuit

4. Control

The control consists of two loops; the loop of the voltage or the fast loop and the loop of he PEM or the slow loop. The voltage controller is responsible of controlling the output voltage of the converter, keeping this in a constant value defined by the user even for load variations. The PEM controller is responsible of controlling the operation of the PEM, keeping it in its optimal point that is, producing the electrical power requested by the load with a minimum current and consequently with a minimum of hydrogen consumption. The control structure of the converter as described is represented in Figure 26.

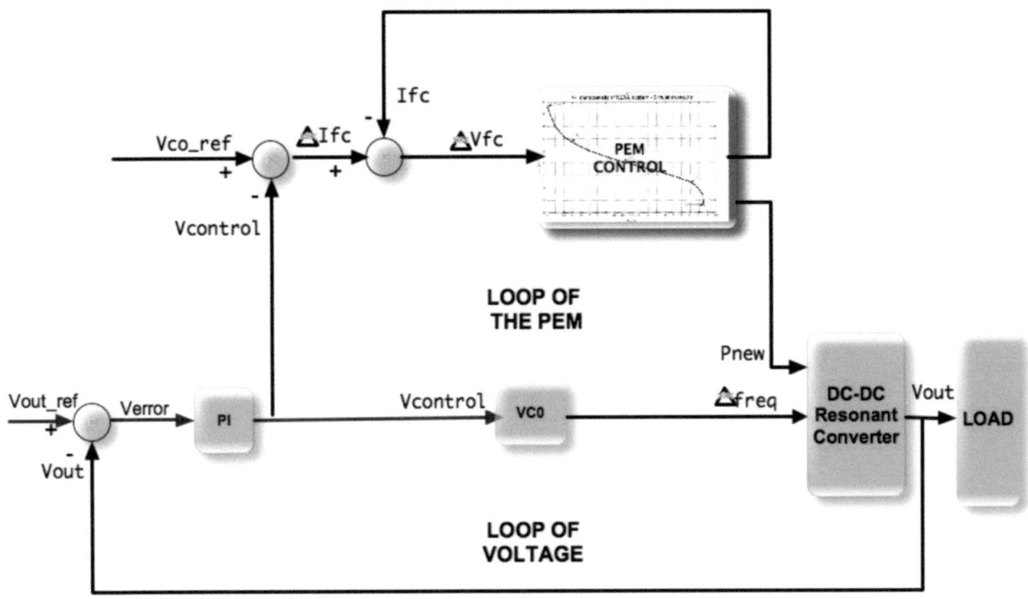

Figure 26. Control structure of the SRC.

http://dx.doi.org/10.5772/54674

5. Experimental setup and results

The experimental setup is represented in Figure 27. It is used to test the all system composed by the PEM Mark 1020, the SRC and the load. The load is composed by a set of several resistors connected in series, whose variation is performed by a manual switch. The fuel pressure that provides a PEM stack is monitored by a standard dial pressure gauge, which maintains it constant in the range of 0.3 to 0.5bar. The ventilator is used to inject the oxidant flow necessary into the stack in order to produce the electrochemical reaction. The voltage of 26.06V represented by the multimetter corresponds to the open-circuit voltage of the PEM.

The experimental results corresponding to the output voltage and current, the PI controller and the resonant current, are presented in this section to validate experimentally the SRC in its stability and dynamics.

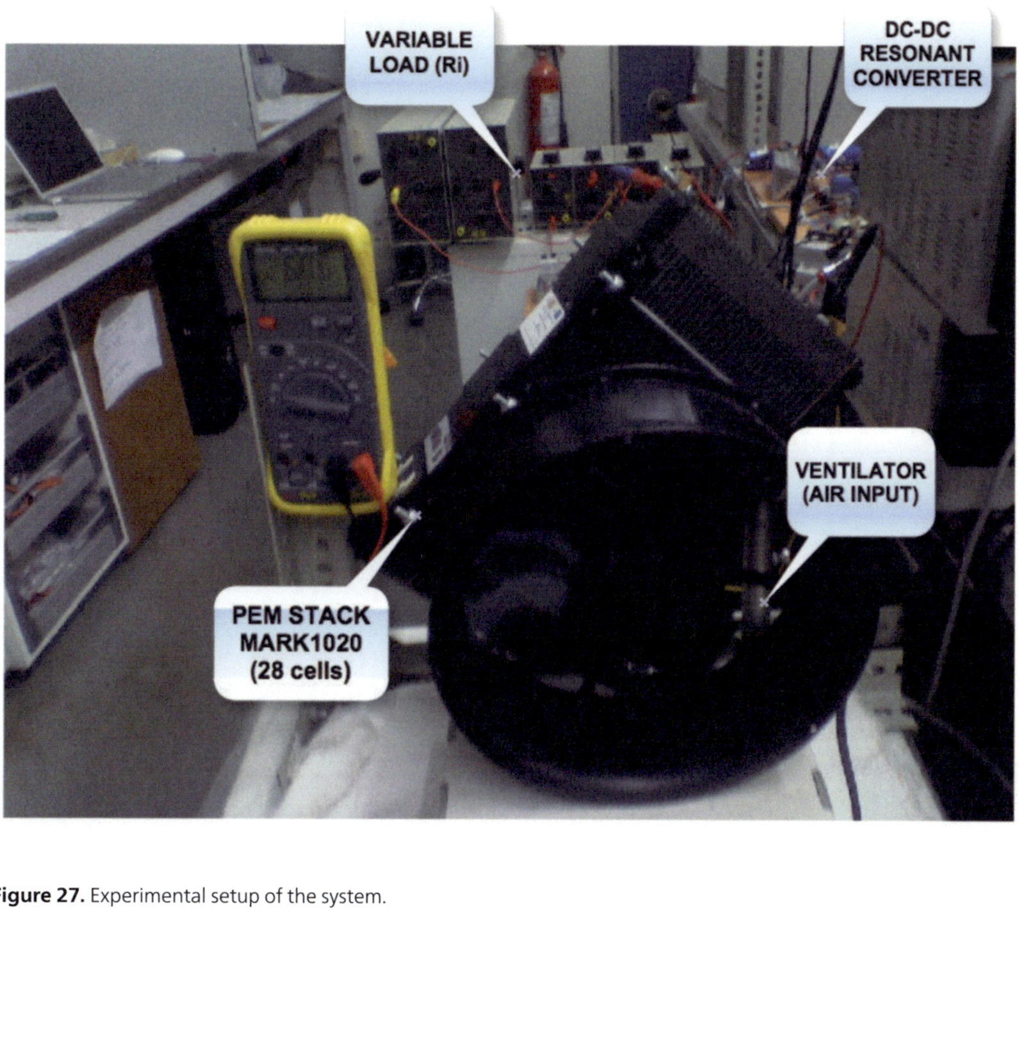

Figure 27. Experimental setup of the system.

5.1. Output voltage and current

Figure 28 below allows validating the stability of the voltage control loop of the converter that is, it can be seen that the output voltage, vout remains constant despite variations in the output current that is, in the load. This condition is valid to both situations namelly, the step-up of load corresponding to Figure 28 a) and the step-down of load corresponding to Figure 28 b).

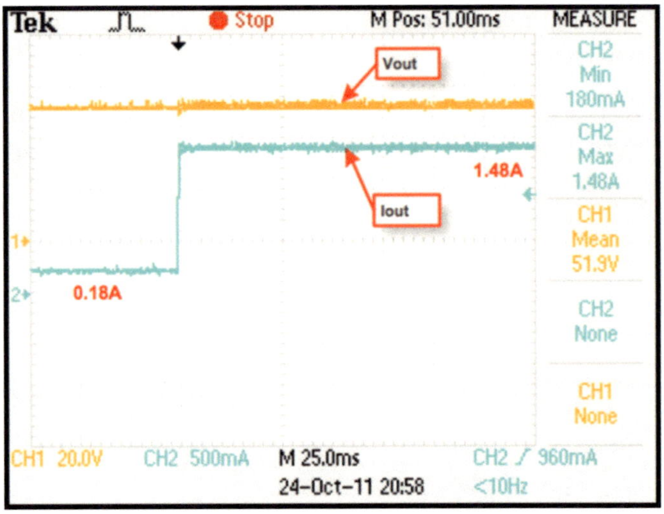

(a) Step-up load condition.

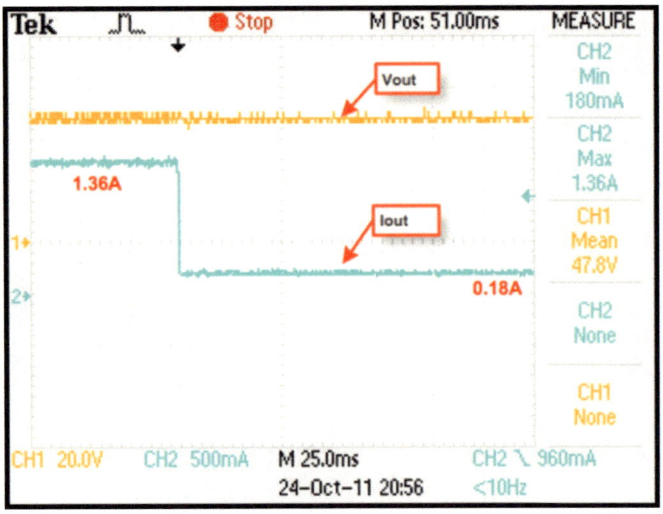

(b) Step-down load condition.

Figure 28. Output voltage (Vout) and current (Iout).

5.2. Proportional integral control (PI)

The dynamics of the system can be evaluated by the analysis of the PI control signal. So, once considered both situations of load variation it appears that the stabilization time of the PI controller is approximately 7ms. In addition it presents a small oscillation which proves that the parameters of the PI control are well adapted to the system Figure 29 a) corresponds to the situation of a step-up load condition while Figure 29 b) corresponds to a step-down of load condition. The error of voltage is given by INA101 such as; $\varepsilon = V_{measured} - V_{reference}$ and accordingly, the objective of the PI controller is to minimize this error for any load variation, as is shown in the two figures below.

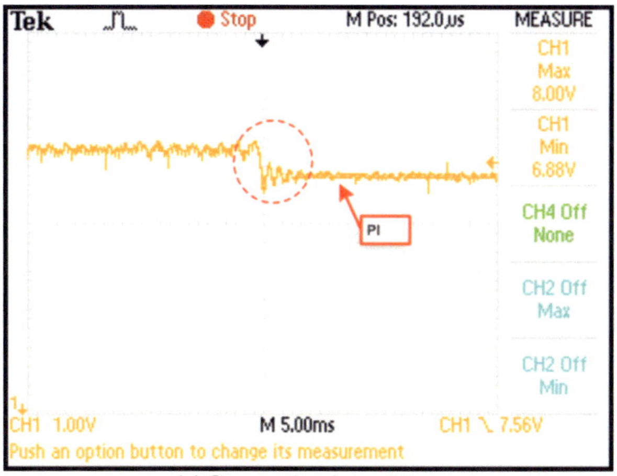

(a) Step-up load condition.

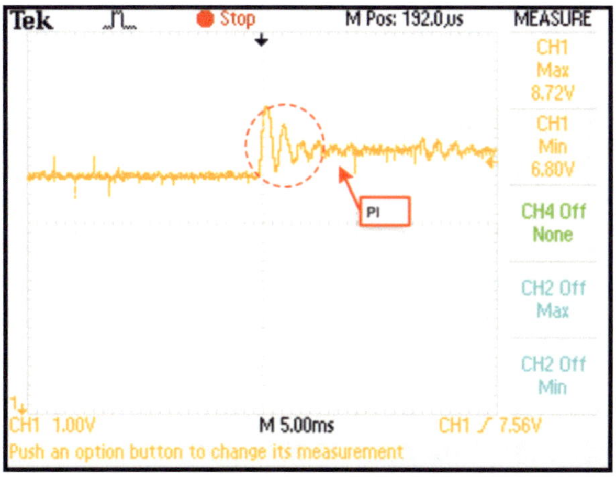

(b) Step-down load condition.

Figure 29. PI Control.

5.3. Resonant circuit operation

Figure 30 corresponds to step-up and step-down load conditions. From its analysis it follows that the converter reacts to the load variation, varying its frequency of operation. Thus, for a small load level (Imin, Rmax) the frequency is low while for a high load (Imax, Rmin) the frequency is high. In dynamic terms it can be seen that the transition in the frequency of operation is instantaneous, hence, we conclude that the system has good dynamic characteristics. It can be also observed that in any of the load variations the output voltage Vout remains constant. This analysis validates the objective defined to the controller, that is it ensures a constant output voltage in order to satisfy the requirements imposed by the power system applications.

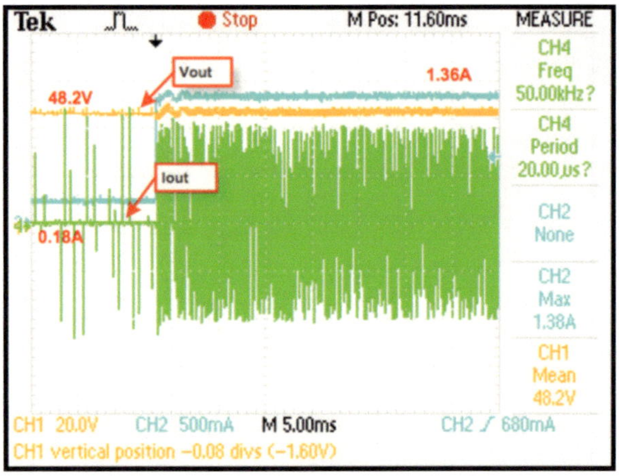

(a) Step-up load condition.

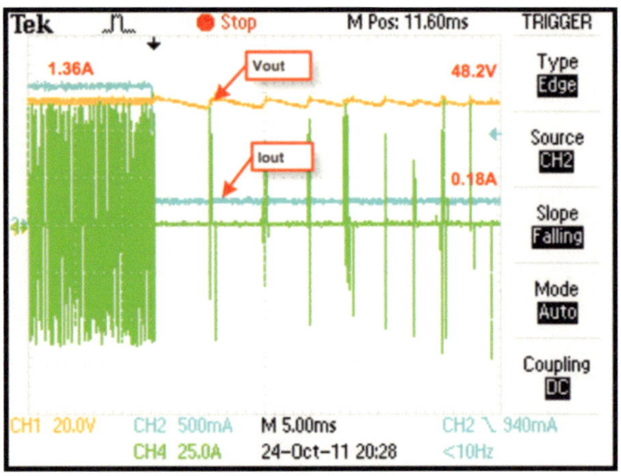

(b) Step-down load condition.

Figure 30. Output voltage and current and resonant circuit operation

http://dx.doi.org/10.5772/54674

6. Conclusions

The main objective of the chapter is to discuss the design and implementation of a power generation system based on fuel cells. Accordingly, a methodology of designing and implementing an efficient high power converter system is presented. Moreover the chapter presents also an electrical equivalent model of the PEM fuel cell, which was validated by experimental tests made with the commercial system MARK 1020.

Authors make considerations on the most suitable topologies of converters for this application type, and satisfying several criterions a series-resonant converter topology is selected, whose principle is based on soft-switching methodology. In this context the design and implementation of the converter consisting of a input filter followed by the full-bridge inverter and the series resonant circuit on the primary side and a diode rectifier and output filter on the secondary side was based on the exploitation of their benefits as compared to other types of converters, namely: low component stresses, high frequency operation and soft-switching commutation. Converter design was done considering the operational constraints of the system MARK 1020.

A particular attention is done to the controller, which ensures a constant output voltage of the converter, in order to satisfy the requirements of the power system application and simultaneously keeps the PEM operating within its optimum operating point. The control implementation was divided into two parts namely: i) the voltage controller, which is responsible for keeping constant the output voltage of the converter even under loading variations and ii) the PEM controller, which is responsible for improving its performance by keeping the PEM fuel cell in its optimal operating point.

Due to significance of the PEM cell behavior the results are firstly presented for the PEM fuel cell model and then for the whole system with load.

The results demonstrate that the converter selected is a good solution to support the approach of improving the efficiency of PEM fuel cells because it allows an appropriated control of the power delivered by the fuel cell as it satisfies the requirements imposed by the load regulation with minimum of losses due to adoption of soft switching commutation.

Nomenclature

PEM – Proton exchange membrane fuel cell

IGBT – Insulated Gate Bipolar Transistor

SRC – Series resonant converter

ZVS – Zero voltage switching

ZCS – Zero current switching

A – Cell active area (cm2)

C – Equivalent electrical capacitance (F)

ENernst – Thermodynamic potential

Jmax – Maximum current density (A/cm2)

n – Number of cells in stack

PO2 – Oxygen partial pressure (atm)

PH2 – Hydrogen partial pressure (atm)

RC – Contact resistance (Ω)

RM – Equivalent membrane resistance (Ω)

T – Cell operating temperature (K)

Vact – Activation voltage drop (V)

Vohmic – Ohmic voltage drop (V)

Vcon – Concentration voltage (V)

ξi, ψ – Parametric coefficients

λ – Membrane thickness (μm)

fr – Resonant frequency (Hz)

wr – angular frequency (rad/s)

Zr – Resonant impedance (Ω)

Lr – Resonant tank inductor (H)

Cr – Resonant tank capacitor (F)

Q – Quality factor

Acknowledgements

The authors would like to thank the Institute of Systems and Robotics of Porto University (ISR-Porto) and Foundation for Science and Technology (FCT), which provided them the conditions and financial support needed to conduct this research.

http://dx.doi.org/10.5772/54674

Author details

Maria Teresa Outeiro[1*] and Adriano Carvalho[2]

*Address all correspondence to: touteiro@isec.pt

1 Department of Electrical Engineering, Institute of Engineering of Coimbra, Institute of Systems and Robotics – Porto, Portugal

2 Department of Electrical and Computer Engineering, University of Porto, Institute of Systems and Robotics – Porto, Portugal

References

[1] Corrêa, J.M., et al. An electrochemical-based fuel cell model suitable for electrical engineering automation approach. IEEE Transactions on Industrial Electronics, 2004; 5(51) 1103-1112.

[2] Yu, D. and S. Yuvarajan. A novel circuit model for pem fuel cells. Proceedings of IEEE Applied Power Electronics Conference and Exposition. 2004.

[3] Friede, W., S. Raël, and B. Davat. Mathematical model and characterization of the transient behavior of a PEM fuel cell. IEEE Transactions on Power Electronics 2004; 5(19) 1234-1241.

[4] Forrai, H., Yanagita Y., and Kato Y. Fuel-cell parameter estimation and diagnostics. IEEE Transactions on Energy Conversion 2005; 3(20): 668-675.

[5] Outeiro, M.T., et al., A Parameter Optimized Model of a PEM Fuel Cell Including Temperature Effects. Journal of Power Sources 2008; 185(2) 952-960.

[6] Outeiro, M.T., et al. Dynamic Modeling and Simulation of an Optimized Proton Exchange Membrane Fuel Cell System. Proceedings of International Mechanical Engineering Congress and Exposition. Seattle, USA. ASME; 2007.

[7] Kristina Haraldsson, K.W. Evaluating PEM fuel cell system models. Journal of Power Sources 2004; 126(1-2) 88-97.

[8] Bernardi, D.M. and Verbrugge M.W. A mathematical model of the solid-polymer-electrolyte fuel cell. Journal of Electrochemical Society 1992; 2477–2491.

[9] Rowe, A. and Li X. Mathematical modelling of proton exchange membrane fuel cells. Journal of Power Sources, 2001; 82-96.

[10] Mann, R.F., et al. Development and application of a generalised steady-state electrochemical model for a PEM fuel cell. Journal of Power Sources 2000; 173-180.

[11] Amphlett, J.C., et al. A model predicting transient responses of proton exchange membrane fuel cells. Journal of Power Sources 1996; 183-188.

[12] Wang, C. and Nehrir M. Dynamic models and model validation for PEM fuel cells using electrical circuits. IEEE Transactions on Energy Conversion 2005; 2(20) 442-451.

[13] Andersen, G.K., et al. A new power converter for fuel cells with high system efficiency. International. Journal of Electronics 2003; 90(11-12) 737-750.

[14] Xu, H., Kong L., and Wen X. Fuel Cell Power System and High Power DC–DC converter. IEEE Transactions on Power Electronics 2004; 19(5) 1250-1255.

[15] Shiju, W. Design and hardware implementation of a soft-switched converter for fuel cell applications. MsC thesis. University of Texas; 2006

[16] Krykunov, O. Comparison of the DC/DC-Converters for Fuel Cell Applications. International Journal of Electrical Computer and Systems Engineering 2007; 1(1) 71-79.

[17] Rathore, A., Bhat A., and Oruganti R. A Comparison of Soft-Switched DC-DC Converters for Fuel Cell to Utility Interface Application. IEEE Proceedings 2007; 588-594.

[18] Rong-Jong, W. and Rou-Yong D. High-Efficiency Bidirectional Converter for Power Sources With Great Voltage Diversity. IEEE Transactions on Power Electronics 2007; 22(5) 1986-1996.

[19] Ying-Chun Chuang, Y.-L.K., Hung-Shiang Chuang, Hung-Kun Chen Implementation and Analysis of an Improved Series-Loaded Resonant DC–DC Converter Operating Above Resonance for Battery Chargers. IEEE Transactions on Industry Applications 2009; 45(3) 1052 – 1059.

[20] Cancelliere P., et al. Modeling and Control of a Zero-Current-Switching DC/AC Current-Source Inverter. IEEE Transactions on Industrial Electronics 2007; 54(4) 2106-2119.

[21] Lee Y.-S., Chiu Y.-Y. and Cheng M.-W. Inverting ZCS Switched-Capacitor Bi-directional Converter, 37th IEEE Power Electronics Specialists Conference, Jeju, Korea, PESC; 2006.

[22] Yuang-Shung, L., et al. Multiple Output Zero-Current Switching Bi-directional Converter. 33rd Annual Conference of the IEEE Industrial Electronics Society, Taiwan Japan, IECON; 2007.

[23] Mohan N. and Underland T. Power Electronics Converters Applications and Design. John Wiley and Sons, Inc. 2003.

[24] Peng, F.Z., et al. A New ZVS Bidirectional DC–DC Converter for Fuel Cell and Battery Application. IEEE Transactions on Power Electronics 2004; 19(1) 54-65.

[25] Krykunov O. Comparison of the DC/DC-Converters for Fuel Cell Applications. International Journal of Electrical Computer and Systems Engineering 2007; 1(1) 71-79.

[26] Lee J.-Y., Jeong Y.-S. and Han B.-M. An Isolated DC/DC Converter Using High-Frequency Unregulated LLC Resonant Converter for Fuel Cell Applications. IEEE Transactions on Industrial Electronics 2011; 58(7) 2926 - 2934.

[27] Xin Kong, A.M.K., Analysis and Implementation of a High Efficiency Interleaved Current-Fed Full Bridge Converter for Fuel Cell System. IEEE Transactions on Power Electronics 2007; 22(2) 543-550.

[28] Averberg A., Meyer K.R. and Mertens A. Current-fed full bridge converter for fuel cell systems. IEEE Power Electronics Specialists Conference, Rhodes, Greece. PESC; 2008.

[29] Changchien S., Liang T., Chen J. and Yang L. Novel High Step-Up DC–DC Converter for Fuel Cell Energy Conversion System. IEEE Transactions on Industrial Electronics 2010; 57(6) 2007 – 2017.

[30] Blaabjerg F., Chen Z. and Kjaer S.B. Power Electronics as Efficient Interface in Dispersed Power Generation Systems. IEEE Transactions on Power Electronics 2004; 19(5) 1184-1194.

Section 8

Integrated System

Chapter 15

Wind Diesel Hybrid Power System with Hydrogen Storage

Mamadou Lamine Doumbia, Karim Belmokhtar and Kodjo Agbossou

Additional information is available at the end of the chapter

http://dx.doi.org/10.5772/52341

1. Introduction

By 2050 the demand for energy could double or even triple as the global population grows and developing countries expand their economies. Energy prices, supply uncertainties, and environmental concerns are driving many countries to rethink their energy mix. The International Energy Agency's Energy Technology Perspectives2008publication projects that energy sector emissions of greenhouse gases (GHGs) will increase by 130% over 2005 levels, by 2050, in the absence of new policies (IEA, 2008).

Renewable energy is part of the solution for the energy problem, and wind energy is one of the cost-effective options for the generation of electricity. The main applications are the generation of electricity and water pumping.

By the end of 2007, in the world, there were around 100,000 wind turbines installed in wind farms, with an installed capacity of 94,000 megawatts, which generated around 300 TWh/year. Wind energy is now part of national policies for generation of electricity in many countries (Vaughn Nelson, 2009).

In 2007, in Europe there were 57,000 MW installed wind power, which generated 3.7% of the electrical demand. The European goal is 20% of electricity generated by renewables by 2020, of which 12–14% would be from wind. In 2010, wind energy provided for nearly 26% of electricity consumption in Denmark, more than 15% in Portugal and Spain, 14% in Ireland and nearly 9% in Germany, over 4% of all European Union (EU) electricity, and nearly 2.9% in the United States (US Department of Energy, 2011). In the future, many countries around the world are likely to experience similar penetration levels as wind power is increasingly considered not only a means to reduce CO_2 emissions but also an interesting economic alternative in areas with appropriate wind speeds. Since 2000, cumulative installed capacity has grown at an average rate of around 30% per year. In 2008, more than 27 GW of capacity were installed in more than 50 countries, bringing global capacity onshore and offshore to 121 GW. Wind energy in 2008 was estimated by the Global Wind Energy Council to have generated some 260 million megawatt hours of electricity. Applications for generation of electricity are divided into the following categories: utility-scale wind farms and small wind turbines (less than 100 kW).

Currently, for remote communities and rural industry the standard is diesel generators. Remote electric power is estimated at over 11 GW, with 150,000 diesel gensets, ranging in size from 5 to 1,000 kW. In Canada, there are more than 800 diesel gensets, with a combined installed rating of over 500 MW in more than 300 remote communities (Vaughn Nelson, 2009). Diesel generators are inexpensive to install; however, they are expensive to operate and maintain, and major maintenance is needed from every 2,000 to 20,000 hours, depending on the size of the diesel genset.

Wind–diesel is considered because of the high costs for generating power in isolated systems. In near future, the market of wind–diesel systems will grow up because of the high cost of diesel fuel. Wind–diesel power systems can vary from simple designs in which wind turbines are connected directly to the diesel grid, with a minimum of additional features, to more complex systems.

There are a number of problems in integrating a wind turbine to an existing diesel genset: voltage and frequency control, frequent stop–starts of the diesel, utilization of surplus energy, and the use and operation of a new technology. These problems vary by the amount of penetration. Wind turbines at low penetration can be added to existing diesel power without many problems, as it is primarily a fuel saver. However, for high wind penetration, storage is needed. Moreover, one of the major drawbacks of wind energy is its unpredictability and intermittency. So, to supply better consumers' energy needs, wind systems have to operate with storage devices. Several energy storage methods have been in development over the past several years. This includes compressed air, pumped hydro, flow battery flywheel, hydrogen storage, etc. It has been proved (E.I. Zoulias, N. Lymberopoulos, 2008; Nelson et al., 2006) that hydrogen can be effectively used as storage medium for intermittent renewable energy sources (RES)-based autonomous power systems. More specifically, excess of RES energy produced from such systems at periods of low demand can be stored in the form of hydrogen, which will be used upon demand during periods when the wind energy is not available.

For many years, Hydrogen Research Institute (HRI) has developed a renewable photovoltaic/wind energy system based on hydrogen storage(M. L. Doumbiaet al., 2009; K. Agbossou et al., 2004). The system consists of a 10 kW wind turbine generator (WTG) and a 1 kW solar photovoltaic (PV) array as primary energy sources, a battery bank, an 5 kW electrolyzer, a 5 kW fuel cell stack, different power electronics interfaces for control and voltage adaptation purposes, a measurement and monitoring system. This renewable energy system is scaled for residential applications size and can be operated in stand-alone or grid-connected mode and different control strategies can be developed.

http://dx.doi.org/10.5772/52341

Numerical modelling is an important part of the design, assessment, implementation and evaluation of autonomous power systems with wind power.

This chapter is devoted to a large scale wind-diesel Hybrid Power System (HPS) applications. It presents theoretical analysis, modelling and control of Wind Energy Conversion Systems (WECS) connected to an autonomous power system with hydrogen storage. The wind generator under study is a Doubly Fed Induction Generator (DFIG) type. The models of the main components (mainly wind turbine, generator, diesel genset, electrolyzer) will be derived. The wind turbine's maximum power point tracking technique will be presented and a power transfer strategy in the interconnected system will be analysed. Performance of the control method is validated to maintain the hybrid power system's frequency. The effectiveness of the proposed hybrid system is validated by simulation using Matlab/ Simulink/ SimPowerSystems environment. The Hybrid Power System (HPS) is composed of a 600 kW diesel genset, three 190 kW wind turbines, three 80 kW alkaline electrolyzers and a 610 kW maximum load.

2. Wind-diesel power system with hydrogen storage

The structures of Hybrid Power System (HPS) can be classified into two categories: AC coupled and DC-coupled (T. Zhou, 2009).

In an AC-coupled HPS, all sources are connected to a main AC-bus before being connected to the grid. In AC-coupled structure, different sources can be located anywhere in the microgrid with a long distance from each other. However, the voltage and the frequency of the main AC bus should be well controlled in order to ensure the stability of the system and the compatibility with the utility network.

In a DC-coupled HPS, all sources are connected to a main DC-bus before being connected to the grid through a main inverter. In a DC-coupled structure, the voltage and the frequency of the grid are independent from those of each source.

However, not all HPSs can be classified into AC or DC-coupled system, since it is possible to have both coupling methods, then a Mixed HPS is obtained. In this case, some advantages can be taken from both structures.

The wind-diesel HPS configuration studied in this work is represented in Fig. 1.

2.1. Wind turbine

Wind turbines come in different sizes and types, depending on power generating capacity and the rotor design deployed. Small wind turbines with output capacities below 10 kW are used primarily for residences, telecommunications dishes, and irrigation water pumping applications. Utility-scale wind turbines have high power ratings ranging from 100 kW to 5 MW. Current wind farms with large capacity wind turbine installations are capable of generating electricity in excess of 500M MW for utility companies (Vaughn Nelson, 2009).

Modern wind turbines are classified into two configurations: horizontal-axis wind turbines (HAWTs) and vertical-axis wind turbines (VAWTs), depending on rotor operating principles. The VAWT configuration employs the Darrieus model named for the famous French inventor.

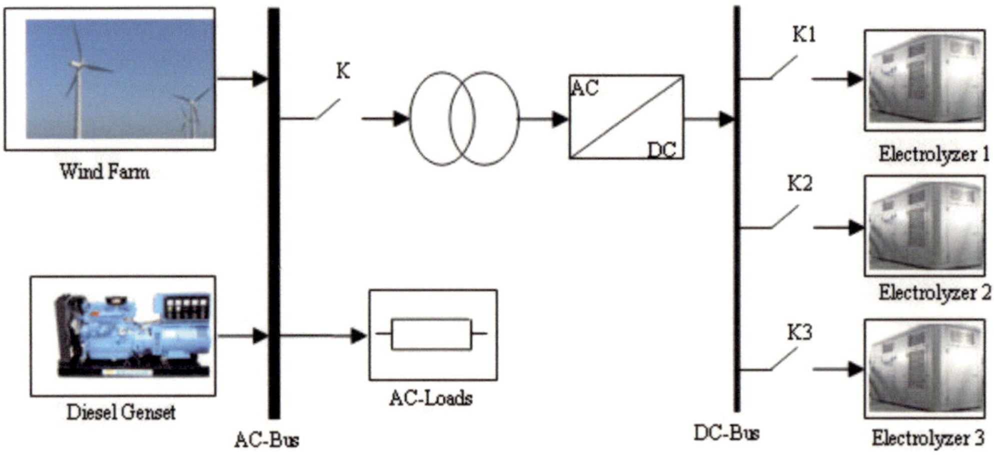

Figure 1. Wind-diesel Hybrid Power System with hydrogen production

HAWTs with two or three blades are the most common. Wind blowing over the propeller blades causes the blades to "lift" and rotate at low speeds. Wind turbines using three blades are operated "upwind" with rotor blades facing into the wind. The tapering of rotor blades is selected to maximize the kinetic energy from the wind. Optimum wind turbine performance is strictly dependent on blade taper angle and the installation height of the turbine on the tower (Vaughn Nelson, 2009).

According to Albert Betz, the mechanical power P_m captured by the turbine from the wind for a given wind speed v_w is computed by the following expression (I. Munteanu et al., 2008; N.M. Miller et al., 2008).

$$P_m = \frac{1}{2}\rho A C_p(\lambda, \beta) v_w^3 \quad (1)$$

$\rho$is the air density in kg/m^3; $A=\pi R^2$is the area in m^2 swept by the blade; $R$is the radius of the blade in m.

The aerodynamic model of a wind turbine can be determined by the $C_p(\lambda,\beta)$ curves. C_p is the power coefficient, which is function of both tip speed ratio λ and the blade pitch angle β. The tip speed ratio is given by:

http://dx.doi.org/10.5772/52341

$$\lambda = \frac{\Omega_r R}{v_w} \tag{2}$$

Ω_r represents the rotational speed of the wind turbine in *rad/sec*.

The maximum power coefficient C_p is determined by Betz as follows (S. Heier, 1998):

$$C_p^{\max}(\lambda, \beta) = \frac{16}{27} \approx 0.593 \tag{3}$$

Hence, even if power extraction without any losses were possible, only 59% of the wind power could be utilized by a wind turbine.

The power coefficient versus the ratio speed as shown in Fig. 2.

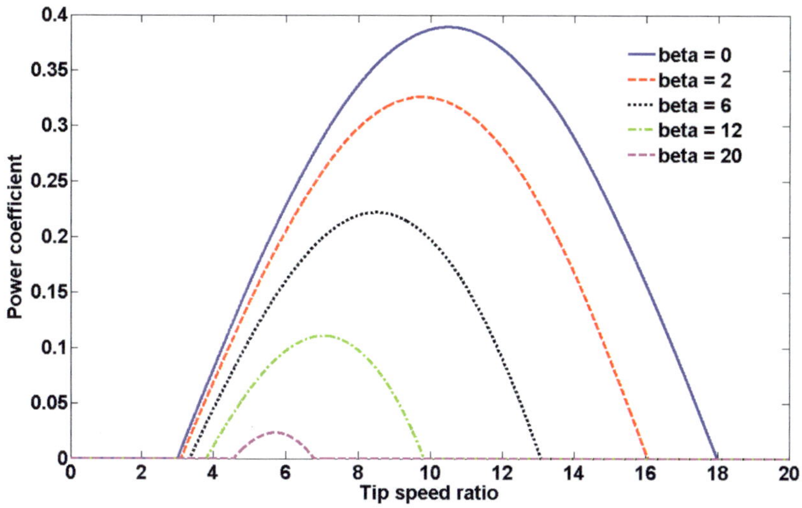

Figure 2. Coefficient of power versus ratio of speed

In our study, the mathematical representation of the power coefficient used for a wind turbine is given by:

$$C_p = 0.398.\sin\left(\frac{\pi(\lambda - 3)}{15 - 0.3\beta}\right) - 0.00394(\lambda - 2)\beta \tag{4}$$

In the aim to extract the maximum active power, the speed of the wind turbine must be adjusted to achieve the optimal value of the tip speed ratio. The block diagram of Fig. 3 shows the Maximum Power Point Tracking (MPPT) technique applied to the generator to produce maximum power. If the wind speed is below the rated value, the WTG operates in the variable speed mode, and C_p is keep at its maximum value. In this operating mode, the pitch control is deactivated. When the wind speed is above the rated value, the pitch control is activated, in the aim to reduce the generated mechanical power (W. Qiao, W. Zhou et al., 2008).

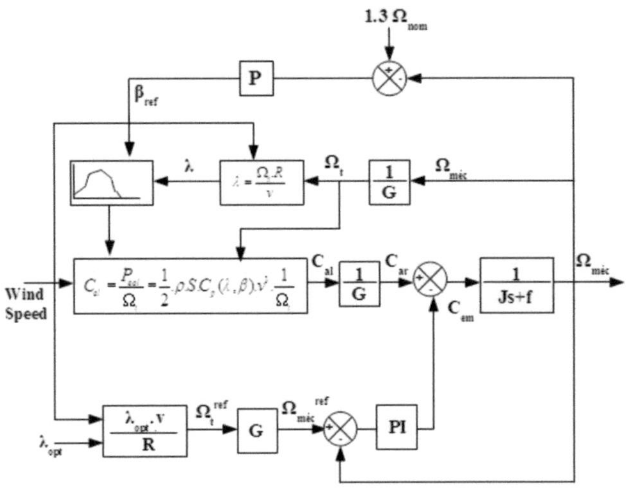

Figure 3. Block diagram of the control of the velocity of the DFIG with MPPT

2.2. Doubly-fed induction generator and its control

a. Doubly-Fed induction Generator

Today, the wind turbines on the market mix and match a variety of innovative concepts with proven technologies both for generators and for power electronics. Wind turbines can operate either with a fixed speed or a variable speed. Most commonly used types of wind turbines generators are asynchronous (induction) and synchronous generators. Among these technologies, asynchronous Doubly Fed Induction Generator (DFIG) has received much attention as one of preferred technology for wind power generation (Fig.4). The DFIG consists of a Wound Rotor Induction Generator (WRIG) with the stator windings directly connected to the constant-frequency three-phase grid and with the rotor windings mounted to a bidirectional back-to-back IGBT voltage source converter.The converter compensates the difference between the mechanical and electrical frequency by injecting a rotor current with a variable frequency.The power converter consists of two converters, the rotor-side converter and grid-side converter, which are controlled independently of each other.The main idea is that the rotor-side converter controls the active and reactive power by controlling the rotor current components, while the line-side converter controls the DC-link voltage and ensures a converter operation at unity power factor (i.e. zero reactive power). Compared to a full rated converter system, the use of DFIG in a wind turbine offers many advantages, such as reduction of inverter cost, the potential to control torque and a slight increase in efficiency of wind energy extraction. Depending on the operating condition of the drive, power is fed into or out of the rotor: in an oversynchronous mode, it flows from the rotor via the converter to the grid, whereas it flows in the opposite direction in a subsynchronous mode. In both cases – subsynchronous and oversynchronous – the stator feeds energy into the grid (T. Ackermann, 2005).

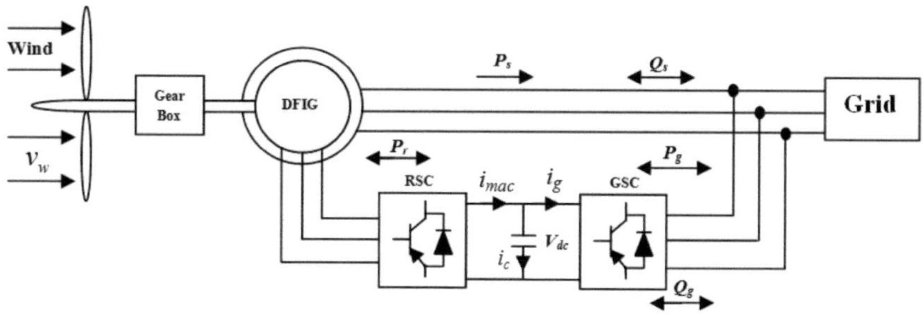

Figure 4. Structure of the DFIG based wind system

The stator and rotor voltages of the DFIG are given by the following expression (Y. Ren, H. Li and J. Zhou, 2009; R. G. De Almeida et al., 2004).

$$\begin{cases} v_{ds} = r_s i_{ds} + \dfrac{d\lambda_{ds}}{dt} - \omega_s \lambda_{qs} \\ v_{qs} = r_s i_{qs} + \dfrac{d\lambda_{qs}}{dt} + \omega_s \lambda_{ds} \\ v_{dr} = r_r i_{dr} + \dfrac{d\lambda_{dr}}{dt} - \omega_r \lambda_{qr} \\ v_{qr} = r_r i_{qr} + \dfrac{d\lambda_{qr}}{dt} + \omega_r \lambda_{dr} \end{cases} \tag{5}$$

r_s and r_r are respectively the resistance of the stator and rotor windings, and ω_s is the rotational speed of the synchronous reference frame.

$$\begin{cases} \lambda_{ds} = L_s i_{ds} + L_m i_{dr} \\ \lambda_{qs} = L_s i_{qs} + L_m i_{qr} \\ \lambda_{dr} = L_r i_{dr} + L_m i_{ds} \\ \lambda_{qr} = L_r i_{qr} + L_m i_{qs} \end{cases} \tag{6}$$

L_s and L_r represent respectively the self-inductance of the stator and the rotor windings, and L_m is the mutual inductance between the stator and the rotor windings.

The electromagnetic torque of the DFIG can be expressed as follow:

$$T_{em} = P(\lambda_{ds} i_{qs} - \lambda_{qs} i_{ds}) \tag{7}$$

P is the number of pole pairs.

To achieve independent control of the stator active power and stator reactive power, a vector-control approach is used. d-q axis connected to the stator's rotating field is chosen and quadratic component of the stator flux is set to zero. The power control is performed through the back to back converter connected to the rotor. Then, the stator voltages can be given according the rotor currents as (D. Aouzellag et al. 2006).

The stator's flux and current equations are:

$$\begin{cases} \lambda_{ds} = L_s i_{ds} + L_m i_{dr} = \Phi_s \\ \lambda_{qs} = L_s i_{qs} + L_m i_{qr} = 0 \end{cases} \tag{8}$$

$$\begin{cases} i_{ds} = \dfrac{\Phi_s}{L_s} - \dfrac{L_m}{L_s} i_{dr} \\ i_{qs} = -\dfrac{L_m}{L_s} i_{qr} \end{cases} \tag{9}$$

Fom equations (5) et (6), rotor's voltages can be rewritten as:

$$\begin{cases} v_{dr} = r_r i_{dr} + \dfrac{d\left(L_r i_{dr} + L_m i_{ds}\right)}{dt} - \omega_r \left(L_r i_{qr} + L_m i_{qs}\right) \\ v_{qr} = r_r i_{qr} + \dfrac{d\left(L_r i_{qr} + L_m i_{qs}\right)}{dt} + \omega_r \left(L_r i_{dr} + L_m i_{ds}\right) \end{cases} \tag{10}$$

$$\begin{cases} v_{dr} = r_r i_{dr} + L_r \dfrac{di_{dr}}{dt} + L_m \dfrac{di_{ds}}{dt} - \omega_r L_r i_{qr} + \omega_r L_m i_{qs} \\ v_{qr} = r_r i_{qr} + L_r \dfrac{di_{qr}}{dt} + L_m \dfrac{di_{qs}}{dt} + \omega_r L_r i_{dr} + \omega_r L_m i_{ds} \end{cases} \tag{11}$$

$$\begin{cases} v_{dr} = r_r i_{dr} + L_r \dfrac{di_{dr}}{dt} - L_m \dfrac{L_m di_{dr}}{L_s dt} - \omega_r L_r i_{qr} - \omega_r L_m \dfrac{L_m}{L_s} i_{qr} \\ v_{qr} = r_r i_{qr} + L_r \dfrac{di_{qr}}{dt} - L_m \dfrac{L_m di_{qr}}{L_s dt} + \omega_r L_r i_{dr} + \omega_r L_m \dfrac{\Phi_s}{L_s} - \omega_r L_m \dfrac{L_m}{L_s} i_{dr} \end{cases} \tag{12}$$

http://dx.doi.org/10.5772/52341

$$\begin{cases} v_{dr} = r_r i_{dr} + \left(L_r - \frac{L_m{}^2}{L_s} \right) \frac{di_{dr}}{dt} - \omega_r L_r i_{qr} - \omega_r \frac{L_m{}^2}{L_s} i_{qr} \\ v_{qr} = r_r i_{qr} + \left(L_r - \frac{L_m{}^2}{L_s} \right) \frac{di_{qr}}{dt} + \omega_r L_r i_{dr} + \omega_r L_m \frac{\Phi_s}{L_s} - \omega_r \frac{L_m{}^2}{L_s} i_{dr} \end{cases} \tag{13}$$

$$\begin{cases} v_{dr} = r_r i_{dr} + \left(L_r - \frac{L_m{}^2}{L_s} \right) \frac{di_{dr}}{dt} - \omega_r \left(L_r - \frac{L_m{}^2}{L_s} \right) i_{qr} \\ v_{qr} = r_r i_{qr} + \left(L_r - \frac{L_m{}^2}{L_s} \right) \frac{di_{qr}}{dt} + \omega_r \left(L_r - \frac{L_m{}^2}{L_s} \right) i_{dr} + \omega_r L_m \frac{\Phi_s}{L_s} \end{cases} \tag{14}$$

$\omega_r$and ω_s are respectively rotor and stator parameters frequencies; $s$is the machine's slip.

As

$$\omega_r = s\omega_s$$

$$\begin{cases} v_{dr} = r_r i_{dr} + \left(L_r - \frac{L_m{}^2}{L_s} \right) \frac{di_{dr}}{dt} - s\omega_s \left(L_r - \frac{L_m{}^2}{L_s} \right) i_{qr} \\ v_{qr} = r_r i_{qr} + \left(L_r - \frac{L_m{}^2}{L_s} \right) \frac{di_{qr}}{dt} + s\omega_s \left(L_r - \frac{L_m{}^2}{L_s} \right) i_{dr} + g\omega_s L_m \frac{\Phi_s}{L_s} \end{cases} \tag{15}$$

From vector control conditions, the stator voltage can be expressed as:

$$V_s = \omega_s . \Phi_s \tag{16}$$

Then, the final vector control equations of the rotor's voltage are:

$$\begin{cases} v_{dr} = r_r i_{dr} + \left(L_r - \frac{L_m{}^2}{L_s} \right) \frac{di_{dr}}{dt} - s\omega_s \left(L_r - \frac{L_m{}^2}{L_s} \right) i_{qr} \\ v_{qr} = r_r i_{qr} + \left(L_r - \frac{L_m{}^2}{L_s} \right) \frac{di_{qr}}{dt} + s\omega_s \left(L_r - \frac{L_m{}^2}{L_s} \right) i_{dr} + s\omega_s \frac{L_m V_s}{\omega_s L_s} \end{cases} \tag{17}$$

2. DFIG's Control

The Rotor Side Converter (RSC) is used to control both active and reactive powers provided by the stator of the DFIG. The control strategy of the RSC is based on the power vector control of the DFIG, and the principle of this control is illustrated by Fig. 5. Different controllers can be used for this purpose.

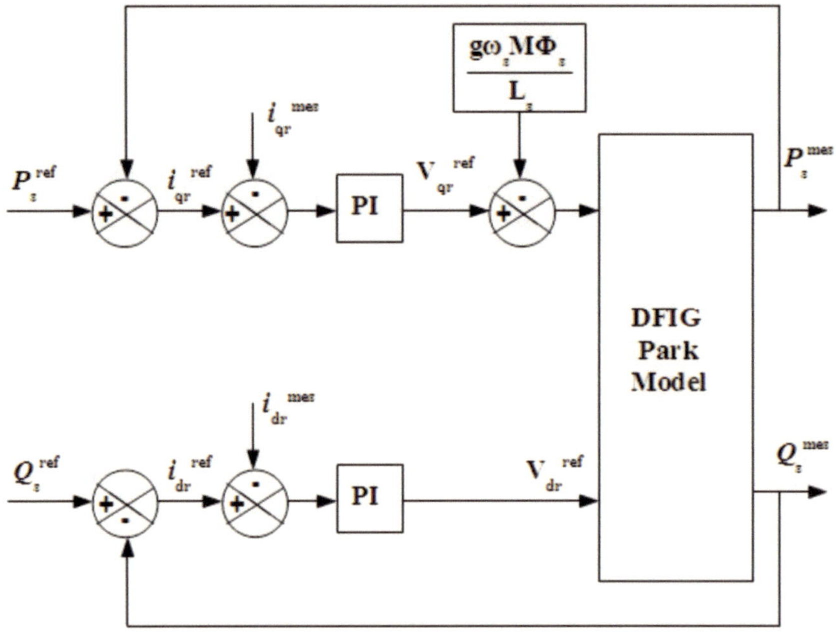

Figure 5. Scheme of the power control vector of the DFIG

The Grid Side Converter (GSC) is used to regulate the DC-link voltage and to adjust the power factor. The GSC is a bidirectional converter which operates as a rectifier when the slip (g) is positive (subsynchronous mode) and as an inverter when the slip is negative (oversynchronous mode).

The active and reactive powers on the grid side are written respectively as follows (X. Yao et al. 2008):

$$\begin{cases} P = \frac{3}{2} V_m . i_d \\ Q = -\frac{3}{2} V_m . i_q \end{cases} \tag{18}$$

$V_m$isthe magnitude of voltage of the grid. The principle of control of GSC is illustrated by Fig. 6.

http://dx.doi.org/10.5772/52341

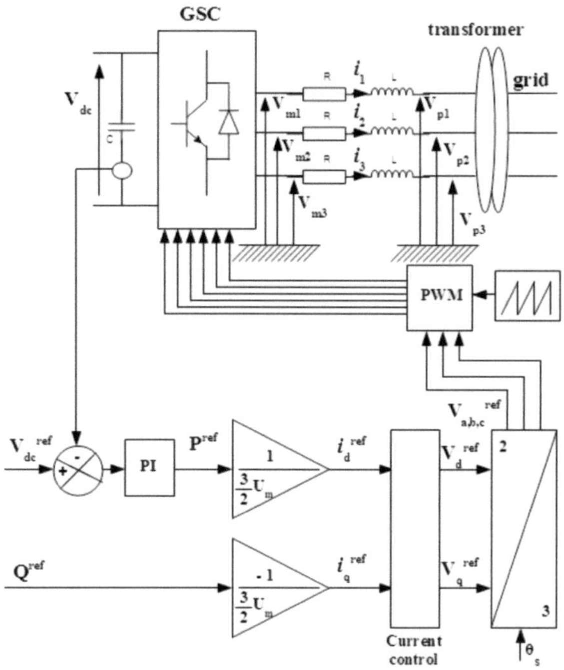

Figure 6. Scheme of the GSC converter control

2.3. Diesel generator modelling

The diesel ge4narator is composed of the diesel engine and Wound Rotor Synchronous Generator (WRSG).

a. Diesel Engine

The model of the diesel engine is shown in Fig. 7 (R. Dettmer, 1990; R. Pena et al., 2002; S. Roy et al., 1993). The dynamic of the actuator is modeled by a first order model with time constant $\tau_1$and gain K_1 (R. Pena et al., 2008; S. Roy et al., 1993). The combustion bloc is represented with gain K_2 and delay τ_2(R. Dettmer, 1990).

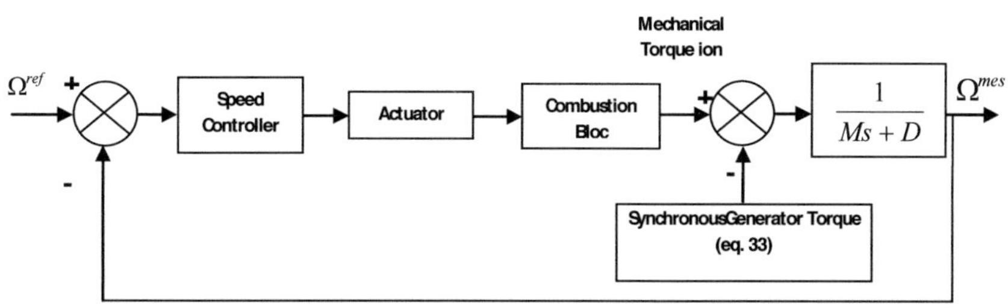

Figure 7. Block diagram of diesel generator model

The actuator is modelled as:

$$\frac{K_1}{1+s\tau_1} \tag{19}$$

The model of the combustion bloc is given by:

$$K_2 e^{-s\tau_2} \tag{20}$$

The delay can be expressed as (R. Pena et al., 2002; R. Pena et al., 2008):j

$$\tau_2 = \frac{60h}{2Nn_c} + \frac{60}{4N} \tag{21}$$

h represents the strokes number, n_c the number of cylinders and N the speed of diesel generator (rpm), Φ is the fuel consumption rate (kg/sec) (F. Jurado and J. R. Saenz, 2002). In the order to maintain constant the frequency of the grid (AC-bus), the speed of the diesel engine must be kept constant when the load varies.

2. Synchronous Generator

The simplified model of the Wound RotorSynchronous Generator (WRSG) can be obtained in *dq* frame (conversion between *abc* and *dq* can be realized by means of the Park Transform) (T. Burton et al., 2001).

The stator armature windings voltages are:

$$\begin{cases} v_d = -R_s i_d + \dfrac{d\lambda_d}{dt} - \omega\lambda_q \\ v_q = -R_s i_q + \dfrac{d\lambda_q}{dt} + \omega\lambda_d \end{cases} \tag{22}$$

$R_s$is the stator winding resistance

The stator fluxes are

$$\begin{cases} \lambda_d = -L_d i_d + L_{md}\left(i_f + i_D\right) \\ \lambda_q = -L_q i_q + L_{mq} i_Q \end{cases} \tag{23}$$

http://dx.doi.org/10.5772/52341

Rotor armature winding voltage is

$$v_f = -R_f i_f - L_d \frac{di_d}{dt} + L_f \frac{di_f}{dt} + L_{md} \frac{di_D}{dt} \tag{24}$$

Damper windings are characterized by

$$\begin{cases} 0 = R_D i_D - L_{md} \dfrac{di_d}{dt} + L_{md} \dfrac{di_f}{dt} + L_D \dfrac{di_D}{dt} \\ 0 = R_Q i_Q - L_{mq} \dfrac{di_q}{dt} + L_Q \dfrac{di_Q}{dt} \end{cases} \tag{25}$$

To eliminate v_d and v_q in the expression of the stator voltages, we introduced the R_cL_c load which is supplied by the synchronous generator.

$$\begin{cases} v_d = R_c i_d + L_c \dfrac{di_d}{dt} - \omega L_c i_q \\ v_q = R_c i_q + L_c \dfrac{di_q}{dt} + \omega L_c i_d \end{cases} \tag{26}$$

By using the equations above, the state space model of the Wound Rotor Synchronous Generator (WRSG) can be written as follow (Belmokhtar et al., 2012a, Belmokhtar et al., 2012b)):

$$[\dot{X}] = [A].[X] + [B].[U] \tag{27}$$

Where

$[A]$: State matrix; $[X]$: State vector; $[B]$: Control matrix; $[U]$: Control vector.

$[X]=[i_d i_q\ i_f i_D i_Q]^T$ and $[B]=[0\ 0\ 0\ v_f\ 0\ 0]^T$

$$\begin{aligned} \frac{di_d}{dt} &= \left(\frac{1}{L_c + \left(\alpha - \dfrac{\alpha\beta}{\chi} \right)} \right) \begin{pmatrix} -(R_s + R_c) i_d + \omega (L_q + L_c) i_q + \dfrac{\beta}{\chi} V_f + \\ \dfrac{\beta}{\chi} R_f i_f + \left(\dfrac{\beta}{\chi} - 1 \right) \dfrac{L_{md}}{L_D} R_D i_D - \omega L_{mq} i_Q \end{pmatrix} \\ \frac{di_q}{dt} &= \left(\frac{1}{L_c + \gamma} \right) \begin{pmatrix} -\omega (L_d + L_c) i_d - (R_s + R_c) i_q + \omega L_{md} i_f + \\ \omega L_{md} i_D - \dfrac{L_{mq}}{L_Q} R_Q i_Q \end{pmatrix} \end{aligned} \tag{28}$$

$$\frac{di_f}{dt} = -\frac{\alpha(R_s+R_c)}{\chi\delta}i_d + \frac{\omega\alpha(L_q+L_c)}{\chi\delta}i_q + \left(\left(\frac{\alpha\beta}{\chi^2\delta}R_f\right)+\left(\frac{R_f}{\chi}\right)\right)i_f +$$
$$\left(\frac{1}{\chi}+\frac{\alpha\beta}{\chi^2\delta}\right)V_f + \left(\left(\frac{\beta}{\chi}-1\right)\frac{\alpha}{\delta\chi}+\frac{1}{\chi}\right)\frac{L_{md}}{L_D}R_D I_D - \frac{\omega\alpha L_{mq}}{\chi\delta}i_Q$$

$$\frac{di_D}{dt} = \frac{L_{md}(R_s+R_c)}{L_D\delta}\left(\left(\frac{\alpha}{\chi}\right)-1\right)i_d + \frac{\omega L_{md}(L_q+L_c)}{L_D\delta}\left(1-\left(\frac{\alpha}{\chi}\right)\right)i_q -$$
$$\frac{L_{md}R_f}{L_D\chi}\left(1+\left(\frac{\alpha\beta}{\chi\delta}\right)+\frac{\beta}{\delta}\right)i_f - \frac{L_{md}}{L_D}\left(\frac{1}{\chi}+\left(\frac{\alpha\beta}{\chi^2\delta}\right)+\frac{\beta}{\chi\delta}\right)V_f -$$
$$\left(\left(\left(\frac{\beta}{\chi}-1\right)\left(\frac{\alpha}{\delta\chi}-\frac{1}{\delta}\right)+\frac{1}{\chi}\right)\frac{L_{md}^2}{L_D^2}+\frac{1}{L_D}\right)R_D i_D + \left(\frac{L_{md}}{L_D}\frac{\omega\alpha L_{mq}}{\chi\delta} - \left(\frac{\omega L_{md}L_{mq}}{\delta L_D}\right)\right)i_Q \tag{29}$$

$$\frac{di_Q}{dt} = -\omega\frac{L_{mq}(L_d+L_c)}{L_Q(L_c+\gamma)}i_d - \frac{L_{mq}(R_s+R_c)}{L_Q(L_c+\gamma)}i_q + \omega\frac{L_{mq}L_{md}}{L_Q(L_c+\gamma)}i_f +$$
$$\omega\frac{L_{md}L_{mq}}{L_Q(L_c+\gamma)}i_D - \left(\frac{L_{mq}^2}{L_Q^2(L_c+\gamma)}+\frac{1}{L_Q}\right)R_Q i_Q$$

$$T_{em} = p\left(\left(L_d-L_q\right)i_d i_q + L_{md}i_f i_q + L_{md}i_q i_D - L_{mq}i_d i_Q\right)$$

P is the number of the poles

In order to improve the efficiency and avoid wet stacking, a minimum load of about 30% to 40% is usually recommended by the manufacturers (J. B. Andriulli et al., 1999). To achieve this goal, the values of R_c and L_c are chosen in the aim to give 35% of the rated power of the diesel generator when it is switched on. Then the partial values of the stator currents i_{d0} and i_{q0} in the dq frame are calculated. The additional values of the stator currents i_{d1} and i_{q1} are computed respectively as follow:

$$\begin{cases} i_{d1} = \dfrac{v_d P_{diesel}^{Add} - v_q Q_{diesel}^{Add}}{\frac{3}{2}\left(v_d^2+v_q^2\right)} \\ i_{q1} = \dfrac{v_q P_{diesel}^{Add} + v_d Q_{diesel}^{Add}}{\frac{3}{2}\left(v_d^2+v_q^2\right)} \end{cases} \tag{30}$$

The active additional power and the reactive power of the diesel generator are expressed respectively as:

http://dx.doi.org/10.5772/52341

$$\begin{cases} P_{diesel}^{Add} = \frac{3}{2}\left(v_d.i_{d1} + v_q.i_{q1}\right) \\ Q_{diesel}^{Add} = \frac{3}{2}\left(v_d.i_{q1} - v_q.i_{d1}\right) \end{cases} \tag{31}$$

The total stator currents of the diesel generator are:

$$\begin{cases} i_{dt} = i_{d0} + i_{d1} \\ i_{qt} = i_{q0} + i_{q1} \end{cases} \tag{32}$$

Then, (27) is expressed as follow:

$$T_{em} = p\left(\left(L_d - L_q\right)i_{dt}i_{qt} + L_{md}i_f i_{qt} + L_{md}i_{qt}i_D - L_{mq}i_{dt}i_Q\right) \tag{33}$$

The diesel generator will operate with minimum load of 35% of the rated power. The alkaline electrolyzers are used as dump load. The electrolyzers are supplied by the surplus power, and then contribute to balance the load demand and power production.

2.4. Alkaline electrolyzer

The decomposition of water into hydrogen and oxygen can be obtained by passing a direct electric current (DC) between two electrodes separated by a membrane and containing an aqueous electrolyte with good ionic conductivity. The electrodes are immersed in an alkaline aqueous solution.

The electrolyzer model is composed of several modules (F. J. Pino et al., 2011). Powered by the DC electrical sources and pure water, an electrolyzer can effectively split water into hydrogen and oxygen. Since it is difficult to obtain analytically the inverse of the equation (34), linear models are used generally in literature (R.Takahashi et al. 2010).

In this paper, the electrolyzer model takes into account the ohmic resistances and cell overvoltages (equation 34) (O. Ulleberg, 1998).

$$V_{cell} = E_0 + \frac{r_1 + r_2 T_{ele}}{A_{ele}} I_{ele} + s_0 \log\left(\frac{t_1 + t_2 / T_{ele} + t_3 / T_{ele}^2}{A_{ele}} + 1\right) \tag{34}$$

$$s_0 = s_1 + s_2 T_{ele} + s_3 T_{ele}^2 \tag{35}$$

$V_{rev}$is the reversible voltage, r_i, $s_i$and t_i are the empirical parameters whose values are determined from experiments (N. Gyawali and Y. Oshsawa, 2010).

The electrolyzer's voltage is expressed as:

$$V_{ele} = N_c . V_{cell} \tag{36}$$

$V_{cell}$is the voltage of electrolyzer cell and N_c is the number of cells of the electrolyzer.

3. Hybrid power system simulation results

The load's power, diesel genset's power and wind generators power are shown respectively in Fig. 8, Fig. 9 and Fig. 10. The diesel genset operates with minimum load of 35% of the rated power (Fig. 9). Three 80 kW alkaline electrolyzers are used as dump load. The surplus power is absorbed bythe electrolyzers (Fig. 11, Fig. 12, Fig. 13) which contribute to balance the load demand and power production (Fig. 14). Electrolyzers contribute to maintain the frequency of the autonomous power system (Fig. 15).

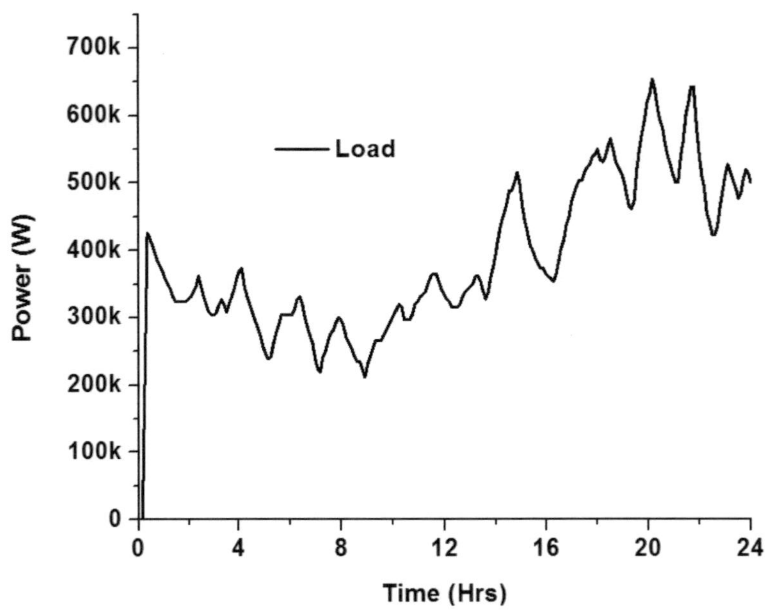

Figure 8. Load profile

http://dx.doi.org/10.5772/52341

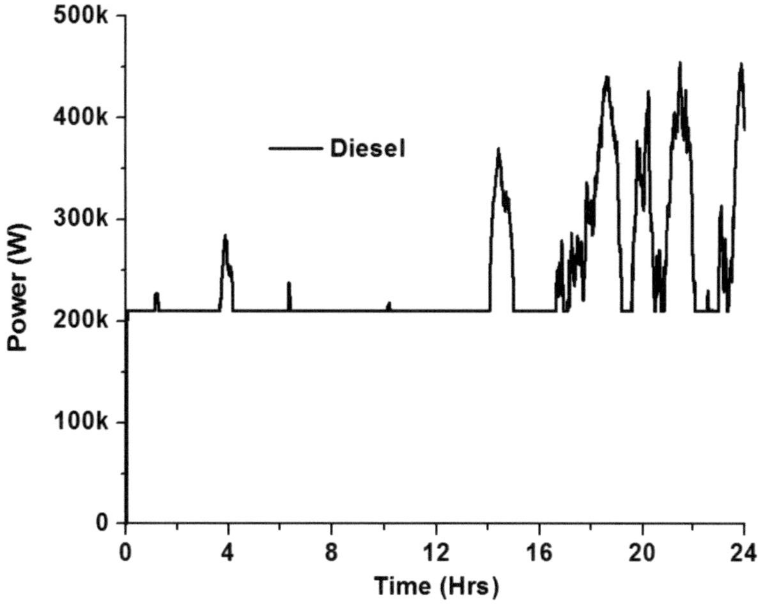

Figure 9. Diesel genset power

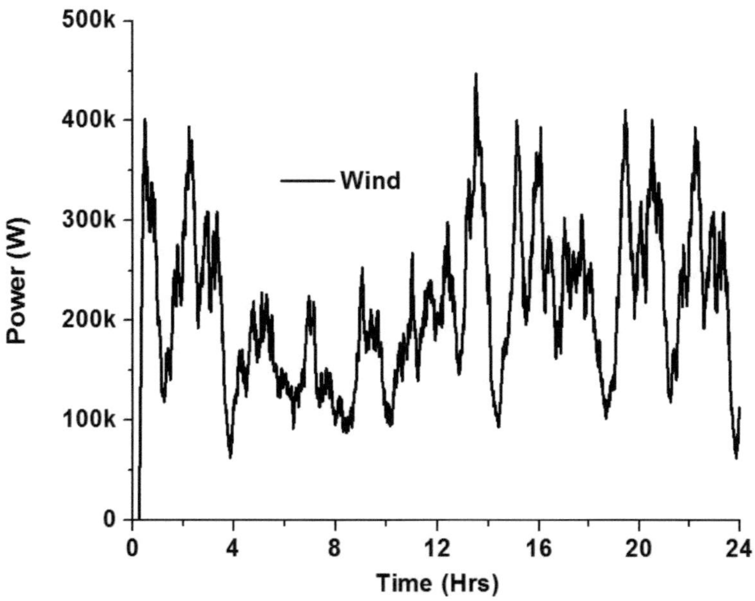

Figure 10. Wind farm power

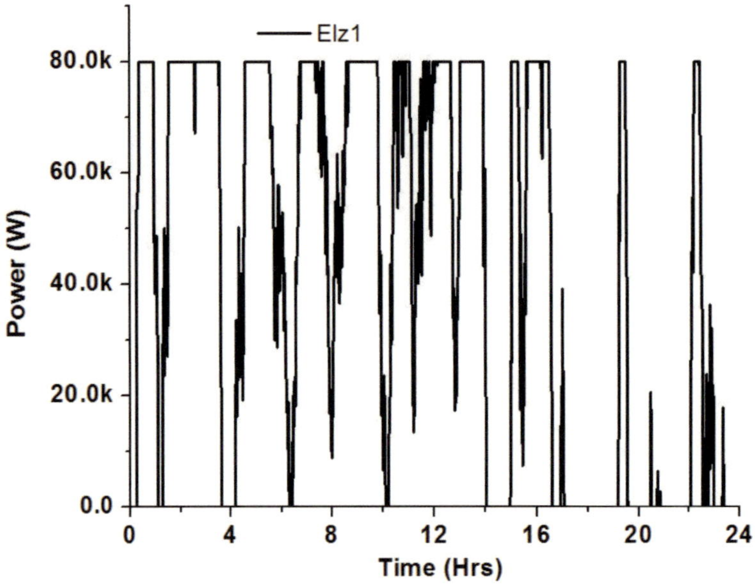

Figure 11. Power absorbed by Electrolyzer 1

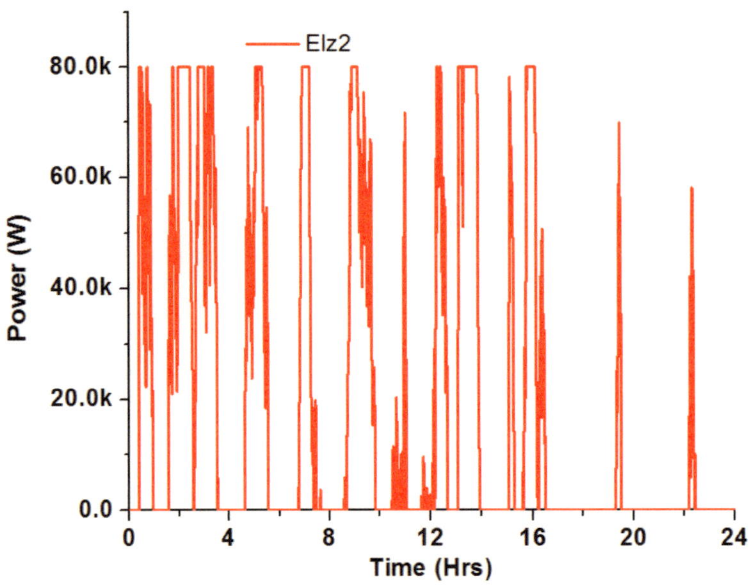

Figure 12. Power absorbed by Electrolyzer 2

http://dx.doi.org/10.5772/52341

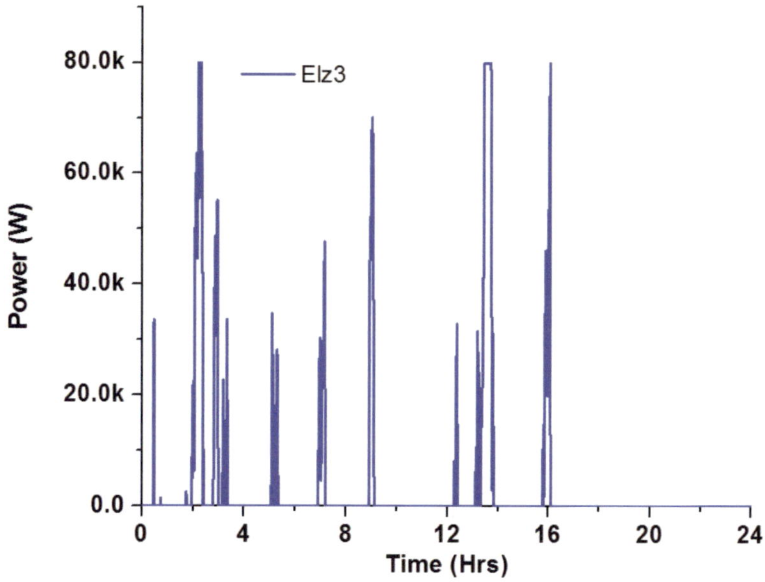

Figure 13. Power absorbed by Electrolyzer 3

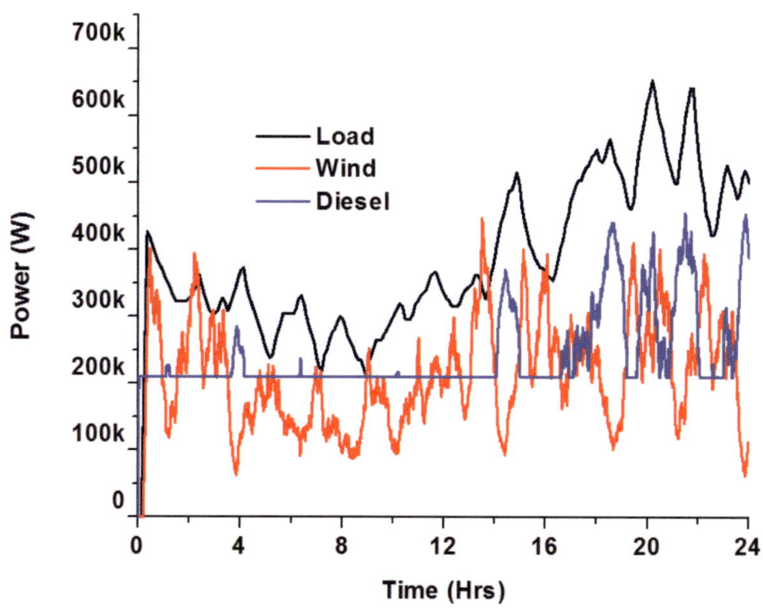

Figure 14. wind, diesel and Load powers

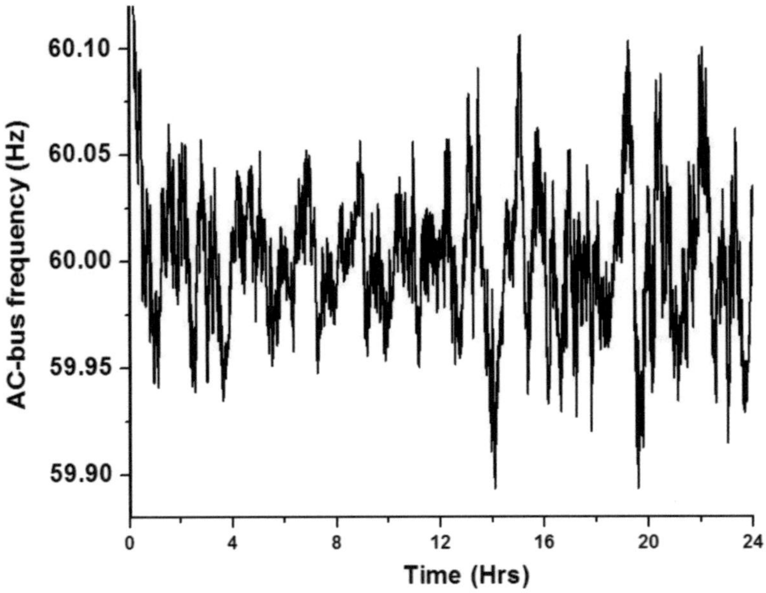

Figure 15. AC-bus frequency

4. Conclusion

This chapter is devoted to a large scale wind diesel Hybrid Power System (HPS). It presents theoretical analysis, modelling and control of Wind Energy Conversion Systems (WECS) connected to an autonomous power system with hydrogen storage. The wind generator under study is a Doubly Fed Induction Generator (DFIG) type. The models of the main components (mainly wind turbine, generator, diesel generator, electrolyzer) are derived. Power transfer strategy in the interconnected system was presented. The effectiveness of the proposed control strategy was validated by simulation using Matlab/ Simulink/ SimPowerSystems environment.

Author details

Mamadou Lamine Doumbia, Karim Belmokhtar and Kodjo Agbossou

Hydrogen Research Institute, Department of Electrical and Computer Engineering, Université du Québec à Trois-Rivières, Québec, Canada

References

[1] International Energy Agency (IEA)(2008). Energy Technology Perspectives 2008, , 5.

[2] Vaughn Nelson,(2009). WIND ENERGY: Renewable Energy and the Environment, CRC Press, 2009.

[3] US Department of Energy(2011). Wind Technologies Market Report 2010, June 2011.

[4] Zoulias, E. I., & Lymberopoulos, N. (2008). Hydrogen-based Autonomous Power Systems, Springer-Verlag.

[5] Nelson et al.(2006). Unit sizing and cost analysis of stand-alone hybrid wind/PV/fuel cell power generation system. Renewable Energy , 31, 1641-1656.

[6] MamadouLamineDoumbia and KodjoAgbossou,"Photovoltaic/Wind energy system with hydrogen based",Renewable Energy, ISBN 978-953-7619-52-7, 2009.

[7] K. Agbossou, M. Kolhe, J. Hamelin and T.K. Bose, « Performance of Stand-alone Renewable Energy System based on Energy Storage as Hydrogen », IEEE Transactions on Energy conversion, vol. 19, no 3, p. 633-640, 2004.

[8] Tao Zhou,(2009). Commande et Supervision Energétique d'un Générateur Hybride Actif Eolien incluant du Stockage sous forme d'Hydrogène et des Super-Condensateurs pour l'Intégration dans le Système Electrique d'un Micro Réseau, thèse de doctorat, 2009, École Centrale de Lille, France.

[9] Munteanu, I., Bratcu, A. I., -A, N., Cutululis, , Ceanga, E., & (2008).", . (2008). Optimal Control of Wind Energy Systems".Springer 2008.

[10] Miller, N. M., Price, W. W., Sanchez-Gasca, J. J., & (2003).", . (2003). Dynamic Modelling of GE 1.5 and 3.6 Wind Turbine-Generators".GE-Powwer Systems Energy Consulting, General Electric International, Inc, Schenectady, NY, USA, Oct, 27, 2003.

[11] Heier, S. (1998). Grid Integration of Wind Energy, Conversions Systems", John Wiley & Sons Ltd, New-York 1998.

[12] Qiao, W., Zhou, W., Aller, J. M., & Harley, R. G. (2008). Wind Speed Estimation Based Sensorless Output Maximization Control for a Wind Turbine Driving a DFIG". IEEE Trans. Power Electron., May 2008., 23(3), 1156-1169.

[13] Thomas Ackermann(2005). Wind Power in Power Systems, John Wiley & Sons, Ltd.

[14] Ren, Y., Li, H., & Zhou, J. (2009). Dynamic Performance of Grid-Connected DFIG Based on Fuzzy Logic Control", Proceeding of the 2009, International Conference on Mechatronics and Automation, August, China, 2009., 9-12.

[15] De Almeida, R. G., Lopes, J. A. P., & Barreiros, J. A. L. (2004). Improving Power System Dynamic Behavior Through Doubly Fed Induction Machines Controlled by Static Converter Using Fuzzy Control", IEEE Trans. on Power Systems. November 2004., 19(4), 1942-1950.

[16] Aouzellag, K., Ghedamsi, , & Berkouk, E. M. (2006). Power Control of a Variable Speed Wind Turbine Driving an DFIG", in ICREPQ06, (220)

[17] Yao, X., Yi, C., Ying, D., Guo, J., & Yang, L. (2008). The grid-side PWM Converter of the Wind Power Generation System Based on Fuzzy Sliding Mode Control", AdvancedIntelligent Mechatronics, IEEE 2008, Xian (Chine), , 973-978.

[18] Dettmer, R. (1990). Revolutionary energy-A wind/diesel generator with flywheel storage".Inst. Electr. Eng. Rev., Apr. 1990., 36, 149-151.

[19] Pena, R., Cardenas, R., Clare, J., & Asher, G. (2002). Control strategy of doubly fed generators for a wind diesel energy system". IECON 02, , 4, 3297-3302.

[20] Roy, S., Malik, O. P., & Hope, G. S. (1993). A k-Step Predictive Scheme for Speed Control of Diesel Driven Power Plants". IEEE Trans. Ind. Appl., , 29(2), 389-396.

[21] Pena, R., Cardenas, R., Proboste, J., Clare, J., & Asher, G. (2008). Wind-Diesel Generation Using Doubly Fed Induction Machines". IEEE Trans. Energy. Convers., March. 2008., 23(1), 202-213.

[22] Jurado, F., & Saenz, J. R. (2002). Neuro-fuzzy control for autonomous wind-diesel systems using biomass".Renewable.Energy 27, , 39-56.

[23] Burton, T., Sharpe, D., Jenkins, N., & Bossanyi, E. (2001). Wind Energy Handbook".John Wiley & Sons, Ltd, 2001.

[24] KarimBelmokhtar et al.(2012a). Modelling and Fuzzy Logic Control of DFIG Based Wind Energy Conversion Systems". IEEE SPEEDAM Conference, Sorrento, Italy, 2012.

[25] KarimBelmokhtar et al.(2012b). New Sensorless Fuzzy Logic Control Strategy of Wind Energy Conversion Systems".IEEE ISIE Conference, Hangzhou, China, 2012.

[26] Andriulli, J. B., Gates, A. E., Haynes, H. D., Klett, L. B., Matthews, S. N., Nawrocki, E. A., Otaduy, P. J., Scudiere, M. B., Theiss, T. J., Thomas, J. F., Tolbert, L. M., Yauss, M. L., & Voltz, C. A. (1999). Advanced power generation systems for the 21st century: Market survey and recommendations for a design philosophy," Oak Ridge National Laboratory, Oak Ridge, TN, Tech. Rep. Nov. 1999.

[27] Pino, F. J., Valverde, L., & Rosa, F. (2011). Influence of wind turbine power curve and electrolyser operating temperature on hydrogen production in wind-hydrogen systems". Journal of Power Sources, , 196(9), 4418-4426.

[28] Takahashi, R., Kinoshita, H., Murata, T., Tamura, J., Sugimasa, M., Komura, A., Futami, M., Ichinose, M., & Ide, K. (2010). Output Power Smoothing and Hydrogen Production by Using Variable Speed Wind Generators".IEEE Transactions on Industrial Electronics, , 57(2), 485-493.

[29] Ulleberg, O. (1998). Stand-alone power systems for the future: optimal design, operation & control of solar-hydrogen energy systems", PhD Thesis, (1998).

[30] Gyawali, N., & Ohsawa, Y. (2010). Integrating Fuell Cell/ Electrolyzer/ Ultracapacitor System Into a Stand-Alone Microhydro Plant". IEEE Transactions on Energy Conversion, ., 25(4), 10921-11101.

Chapter 16

Sustainable Power Generation Through Co-Combustion of Agricultural Residues with Coal in Existing Coal Power Plant

Wan Azlina Wan Ab Karim Ghani and
Azil Bahari Alias

Additional information is available at the end of the chapter

http://dx.doi.org/10.5772/52566

1. Introduction

Waste-to-energy is gaining more and more attention as landfill costs and environmental concerns increase in many developed countries including Malaysia. Biomass from agricultural residues is one of the most important sources of renewable energy in Malaysia.The National Biofuel Policy, launched in 2006 encourages the use of environmentally friendly, sustainable and viable sources of biomass energy. Under the Five Fuel Policy, the government of Malaysia has identified biomass as one of the potential renewable energy. Malaysia produces at least 168 million tonnes of biomass, including timber and oil palm waste, rice husks, coconut trunk fibres, municipal waste and sugar cane waste annually. Being a major agricultural commodity producer in the region Malaysia is well positioned amongst the ASEAN countries to promote the use of biomass as a renewable energy source.

Combustion of agricultural residues is commonly used in industries for energy recovery. However, many researchers found that stand alone biomass firing is difficult to get high efficiency [1-5]. Thus, co-firing biomass with coal in industrial and utility boilers could offer an alternative approach with improved combustion efficiency, lower-cost and reduced risk technology. Significant co-combustion potential for biomass and waste materials exists in all European Union (EU) countries and this is mirrored on a worldwide basis, creating a significant market for equipment and services. Fluidized bed combustion (FBC) technology has already proven highly efficient, economic and environmentally sound combustion of a wide variety of fuels in comparison conventional combustors. Hence, with the current demands in electricity and with the recent developments in biomass energy, co-combustion of biomass with coal must be recognized as one of the most important source of energy for the foreseeable future.

1.1. Biomass as potential renewable resources

A recent study shows that Malaysia has been one of the world's largest producers and exporters of palm oil for the last forty years. The Palm Oil industry, besides producing Crude Palm Oil (CPO) and Palm Kernel Oil, produces Palm Shell, Press Fibre, Empty Fruit Bunches (EFB), Palm Oil Mill Effluent (POME), Palm Trunk (during replanting) and Palm Fronds (during pruning). Almost 70% of the volume from the processing of fresh fruit bunches (FFB) is removed as waste.Malaysia has approximately 4 million hectares of land under oil palm plantation. Over 75% of total area planted is located in just four states, Sabah, Johor, Pahang and Sarawak, each of which has over half a million hectares under cultivation. The total amount of processed FFB was estimated to be 75 million tons while the total amount of EFB produced was estimated to be 16.6 million tons. Around 58 million tons of POME is produced in Malaysia annually, which has the potential to produce an estimated 15 billion m^3 of biogas can be produced each year [6].

Rice husk is another important agricultural biomass resource in Malaysia with good potential for power cogeneration. An example of its attractive energy potential is biomass power plant in the state of Perlis which uses rice husk as the main source of fuel and generates 10 MW power to meet the requirements of 30,000 households. The US$15 million project has been undertaken by Bio-Renewable Power Sdn. Bhd in collaboration with the Perlis state government, while technology provider is Finland's Foster Wheeler EnergiaOy. Under the EC-ASEAN Cogeneration Program, there are three ongoing Full Scale Demonstration Projects (FSDPs) – Titi-Serong, Sungai Dingin Palm Oil Mill and TSH Bioenergy – to promote biomass energy systems in Malaysia. The 1.5MW TitiSerong power plant, located at Parit Buntar (Perak), is based on rice husk while the 2MW Sungai Dingin Palm Oil Mill project make use of palm kernel shell and fibre to generate steam and electricity. The 14MW TSH Bioenergy SdnBhd, located at Tawau (Sabah), is the biggest biomass power plant in Malaysia and utilizes empty fruit bunches, palm oil fibre and palm kernel shell as fuel resources [7].

1.2. Fluidised Bed Combustion (FBC) technology

Fluidised bed combustion technology is one of the most significant recent developments in both coal and biomass incineration over conventional mass burning incinerator designs. This technology has been accepted by many industries because of its economic and favourable environmental consequences.

The major advantages of fluidized bed combustors are [8,9]:

- Uniform temperature distribution due to intense solid mixing (no hot spots even with
- strongly exothermic reactions
- High combustion efficiencies
- FBC systems have a very short residence time for their fuels (making these systems

http://dx.doi.org/10.5772/52566

- highly responsive to rapid changes in heat demand).
- Large solid–gas exchange area by virtue of the small solids grain size
- High heat-transfer coefficients between bed and the heat exchanging surfaces; the
- intense motion of the fluidized bed makes it possible to combust a wide range of fuels
- having different sizes, shapes, moisture contents and heating values.
- The fuel supplied can be either wet or dry
- The high heat capacity of the fluidized bed permits stable combustion at low temperatures (i.e. 850°C), so that the formation of thermal and prompt nitrogen oxides is suppressed

Sets against these advantages are the following disadvantages:

- Solid separation equipment required because of solids entrained by fluidizing gas resulting in a high dust load in the flue gas;
- Possibility of de-fluidization due to agglomeration of solids;

1.3. Combustion studies

Fluidized bed combustion of alternative solid fuels (including biomass) are attractive as a result of the constantly increasing price of fossil fuels, the presence of high quantities of wastes to be disposed of and global warming issues. Extensive experimental investigation has been carried out to date on the feasibility and performance of different biomass fuels FB combustion such as rice husk [10-13], animal waste [14-15], municipal solid waste (MSW) [16-19] and Refuse Derived Fuel (RDF) [5]. In whatever form biomass residues are fired (loose, baled, briquettes, pellets), a deeper understanding of the combustion mechanisms is required in order to achieve high combustion efficiency and to effectively design and operate the combustion systems. The combustion properties and their effect on combustion mechanisms are all important information required to understand the combustion characteristics of biomass residues and their co-combustion with coal in FBC.

In general when a single coal or biomass particle enters a fluidised bed furnace, then three phenomena occur namely [20-22]:

- Heating up and drying – the fuel particle temperature will rise to its ignition temperature and beyond
- Devolatilization (pyrolysis) – for a short period of time (<10 second), volatile matter will be evolved and can be burnt at or beyond the particle.
- Char oxidation – the remaining solid combustible matter (mostly carbon), will be oxidised relatively slowly with the evolution of heat until only incombustible ash remains.

Fuel properties and combustion operating parameters significantly influenced the combustion efficiency and emissions. Armesto et al (2002) has stated that the bed temperature has an effect on combustion efficiency, which improves from 97% to 98% as bed temperature increased from 840 to 880°C. Also, they found that the efficiency increased with decreasing fluidising velocity. They claimed that when fluidisation velocity increased above 1.0 m/s, the combustion efficiency decreased. This behaviour was attributed to an increase in the elutriation of unburned carbon. On the contrary, Suthum (2000) found that the combustion efficiency increased from 88% to 92% with increasing excess air up to 30% during combustion of oil palm waste in a 10 kW FBC with over-bed feeding. Saxena et al., 1993 reported similar results. It was suggested that there is an optimum balance between the carbon to CO conversion rate and increased elutriation with high excess air [22]. Furthermore, Fahlstedt et al (1997) performed a series of tests on co-firing wood chips, olive pit and palm nut shell with coal in 1MW FBC facility. It was noted that the co-combustion had a slightly higher carbon combustion efficiency based on flue gas emissions (97.2 -98.1%) than coal-only combustion (97.1%) [22]. The reason is likely due to the higher volatile matter content of the biomass fuels. Increased volatile matter will also increase the fuel reactivity and hence reduce the unburned carbon. In contrast, a decrease in combustion efficiency was obtained by Armesto et al (2003) and Suksankraison et al (2004) during co-combustion of Lignite-olive waste and Lignite-MSW mixture, respectively, even though the volatility of the fuel used quite similar (60-70% VM). The decrease was mainly attributed to a drop in the bed temperature. Since most fixed carbon generally burns in the bed while the volatile gas burns in the freeboard, there is insufficient chance for CO conversion to CO_2. As the freeboard temperature is maintained at a higher value, devolatilisation occurred rapidly and produced more volatile gases. As the biomass fraction increased, the reduced fixed carbon gives more chance for the volatiles to escape combustion.

Significant fluctuations of CO emissions were reported during co-combustion of biomass in a FBC. The value of the CO concentration in the flue gas has been found to depend on the type of fuel, fuel properties (volatility, particle size and density) and the operating conditions (bed and freeboard temperature, excess air, secondary air). In addition to the expected immediate ignition and the high volatile matter contents, the volatiles consist mainly of the combustibles (CO, H_2, CxHy). These factors together indicate that the combustion of the volatiles would be the dominant step during the biomass combustion. At higher temperatures, the combustibles (CO, H_2, CH_4) accounted for more than 70–80 % of the gas components [23]. Saxena et al (1994) found that the hydrodynamic activity in the bed is related to the solid mixing and gas-solids contacting and these in turn are directly related to CO emissions. Higher bed temperature seems to provide optimum conditions for rapid de-volatilisation and hence increased conversion CO to CO_2 [19]. They found that in the turbulent regime, the carbon utilisation efficiency reached a maximum and a further increase in the fluidisation velocity had an insignificant influence on the bed hydrodynamics and hence CO emissions. Similarly, as most of the biomass combustion was observed to take place in the freeboard, the supply of oxygen to this zone in amounts sufficient to achieve satisfactory combustion had to be ensured. Furthermore, Sami et al (2001) found that if the level of CO was within acceptable limits, then approximately 10% excess air and a temperature of 650°C provided optimum conditions for the combustion of manure in a fluidised bed unit [17].

However, there was a significant improvement in CO emissions, particularly when the air to the freeboard was introduced at different heights (air staging). The CO levels were brought down to about 60 mg/N m^3 at 11%O_2 in the flue gases, which is very close to what is permitted by EU directives; 50 mg/N m^3 at 11% O_2.The trends observed during single fuel combustion are reflected also in co-combustion: in the practically important cases with moderate amounts of biomass (an energy fraction of less than 25%) the properties of the base fuels dominate the emission obtained. Suksankraisorn et al (2004) reported that for 100% lignite combustion, CO drops significantly as excess air increases due to the increased CO to CO_2 conversion [16]. The emission of CO is relatively insensitive to changes in excess air and waste fraction, which further strengthen the argument that co-combustion is dominated by the combustion of the volatiles in the freeboard zone.

The feasibility of FBC technologies has been widely demonstrated for the combustion of a variety of fuels. As a drawback, severe problems of agglomeration in the bed as well as fouling and slagging may sometimes occur, especially during combustion of biomass fuels as some agricultural residues have high contents of alkali oxides and salts, the low melting points of which may lead to various problems during combustion. Werther et al (2000) encountered the problems of sintering and agglomeration during the combustion of coffee husks in a 150 mm diameter fluidized bed combustor [24].

1.4. Problem statement

Although there are many potential benefits associated with co-combustion, there are several combustion related concerns associated with the co-combustion of coal and biomass. Utilization of solid biomass fuels and wastes sets new demands for boiler process control and boiler design, as well as for combustion technologies, fuel blend control and fuel handling systems. For example, the different mineral matter composition (high alkali levels) and mode of occurrence (mostly mobile forms) in biomass results in concerns over enhanced fouling and slagging of pulverized coal boilers, particularly when firing agricultural residues or herbaceous materials. The economics of co-combustion in pulverized coal boilers are closely tied to the biomass preparation costs (i.e. drying and milling), so an improved understanding of the effect of biomass particle size and moisture content on combustor performance is needed (e.g. in the areas of flame stability, flame shape, and carbon burnout).Thus, this research was carried out with the objective to characterise biomass properties that affect the co-combustion of biomass with coal, in particular biomass that is available in large quantities in Malaysia.

1.5. Research scope and application

This research was performed with the objective to determine the combustion efficiency of the existing coal-fired combustor during co-firing agricultural residue with coal. The efficiency is calculated mainly based on the carbon monoxide and carbon dioxide emissions. Furthermore, this works also to demonstrate the technical feasibility of a fluidized bed as a clean technology for burning agricultural residues. In addition, the effect of biomass properties such as such as particle size, particle density and volatility as well as influences of operating parameters such as excess air, fluidizing velocity on axial temperature profile, the combustion efficiencies and CO emissions are also being investigated.

2. Materials and experimental

2.1. Raw materials and characterizations

In this study British coal, rice husk and palm kernel shell originated from Perlis were employed as fuel. These fuels were open air dried for 2 to 3 days to remove moisture. The proximate and ultimate analyses were performed on coal and rice husk are summarized in Table 1.

	British Coal	Palm Kernel shell	Rice Husk
Proximate Analysis (wt % dry basis) Volatile matter	58.90	18.60	15.00
Fixed carbon	38.20	72.50	60.70
Ash	2.90	8.90	24.30
Ultimate Analysis (wt % dry basis)			
Carbon	80.10	49.5	36.20
Hydrogen	5.30	6.74	5.71
Nitrogen	0.90	1.85	0.10
Sulfur	0.70	0.00	0.00
Oxygen	13.00	41.91	57.99
Calorific value (MJ/kg)	31.1	18.0	13.5
Particle size (mm)	1.4-4.8mm	3x6 mm	0.8x1.00 mm
Particle density (kg/m^3)	1200	435	98

Table 1. Coal and biomass characterization

2.2. Experimental set up and procedures

Figure 1, is a schematic diagram of the Atmospheric Fluidised Bed combustor used in this investigation. The system comprises of a 0.15m diameter and 2.3m high combustion chamber, allows for bed depths of up to 0.3m using 850μm sand, cyclone, screw feeder and gas analyzer. The combustor body is constructed from 1 cm thick 306 stainless steel and covered in Kaowool insulation to prevent excessive heat loss during operation. Fluidising air was introduced at the base of the bed through a nozzle distributor and provided fluidisation and combustion air. Start up of the bed was achieved using an in-bed technique; Propane was introduced directly into the distributor plate by injectors and mixed with air in the nozzles, providing a combustible mixture at the nozzle exit. Bed and freeboard temperatures were measured at 8 different heights above the distributor plate by means of sheathed Ni/Cr-Ni thermocouples (TC) type K. Fuel was fed pneumatically into the bed surface from a sealed hopper through an inclined feeding pipe the flowrate for which flow rate was controlled by a screw-feeder. A cyclone was fitted to the combustor exit and the carryover from the bed was collected for analysis. CO and O_2 were measured using a Xentra 4904 B1 continuous emissions analyzer, whereas CO_2 was measured using a non-dispersive infrared absorption spectrometry analyser. A fly ash sample was collected from the catch pot after finishing the combustion run. The fly ash sample was then weighed and analysed to determine the total amount of unburned carbon of the fuels in the test. The percentage of combustion efficiency was computed using Equation 1. Based on the values of combustion efficiency from experiments where duplicate runs were conducted under almost identical conditions, the combustion efficiency values should be within ±2%.

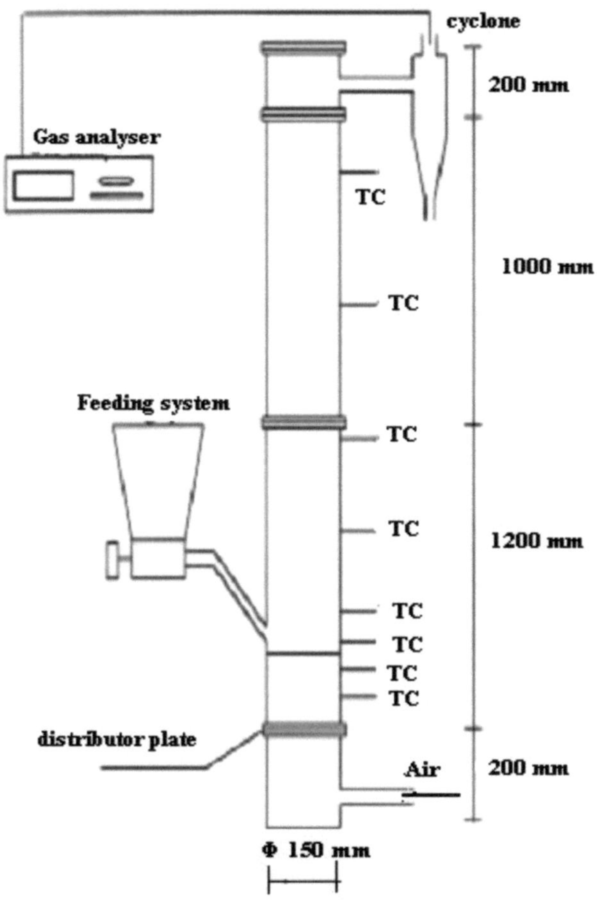

Figure 1. Schematics diagram of the laboratory scale fluidized bed combustor (Tc =Thermocouple).

2.3. Operating conditions

In this experiment, baseline data was first obtained for single combustion of 100% British bituminous coal. Also, single combustion of other biomass fuels was carried out to investigate their combustion characteristics in comparison to coal during the co-combustion study. Co-combustion tests at biomass fractions of 30%, 50%, and 70% were performed. For each biomass fraction, excess air was varied from 30% to 70% at 20% intervals. For each excess air condition, air staging combustion was applied where the total secondary air is maintained at 65 l/min (about 10-20% to total air ratio). In order to study the impact of fuel property changes (volatiles, ash, and combustibles), heat input was fixed at the design value of the experimental rig i.e. 10 kW. The combustion tests were operated in the bed temperature range of 700-950 °C and superficial velocity range of 0.63 – 1.12 m/s.

2.4. Carbon combustion efficiency calculation

The carbon combustion efficiency of a system has been expressed as:

$$\eta CE = \frac{B}{C} x 100\% \tag{1}$$

where$B$ and C are *the mass fractions of burnt and total carbon in the fuel*, respectively. Knowing the flue gas composition, the flue gas composition, fractional excess air and the fuel ultimate analyses of the fuel, B can be determined [20].

This method is particularly appropriate for solid fuels and is described as follows:

Let C, H, O, N, and S be the mass fractions of *carbon, hydrogen, oxygen, nitrogen and sulphur*, respectively, in the feed.

Further, let A and B be the *mass fractions of unburned and burnt carbon*, respectively, in the fuel. Then,

$$A + B = C \tag{2}$$

Further define

$$P = \frac{C\ converted\ to\ CO}{C\ converted\ to\ CO + CO_2} = \frac{C\ converted\ to\ CO}{B} \tag{3}$$

$$C\ converted\ to\ CO = PB \tag{4}$$

$$C\ converted\ to\ CO_2 = (1 - P)B \tag{5}$$

$$Mass\ of\ CO_2\ in\ the\ flue\ gas = \left(\frac{44}{12}\right)(1 - P)B \tag{6}$$

http://dx.doi.org/10.5772/52566

$$Mass\ of\ CO\ in\ the\ flue\ gas = (28/12)PB \tag{7}$$

$$O_2\ consumed\ to\ produced\ CO_2\ +\ CO = (32 - 16P)B/12 \tag{8}$$

Assuming that *H, N,* and *S* present in the fuel are completely converted to H_2O, *NO* and SO_2 respectively,

$$O_2\ consumed = (16/2)H\ +\ (16/14)N\ +\ (32/32)S\ =\ X1 \tag{9}$$

$$SO_2\ produced = (64/32)S \tag{10}$$

$$NO\ produced = (30/14)N \tag{11}$$

Therefore, total O_2 required for stoichiometric combustion of fuel

$$(32/12)C\ +\ (16/2)H\ +\ (16/14)N\ +\ (32/32)S\ -\ O\ =\ X2 \tag{12}$$

Let *Z* be the fractional excess air supplied, which is defined as the excess air divided by the stoichiometric air. Therefore,

$$O2\ supplied\ =\ X2(1+Z) \tag{13}$$

$$Mass\ of\ N2\ in\ the\ flue\ gas = (79/12)(28/32)\ X2(1+Z) \tag{14}$$

$$O2\ consumed\ during\ combustion = (32 - 16P)B/12\ +\ X1 \tag{15}$$

$$Mass\ of\ O2\ in\ the\ flue\ gas = \left(X2(1+Z) - \frac{(32 - 16P)B}{12} + X1\right. \tag{16}$$

Let *F* be the mass of dry flue gas can also be estimated from the flue gas composition. The flue gas flow rate and composition are not appreciably influenced by neglecting the presence of SO_2 and *NO* in the flue gas. Hence the flue gas may be taken as consisting of *CO*, *CO2*, N_2 and O_2.

Let Y be the mass of dry flue gas per unit mass of C burnt in the fuel. Then,

$$Y\ =\ \{44[CO_2]\ +\ 32[O_2]\ +\ 28[N_2]\}/12\{[CO]\ +\ [CO_2]\} \tag{17}$$

The square brackets represent the volume fraction of the particular chemical species in the flue gas and Y can be simplified to

$$Y\ =\ \{4[CO_2]\ +\ [O_2]\ +\ 7\}/3\{[CO]\ +\ [CO_2]\} \tag{18}$$

By substituting

$$[CO] + [N_2] + [CO_2] + [O_2] = 1 \tag{19}$$

mass of dry flue gas per unit mass of the fuel is

$$F = YB \tag{20}$$

Substituting F in (A-6.18) into (A-6.17), then the fraction of C burnt, B, can be written as follows:

$$B = \frac{\left[4.29(1 + Z)\left[\left(\frac{32}{12}\right)C + \left(\frac{16}{2}\right)H + \left(\frac{16}{4}\right)N + \left(\frac{32}{32}\right)S - O\right] - 8H + N + S\right]}{Y - 1} \tag{21}$$

$$\eta CE = (B(21) + Unburned\ carbon\ in\ ash) \ / \ C \times 100\% \tag{22}$$

3. Results and discussion

This section describes the combustion of agricultural residue in a fluidized bed combustor. The influences of fuel properties such as particle size, particle density and volatility as well as influences of operating parameters such as excess air, fluidizing velocity on axial temperature profile, the combustion efficiencies and CO emissions are discussed.

3.1. Carbon combustion efficiencies

The combustion tests were performed using different coal mass fraction; 0, 50 and 100%, corresponding to heat input of 10kW under optimum excess air conditions. Figure 2 shows the effect of different mixtures of rice husk and palm kernel shell with coal on carbon combustion efficiency with the same heat input. Generally, Carbon combustion efficiency for single biomass (rice husk and palm kernel shell) but increases with increasing coal addition and experimental runs. The following carbon combustion efficiencies, from Eqn. (1), range between 67-75% for burning 100% rice husk and 80-83% for burning 100% palm kernel shell, 83-88%, and 86-92% for 50% of coal addition to rice husk and palm kernel shell, respectively. The improved carbon combustion efficiency by co-combustion of rice husk with coal can be attributed to an increase in bed temperature, Figure 3, which is caused by the addition of fixed carbon content in the mixture. This fixed carbon, from coal, burns in the bed while the volatile gas burns in the freeboard region. Thus, there is more chance for fuel conversion carbon to carbon dioxide as the coal fraction increases and less volatile and tend to escape combustion, because of the reduced biomass concentration [16].

In addition, increasing the fluidizing velocity increases the turbulence in the bed leading to better solid mixing and gas-solid contacting and so as the amount of carbon in the bed is burnt at higher rate. Consequently, higher carbon burn out obtained leads to higher carbon

http://dx.doi.org/10.5772/52566

combustion efficiency. However, when the combustion is stabilized, increasing fluidizing velocity contributed to a greater particle elutriation rate than the carbon to CO conversion rate and hence increased the unburned carbon [5]. However, when the combustion is stabilized, increasing the fluidizing velocity contributes to a greater particle elutriation rate than carbon to carbon monoxide conversion rate and increases the amount of unburned carbon. This phenomenon can be seen in Figure 2 where the carbon combustion efficiency is lower than expected for 50% rice husk mixtures when the fluidizing velocity increases beyond the optimum value. Apart from solid mixing, increasing the fluidizing velocity also influences the fuel particle settling time during the combustion process in the FBC. Increasing fluidizing velocity drives the lighter fuel particles upwards and into the freeboard region, which is indicated by higher freeboard temperatures. Thus, the settling time for the biomass to reach the bed will be greater and a significant portion of the combustion will be completed before the particles return to the bed is reached, although this is dependent upon fuel particle size and density. This settling time depends on the fuel particle size and particle density. The greater settling time the higher the freeboard temperature due to greater volatile combustion contributing to higher combustion efficiency providing the bed temperature is maintained within the range of 800-900°C.

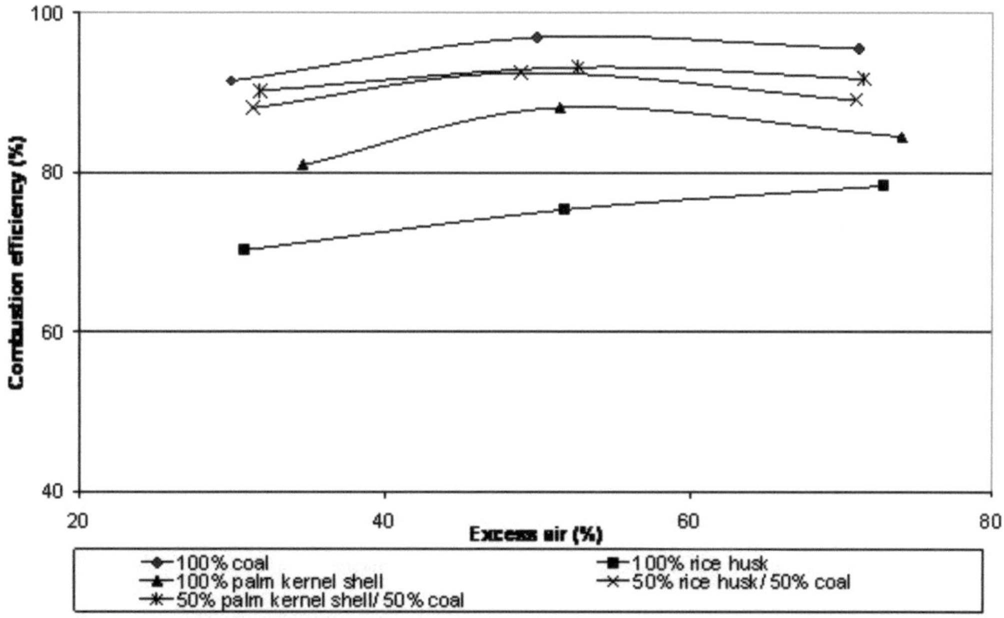

Figure 2. Combustion efficiency during co-combustion of coal with rice husk and palm kernel shell

3.2. Temperature profiles

Figure 3 illustrates the axial temperature distributions along the FBC height for fuel studied at 50% excess air. As can be seen from the figure, coal combustion gives higher bed temperature (y = 0-40cm) but lower freeboard temperature (y = 450-120cm) in comparison to biomass. Then, all the temperatures shows start to fall from 120 cm above distributor plate indicating that most of the combustion was completed. This significant combustion behaviour can be explained by the devolatilization process of the fuel [17]. With high volatility (more than 50%) and low ignition temperature (250-350°C), biomass (rice husk and palm kernel shell) will start to devolatilize upon feeding at 45 cm of the FBC height (freeboard region) and was mostly burned before it reached the bed region. While coal with low volatility (30%) and higher ignition temperature (400-600°C) will travel down to the bed and completed combustion in the bed region. This was also greatly influenced by settling velocity of the fuel particles which correspond to the fuel particle size and fluidizing velocity [5]. Those explain why palm kernel shell has higher bed temperature than rice even though the volatility is almost similar (see Table 1).This was due to the fact that greater particle size contributed to a greater devolatilization time and settling time. The greater settling time, the higher the freeboard temperature due to greater volatile combustion contributing to higher combustion efficiency providing the bed temperature is maintained within the range of 800-900˚C.

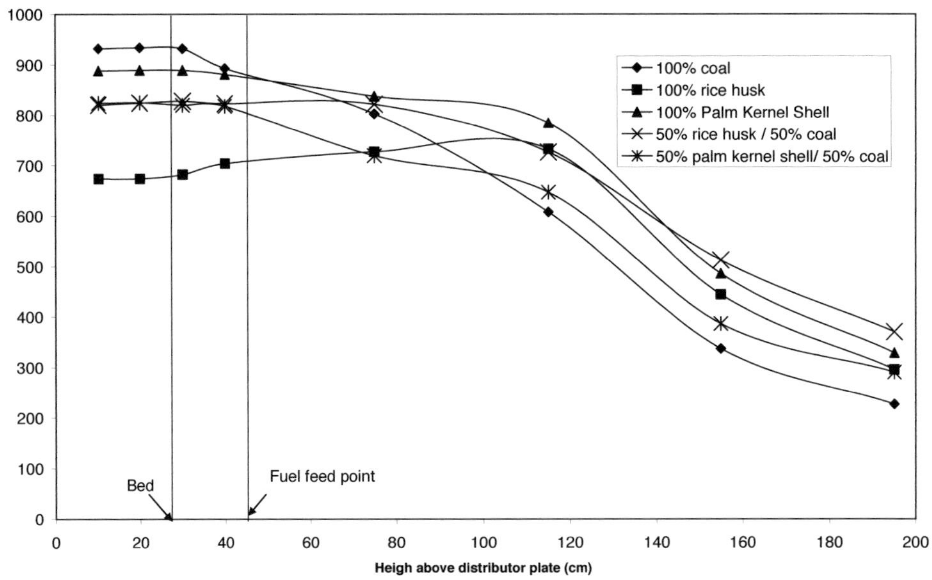

Figure 3. Axial temperature profile for co-combustion of coal with rice husk and palm kernel shell combustion in the case of excess air = 50%

Significant increment of carbon combustion efficiencies was noted with coal addition to biomass fraction (see Figure 2).The improvement can be attributed to an increase in bed temperature, Figure 3, which is caused by the addition of fixed carbon content in the mixture. This fixed carbon, from coal, burns in the bed while the volatile gas burns in the freeboard region. Thus, there is more chance for fuel conversion carbon to carbon dioxide as the coal fraction increases and less volatile and tend to escape combustion, because of the reduced biomass concentration. Furthermore, this can be explain by the fact that biomass fuels with lower density (about half) compared to coal tend to burn in freeboard and coal tends to burn in the bed region. Therefore, the addition of coal in biomass increases the amount of fixed carbon reaching the bed resulting in higher bed temperatures. This observation agrees with the results of Abelha et al. (2003) and Suksankraisorn et al. (2003) who investigated the co-firing of coal and chicken litter and co-firing of lignite with municipal solid waste in a fluidised bed combustor, respectively.

3.3. Carbon monoxide (CO) emissions

In order to enable comparison of CO from all tests were converted to CO emitted 6% flue gas oxygen. Figure 4, it is evident that there are significant fluctuations in CO emissions, which between 200 and 900 ppm under the same conditions. The orders of fluctuation were similar to those observed by Abelha et al. (2003) and W.A.W.A.K. Ghani et al. (2009). The fluctuations are caused by slight variations in feed composition and this effect is reflected in the temperature profiles. It is noted that the addition of coal has no significant influence on CO emissions during all co-combustion cases, except at coal (50%) / rice husk (50%) where emissions tend to be lower than expected in reference to the other rice husk fractions. This phenomenon is due to the synergistic nature of the coal and rice husk mixture, which enhances the fuel reactivity and lowers the CO emissions [10]. In most cases the emission of CO seems relatively insensitive to changes in excess and fluidising air. This insensitivity is due to increased segregation of fuels in the combustor between the feed point and the bed. If the combustor receives a batch with a relatively high amount of fuel pellets, then as burning CO_2 is produced since the pellets need to be heated and dried first. While this occurs, oxygen is not consumed and results in high CO emissions. The decrease in CO levels at low percentages of excess air, not below than 50%, can be attributed to low excess air, relatively high bed temperatures (about 900°C) causing rapid enhances and ignition of volatiles from rice husk. Thus, higher CO to CO_2 conversion rates was observed which enhanced the reactivity of the mixture [13].

3.4. Analyses of carryover

Tables 2 and 3 present the ash collection and unburned carbon analyses during combustion tests. Generally, the mass balance on the ashes particles accounted for over 90% of the ash input from the fuel. The analyses of the ash collected in all tests for unburned carbon demonstrates that with biomass only, there was the least amount of unburned carbon detected in ash collected from the cyclone. However, the unburned carbon content increased when coal was added which suggested that some fine particles were elutriated with the fluidising gases. The amount of unburned carbon was, however, quite low, corresponding to about less than 5% of the total carbon input. Such observations seem to suggest that the large particle size and lower heating value of the biomass fuel did not adversely affect combustor performance, probably due to the higher volatile matter content of the biomass fuel. The volatile matter burns rapidly and the higher volatile matter content of the biomass can also result in a highly porous char, thus accelerating the char combustion as well. In all cases the amount of unburned carbon in the ash increased as the percentage s of coal increased which is due to the low volatility of coal. For the biomass materials the low density of palm fibre and rice husk are also led to increased carbon content in the ash. The initial particle size of the biomass does not appear to be significant.

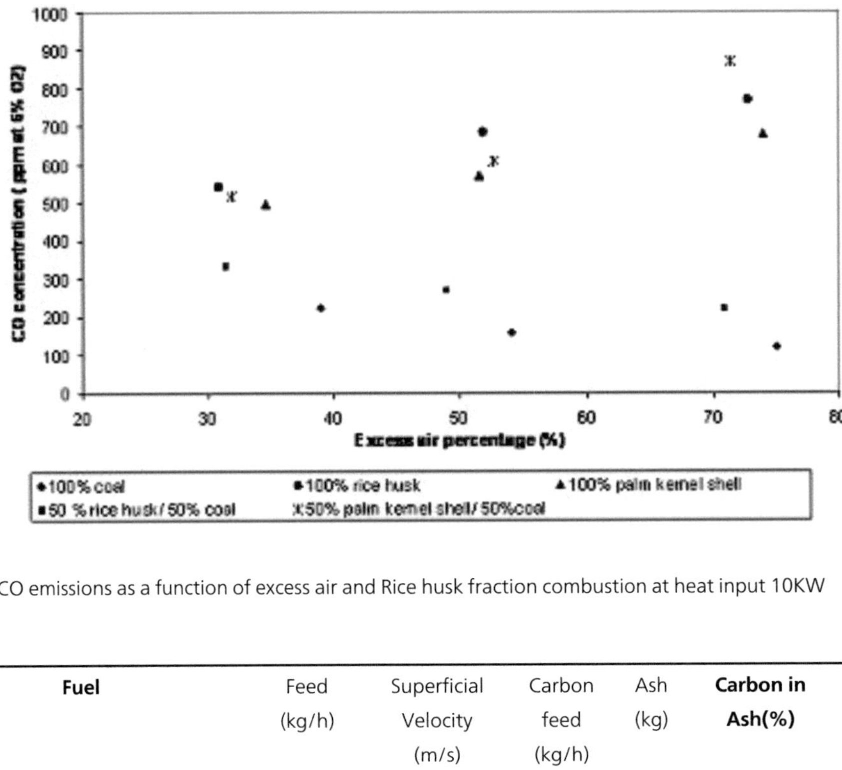

Figure 4. CO emissions as a function of excess air and Rice husk fraction combustion at heat input 10KW

Fuel	Feed (kg/h)	Superficial Velocity (m/s)	Carbon feed (kg/h)	Ash (kg)	**Carbon in Ash(%)**	Efficiency (%)
Coal (100%)	1.20	0.67	0.900	0.039	23.0	90.25
Rice husk (100%)	2.97	0.56	1.038	0.621	14.5	66.62
Coal (30%) : Rice husk (70%)	2.16	0.99	1.149	0.348	20.9	75.33
Coal (50%) : Rice husk (50%)	1.60	0.85	1.159	0.196	28.7	83.24
Coal (70%) : Rice husk (30%)	1.40	0.81	1.094	0.176	26.6	86.07

Table 2. Ash analysis for single and co-combustion of coal and rice husk at varies percentage of excess air

Fuel	Feed (kg/h)	Superficial Velocity (m/s)	Carbon feed (kg/h)	Ash (kg)	**Carbon in Ash(%)**	Efficiency (%)
Coal (100%)	1.20	0.67	0.900	0.039	23.0	90.25
Palm kernel shell (100%)	1.97	0.59	0.898	0.028	5.0	80.67
Coal (30%) : Palm kernel shell (70%)	1.74	0.74	0.949	0.030	11.7	80.73
Coal (50%) : Palm kernel shell (50%)	1.59	0.65	0.962	0.031	14.9	89.86

Table 3. Ash analysis for single and co-combustion of coal and palm kernel shell at varies percentage of excess air.

4. Conclusion

The conclusions obtained in the present investigation on the temperature profile, carbon combustion efficiency and CO emissions in a 10 kW FBC can be summarised as that biomass combustion behaves differently in comparison to coal due to the significant difference in volatile matter content and variations of particle size and particle density. The carbon combustion efficiency was influenced by the operating and fluidising parameters in the following order: a) settling velocity; b) coal mass fraction; c) fluidising velocity; d) excess air and e) bed temperature (Tb). The maximum carbon combustion efficiency increased in the range of 3% to 20% as the coal fraction increased from 0% to 70%, under various fluidisation and operating conditions. Generally, the carbon combustion efficiency increased with increases of excess air and peaks at 50%. The corresponding increasing carbon combustion efficiency with excess air from 30-50% was found to be in the range of 5 – 12 % at 50% coal mass fraction in the biomass mixture. Further increase of excess air to 70% reduced the carbon combustion efficiency. Increasing the fluidising velocity increases the turbulence in the bed leading to better solid mixing and gas-solid contacting and shows as the amount of carbon in the bed is burnt at higher rate. However, when the combustion is stabilised, increasing fluidising velocity contributed to a greater particle elutriation rate than the carbon to CO conversion rate and hence increased the unburned carbon. Apart from solid mixing, increasing fluidising velocity also influenced settling time of fuel particle during the combustion process in FBC. Increasing fluidising velocity brought the lighter fuel particle upward to the freeboard region and completed before they reached the bed surface.The bed temperature had a small effect on carbon combustion efficiency for the biomass fuels. The turbulence created by increasing excess air related with increases in fluidising velocity had a greater influence than reduced bed temperature. Significant fluctuations of CO emissions observed when coal was added into almost all biomass mixtures depending upon excess air ranges between 200-1500 ppm.The analyses of the ash collected in all tests for unburned carbon demonstrates that with biomass only, there was less unburned carbon detected in the ash collected from the cyclone indicating that the combustion of fixed carbon was almost complete. The percentages of unburned carbon increased in the range 3 to 30% of the ash content with the increases of coal fraction in the coal/biomass mixture. This can be explained by the fact that as the coal fraction increased the higher char combustion and less volatiles combustion occurred. Moreover, the elutriated carbon loss increased as fluidising velocity increased resulting in the lower carbon combustion efficiency. On the contrary, it was found that the bed temperature had no strong influence on carbon loss during the tests.As a conclusion, the combination factors of operating parameters attributed to the resulting effects of biomass co-combustion.

5. Further research

As for further research work, the followings are suggested:

a. Modify the combustor to compare in-bed with overbed feeding of coal/biomass mixtures.

b. Investigate the effect of air staging on combustion.

c. Study the effect of bed temperature on combustion efficiency by having a cooling coil in the bed instead of using air flowrate.

d. To investigate the release of NO_x from coal/biomass mixtures. Although the nitrogen content of biomass is generally low the compounds have a lower molecular weight and are more volatile.

e. Investigate co-firing of a wide range of densified biomass fuels that are currently available with coal.

Abbreviations and nomenclature

CO= Carbon monoxide

ASEAN = Association of South East Asian Nations

EU= European Union

FBC= Fluidized bed combustor

MSW= Municipal solid waste

RDF= Refuse Derived Fuel

H_2=Hydrogen

CxHy= hydrocarbons

CH_4=methane

O_2= Oxygen

η_{CE}= Carbon combustion efficiencies

http://dx.doi.org/10.5772/52566

B = burnt carbon

C = Carbon

H = Hydrogen

N= nitrogen

S= sulphur

H_2O= water

NO = Nitrogen Oxide

SO_2= Sulfur dioxide

Z = fractional excess air supplied

F = mass of dry flue gas

Y = mass of dry flue gas per unit mass of C burnt in the fuel

Acknowledgment

This work was performed within the University of Sheffield, a research project supported by the Ministry of Science, Technology and Innovation (MOSTI), Malaysia.

Author details

Wan Azlina Wan Ab Karim Ghani[1] and Azil Bahari Alias[2]

1 Department of Chemical and Environmental Engineering, the Universiti Putra Malaysia, 43400 UPM Serdang, Selangor, Malaysia

2 Faculty of Chemical Engineering, University Technology MARA, 40450 Shah Alam, Selangor, Malaysia

References

[1] Natrajan, R., Nordin, A., and Rao, A. N. Overview of combustion and gasification of rice husk in fluidised bed reactors. Biomass and bioenergy 1998; 14: 533- 546.

[2] Cliffe, K. R., and Sathum, P.Co-combustion of waste from olive production with coal in a fluidised bed', Waste management 2001; 21: 49-53.

[3] Guilin, P. Shigeru, A., Shigekatsu, M., Seiichi, D., Yukihisa, F., Motohiro, K and Masataka, Y. Combustion of refused derived fuel in a fluidised bed.Waste management 1998; 1: 509-512.

[4] Armesto, L., Bahillo, A., Veijonen, K., Cabanillas, A. and Otero, J.Combustion behaviour of rice husk in a bubbling fluidised bed. Biomass and Bioenergy 2002; 23(3): 171-179.

[5] W.A.W.A.K. Ghani, A.B. Alias, R.M. Savory and K.R. Cliffe.Co-combustion of Refuse derived fuel with Coal in a 10 kW Fluidised Bed Combustor. Journal of Engineering Science and Technology 2009; 4: 122-131.

[6] Malaysian Oil Palm Statistics 2002, 22nd Edition, MPOB, Data is for the year 2002.

[7] Cleantechsolution:Biomass Energy Potential in Malaysia: http://www.cleantech-loops.com/biomass-energy-in-malaysia/(accessed 3 August 2012).

[8] Basu, P., and Fraser, S.A., Circulating Fluidized Bed Boilers, Butterworth-Heinemann, USA, 1991.

[9] Bhattacharya, S.C. and Wu, W., 'Fluidised bed combustion of rice husk for disposal and energy recovery', Energy from Biomass and Wastes XII, 1989, 591- 601

[10] Mutanen, K., Circulating Fluidized Bed Technology in Biomass Combustion—Performance, Advances and Experiences, Proceedings, 2nd Biomass Conference of the Americas: Energy, Environment, Agriculture and Industry, Portland, USA, August 1995, pp. 449–460

[11] Atmospheric Fluidized Bed Coal Combustion: Research, development and application (M. Valk, Ed.), Elsevier, The Netherlands, 1995.

[12] Kelleheler, B. P., Leahy, J.J, Henihan, A.M, O'Dwyer, T.F, Sutton, D., Leahy, M.J. Advances in poultry litter disposal-a review.Bioresource Technology 2002; 83: 27-36

[13] Kuprianov, V. I and Pemchart, W. 'Emissions from a conical FBC fired with a biomass fuel', Applied Energy, 74, 2003, pp. 383–392

[14] Abelbha, P., Gulyurthu, I., Boavida, D., Seabra Barros, J., Cabrita, I., Leahy, J.,Kelleher,B.,andLeahy M. Combustion of poultry litter in fluidised bed combustor. Fuel2003;82: 687-692.

[15] Suksankraisorn, K, Patumsawad, S. and Fungtammasan, B. Combustion studies of high moisture content waste in a fluidised bed', Waste Management 2003;23:433–439.

[16] Suthum, P. Co-combustion of high moisture content of MSW with coal in fluidised bed combustor. Ph.D. thesis. The university of Sheffield, 2000.

[17] Cooke, R.B, Goodson, J and Hayhurst, A. N.The combustion of solid wastes as studied in fluidised bed', Trans IchemEPart B 2003; 81: 156-166.

[18] Suksankraisorn,K, Patumsawad, S., Vallikul, P., Fungtammasan, B. and Accary, A. Co-combustion of municipal solid waste and Thai lignite in a fluidized bed.Energy Conversion and Management 2004; 45: 947–962.

[19] Sami M,Annamalai K, and Woodridge M. Co-combustion of coal and biomass fuel blends. Progress Energy Combustion Science 2001; 27:171–214.

[20] Leckner, B., Amand, L. E. Lucke, K. and wether, J. Gaseous emissions from –co-combustion of sewage sludge and coal/wood in a fluidised bed. Fuel2004; 3: 477-486.

[21] Tillman, D. A. Biomass cofiring: the technology, the experience, the combustion consequences. Biomass and Bioenergy2000; 19(6): 365-384.

[22] Saxena SC and Jotshi CK. Fluidized bed incineration of waste materials. Progress in Energy and Combustion Science 1994; 20:281.

[23] Fahlstedt I, Lindman E, Lindberg T, Anderson J. Co-firing of biomass and coal in a fluidised bed combined cycle. Results of pilot plant studies. In: Proceedings of the 14th International Conference on Fluidized Bed Combustion in Vancouver, Canada, vol. 1, 1997. pp. 295–299.

[24] Werther, J., M. Saenger. Combustion of agricultural residues'. Progress in Energy and Combustion Science2000; 26(1):1-27.